Radiation Myelopathy

Timothy Schultheiss

Radiation Myelopathy

 Springer

Timothy Schultheiss
Department of Radiation Oncology
City Of Hope National Medical Centre
Duarte, CA, USA

ISBN 978-3-030-94660-9   ISBN 978-3-030-94658-6   (eBook)
https://doi.org/10.1007/978-3-030-94658-6

This Springer imprint is published by the registered company Springer Nature Switzerland AG
The registered company address is: Gewerbestrasse 11, 6330 Cham, Switzerland

Preface

Radiation myelopathy is a terrifying complication of cancer therapy. It can lead to quadriplegia and death. As with traumatic spinal cord injuries, it is not uncommon for victims of this progressively debilitating injury to take their own lives.

It is no surprise then that hundreds of research papers have been devoted to presenting the clinical data available and to describing experimental results. Experiments in the spinal cord's response to radiation have been performed in virtually every animal model used in radiation studies. These experiments include not only the dose response, but also many aspects of fractionation effects, volume effects, retreatment, dose-response modification, and of course the pathogenesis of injury. The radiation injury is rarely equivocal, and therefore the endpoint of both the experiments and the clinical outcome is rarely in doubt.

Because the spinal cord is small and has a simple geometry, it is ideal for describing radiation responses using biomathematical models. Generally the dosimetry is simple and the dose is uniform. In clinical studies, the dose to the spinal cord can usually be well described by a single dose without the need to resort to a dose-volume histogram.

With the possible exception of the skin, the spinal cord is the organ most studied in the radiation literature. Unlike the skin, the spinal cord's response is nearly always easily dichotomized—myelopathy has or has not occurred. Thus the radiation response of the spinal cord has been a favorite subject of study, not only because of the crucial knowledge that may be gained regarding the clinical complication, but also because it can serve as a testing ground for all biomathematical models of radiation response. If a model cannot be successfully deployed in this area, there is little hope it can describe the results of any other radiation complication.

The purpose of this book is to coalesce in a single volume most of the clinical dose-response data and the experience with the numerous experimental studies. Frankly, some experimental and clinical studies do not seem to fit well with the totality of knowledge in this area. By compiling this information, it may be possible to identify those outliers as one would identify outlying data points.

Reflecting the literature, the most extensively covered topic in this book is the rodent model of radiation myelopathy. The rat has been the proving ground for every class of radiation myelopathy experiment.

Like much of the medical literature, publications in this field have a broad range in the quality of the study design, execution, and analysis, and the overwhelming majority of study conclusions have not been critically assessed by the target consumers of radiation oncology literature. Accordingly, much of the experimental literature that is discussed here has been subjected to re-analysis in an effort to determine to what degree the conclusions are supported by the data. The findings are mixed. In the early literature, results of the data analysis were commonly reported without providing the reader with any direct exposure to the raw data in either table or graphical form. Being given values or graphs of D_{50} without even seeing the dose-response curve fitted to the data leaves the reader without any basis for assessing whether the data and the model agree. In fact, it was also common to use so few animals per dose group that either one or zero data points had responses between 0% and 100%. This issue is addressed in considerable detail.

As this field of study matured, certain mathematical models used to describe the radiation response have become standardized and accepted as being useful, and possibly accurate. However, it has been somewhat unusual for these models to be assessed for their ability to describe the data, that is, goodness-of-fit tests are not the standard. One objective of this book is to re-analyze data from seminal works to determine the degree to which our beliefs are supported by the data. In some cases, this re-analysis broadens or clarifies the current understanding of the subject. This exercise also discloses areas where specific information is missing. The areas of volume effects and retreatment have received particular attention in this work.

This book also contains original work that has not been previously published. A new proposal to describe the final step in the pathogenesis that leads to white matter necrosis is offered. A composite dose-response function for both the cervical and thoracic spinal cord levels is presented, with the new component being the thoracic cord's dose response. A new model of retreatment is put forward that does not equate expression of damage and dose. The last chapter is a presentation of a multi-species dose-response model that has only been published in abstract form.

Although we know a great deal about radiation myelopathy and how to avoid it, the most important aspect of this injury has remained elusive—how to treat it. Some progress has been made and many now believe that the elements of the pathogenesis are sufficiently well understood to make logical attempts in its prevention and treatment.

The clinical occurrence of RM frequently leads to civil cases of malpractice and sometimes product liability where machines fail to perform as anticipated. This area is briefly addressed in the Appendix.

I would like to acknowledge the mentorship of Colin G. Orton, who introduced me to medical physics and Lester J. Peters whose guidance in clinical research and academic medicine was indispensable. Finally, the daily collaborations with L. Clifton Stephens grew to be the most important experience in my research career. In the writing of this book, I have been fortunate to enjoy the enduring patience of my wife, Ginamarie.

Duarte, CA, USA   Timothy Schultheiss

Contents

Contents

About the Author

Timothy Schultheiss Radiation myelopathy has been the primary research area for Timothy Schultheiss for 40 years, but he has numerous publications on the radiation dose response of many organs in addition to the spinal cord. His research contributions have been acknowledged by the awarding of Fellowships in the American College of Radiology, the American Association of Medical Physicists, and the American Society for Radiation Oncology. He is currently Professor Emeritus at City of Hope National Medical Center after having retired as Director of Radiation Physics. Before that he served in the same capacity at Fox Chase Cancer Center.

Abbreviations

AAPM	American Association of Physicists in Medicine
ACE	Angiotensin-converting enzyme
ACR	American College of Radiology
AIC	Akaike information criterion
AQP	Aquaporin
AUC	Area under the curve
AZQ	Aziridinylbenzoquinone
B&S	Bath and shower
BBB	Blood-brain barrier
BED	Biologically effect dose
BIANCA	BIophysical ANalysis of cell death and Chromosome Aberrations
BID	Twice Daily
BN	Brown norway
BNCT	Boron neutron capture therapy
BPA	p-Boronophenylalanine
BSCB	Blood-spinal cord barrier
BSH	Borocaptate Sodium
CDF	Trademark for Fischer rat
CI	Confidence interval
Co	Cobalt
CSF	Cerebro-spinal fluid
CTV	Clinical target volume
DFMO	Difluoromethylornithine
DMF	Dose-modifying factor
DVH	Dose-volume histogram
ED	Effective dose
ERD	Extrapolated response dose
ESD	Equivalent single dose
GLM	Generalized linear model
H&N	Head and neck
HBO	Hyperbaric oxygen
HDR	High dose rate
HRP	Horseradish peroxidase
HYTEC	Hypofractionation treatment effects in the clinic

IGF	Insulin-like growth factor
IMRT	Intensity modified radiation therapy
IR	Incomplete repair
ISF	Interstitial fluid
KM	Kaplan-Meier
KV	Kilovoltage
LDR	Low dose rate
LEM	Local effects model
LENT	Late effect in normal tissues
LET	Linear energy transfer
LL	Log likelihood
LQ	Linear-quadratic
LR	Likelihood ratio
MC	Monte Carlo
MFD	Multiple fractions per day
ML	Maximum likelihood
MLE	Maximum likelihood estimation (or estimate)
MRC	Medical Research Council
MS	Multiple sclerosis
MTX	Methotrexate
MV	Megavoltage
NPC	Neural progenitor cell
NPV	Negative predictive value
NSD	Nominal standard dose
NTCP	Normal tissue complication probability
OAR	Off axis ratio; Organ at risk
OER	Oxygen enhancement ratio
OPC	Oligodendrocyte progenitor cell
OR	Odds ratio
PA	Polyamine
PDGF	Platelet-derived growth factor
PPV	Positive predictive value
PVS	Perivascular space
QD	Daily
QUANTEC	Quantitative Analysis of Normal Tissue Effects in the Clinic
RBE	Relative biological effectiveness
RFP	Request for proposals
ROC	Receiver-operator
RTOG	Radiation Therapy Oncology Group
SAS	Statistical analysis system
SBRT	Stereotactic body radiation therapy
SD	Standard deviation
SEM	Standard error of the mean
SLD	Sublethal damage
SOBP	Spread-out bragg peak

SPF	Specific pathogen-free
SRS	Stereotactic radiosurgery
SSD	Source to skin distance
TDF	Time-dose factor (or fractionation)
TG	Task group
TL	Thoracolumbar
TN	Thermal neutrons
VEGF	Vascular endothelial growth factor
WM	White matter
WMN	White matter necrosis
XRT	X-ray therapy (radiation therapy)

Symbols

$\mid$	Given, as in $P(a\mid b)$ is "probability of a given b"
α	Coefficient of total dose in the LQ model
α/β	Ratio that determines the fractionation sensitivity in the LQ model
β	Coefficient of Dd in the LQ model
d	Dose per fraction
D	Total dose
D_{50}	Median tolerance dose
DMF	Dose modifying factor
$D_{x\%}$	Dose to the hottest $x\%$ of volume
ED_{50}	Median tolerance when dose is expressed as an equivalent dose
k	Coefficient of $\ln(D)$ in a logistic model of dose response
ln	Log-natural logarithm
LR	Likelihood ratio
m_i	Number of subjects with variate pattern i
N	Number of fractions
n	Number of covariate patterns in a data set
P	Probability of radiation myelopathy
p	Probability of event (other than radiation myelopathy)
r_i	Number of responders with variate pattern i
T	Total treatment time
$t_{1/2}$	Half-time of repair
V_D	Volume that receives dose D or more
x	Dummy variable
z	Systematic component of a GLM

List of Figures

List of Tables

History and the Search for the Tolerance Dose

1

1.1 Early Clinical Reports: How We Arrived at the 45-Gy Tolerance Dose

One of the first facts medical students and residents learn about clinical radiation oncology is that the spinal cord tolerance is 45 Gy in 25 fractions. The many reasons this is not correct will be addressed later. For now, it is sufficient to recognize that this statement is a metaphorical stop sign that reads, "Do not exceed 45 Gy to the spinal cord until you know a bit more about what you are doing." How this particular value was reached deserves some exposition.

In 1941, Ahlbom appears to be the first to report radiation myelopathy (Ahlbom 1941). It is useful to note that this report, and others of the 1940s, was generated from studies of relative large populations of patients, rather than case reports. One might reasonably guess that during the early part of the twentieth century, most cases of radiation myelopathy were attributed to cancer progression. Only when a specific set of symptoms with a common pattern of progression were observed in a number of patients was a diagnosis made of injury to the central nervous system. The injury was quickly identified as a myelitis, but Boden's 1948 paper in the British Journal of Radiology was the first to call it "radiation myelitis," presumably to assert clearly that it was a result of the radiation and not tumor progression (Boden 1948). This was also the first paper to be dedicated solely to this injury. The somewhat more accurate term "Radiation myelopathy" was in use by the early 1960s (Dynes 1960; Alajouanine et al. 1961; Innes and Carsten 1961; Pallis et al. 1961; Verjaal 1964).

In the early days of radiation therapy, as now, the most important consideration regarding serious radiation complications was to avoid them, not to study them. Therefore in the early literature one finds a great deal of effort put toward divining doses that could be proffered as safe doses. Boden suggested dose limits that should be respected when irradiating the spinal cord, relevant to the practice patterns of the day. These were extended in a subsequent paper using the Strandquist formula where $D_2 = D_1 \, (T_2/T_1)^{0.22}$ with dose D_i being given in overall time T_i (Strandquist

© Springer Nature Switzerland AG 2022

T. Schultheiss, *Radiation Myelopathy*,

https://doi.org/10.1007/978-3-030-94658-6_1

Table 1.1 Lampe's conversion of Boden's limits using Strandquist equation. Boden's large field limit was 3500 r/17 days, and the small field limit was 4500 r/17 days

Treatment duration (days)	Number of fractions at 5/week	r		Gy	
		Large field	Small field		
21	15	3650 (3667)	4700 (4714)	35.2	45.3
28	20	3900 (3906)	5000 (5022)	37.5	48.2
35	25	4100 (4103)	5300 (5275)	39.4	50.6
41	30	4300 (4248)	5500 (5462)	40.8	52.4

1944). This formula (based on the inaccurate Schwarzschild's law) was very speculative and was certainly wrong as an isoeffect formula for late effects such as radiation myelopathy (Kajanti 1994). In terms of a more modern fractionation, Lampe extended the treatment duration to 42 days and estimated 4300–5500 r as safe for large and small fields, respectively (Lampe 1958). Converting these doses in r to Gy, we get approximately 41 and 53 Gy. The complete set of doses reported by Boden and Lampe (based on Boden) is given in Table 1.1. Also in Table 1.1 is shown the number of fractions that would have been given assuming five treatments per week. Finally, the doses converted to Gy are given. What becomes clear is that all of the small field doses were far too aggressive, and the large field doses were too conservative up to about 25 fractions, and the 30-fraction dose is again too conservative. Only the 25-fraction dose for large fields (defined as encompassing the entire cervical cord) was close to today's accepted practice.

It seems then that serendipity was the initial force resulting in a dose limit of about 50 Gy in 25 fractions. The problems with the other dose limits may have been in part a result of the actual fractionation used in the original data, which was unstated, and certainly partially due to the use of the Strandquist isoeffect equation. In 1963, Buschke further confirms 5000 r as a dose limit in an editorial reflecting a presentation given in 1962 (Buschke 1963). Once Fletcher and Million (1965) reported the same number, with the units now changed to rads, the upper limit of 50 Gy to the cervical cord had been firmly established (Fletcher and Million 1965). However, it would be another 10–20 years before it was realized that the overall time factor is of much less importance than the dose per fraction when it comes to late effects rather than acute effects of radiation.

The progression from an accepted safe dose being 50 Gy in 25 fractions to a 5% incidence for 45 Gy in 25 fractions is an interesting story involving a statistical misadventure and contradictory statements.

In 1972 the proceedings of the (1970) sixth Annual San Francisco Cancer Symposium were published as Volume 6 of Frontiers of Radiation Therapy and Oncology titled Radiation Effects and Tolerance, Normal Tissue (Rubin and Casarett 1972). In this issue, Rubin and Casarett made a very important step away from defining and deploying "safe" doses for treating normal organs and toward the utilization of the concept of dose response of normal organs. They defined minimal and maximal tolerance doses as those which would cause a 1–5% or a 25–50% complication rate in a tissue, respectively. So in truth, they were roughly defining two points on the dose-response curve. The minimal tolerance dose for the spinal cord was given as 50 Gy (in keeping with the above history) and the maximal toler- ance dose given as <60 Gy. No references were cited to support these values.

In 1975, the American College of Radiology published Radiation Biology and Radiation Pathology Syllabus edited by Rubin et al. (1975). In the Preface, Rubin thanked Robert Cooper and Theodore Phillips who shared with him "the responsi- bility of editing the entire syllabus." This is significant because the guidance related to RM was based largely on the work of Philips. Philips' work plays an important role in the new dose limit for the spinal cord.

The first chapter contains tables of minimal and maximal tolerance doses for various organs and tissues. In this volume, 45 Gy was given as the minimal toler- ance dose and 55 Gy as the maximal, assuming conventional fractionation of about 10 Gy per week. (The minimal and maximal tolerance doses had now been rede- fined to be the doses associated with 5% and 50% complication rates, respectively.) The references cited in this table to support these numbers are Philips and Buschke, "Radiation tolerance of the thoracic spinal cord," and Wara, Philips, Sheline, and Schwade, "Radiation tolerance of the spinal cord" (Phillips and Buschke 1969; Wara et al. 1975). In the chapter of this book devoted to radiation injury to the spinal cord, the dose associated with a 5% incidence of complication is given as 50 Gy, but it goes on to say "most treatment plans introduce shielding after 4500 rads at 200 rads per fraction."[1]

The paper by Phillips and Buschke was published shortly after Ellis introduced the Nominal Standard Dose (NSD) concept, discussed below (Ellis 1969). These authors deployed the NSD in their analysis, but just months after its introduction, they were correctly raising the point that it was not sufficiently sensitive to fraction- ation effects, that is, the impact of few fractions and consequently higher doses per fraction was not adequately handled by the NSD formula. They used scattergrams of total dose versus overall treatment time and number of fractions (plotted on a log-log scale) to determine safe dose regimens. Although they quoted some values of the incidence of RM versus dose, this is where they made a wrong turn. They combined their cases and controls from UCSF with nine cases only, no controls, from two other studies, Atkins and Tretter and Locksmith and Powers (Atkins and Tretter 1966; Locksmith and Powers 1968). Obviously, this yields incorrect values of the frequencies of RM versus dose and is an egregious example of selection bias. Nonetheless, the dose limits in this study were based on the log-log plots and not any sort of dose-response analysis.

An important side topic arises here—that of treating all fields every day. The authors noted the importance of the number of fractions and by implication the size of the dose per fraction. However, they did not make the connection that by treating alternate fields on alternate days, large and small doses per fraction were given on alternate days rather than equal dose per fraction every day. Both Atkins and Tretter

---

[1] It is further stated that the thoracic cord is more radiosensitive than the cervical cord, citing Kramer and Philips and Buschke. Although the latter reference does not address the relative sensi- tivity of the cervical and thoracic spinal cord, it may be assumed that Philips must have agreed with this conclusion since he edited (at the very least) this chapter.

and Locksmith and Powers treated their anterior and posterior fields using a single field per day and not both fields each day.

Wara et al. expand the UCSF data by including more years and by adding additional studies from the literature (Wara et al. 1975). Again, only cases without controls were added, but this time they also perform a dose-response analysis, and they modify the constants in the NSD formula based on scattergrams using their data combined with the cases from the literature. Using a "probit analysis," they determined the doses that yield a 1% and 50% complication rate. From these doses one can calculate the dose that gives a 5% complication rate. It is 44.56 Gy. Perhaps this is where Rubin's value of 45 Gy originates. However, it is useful now to note that this value relies on flawed analysis that combined cases and controls from one institution with cases alone from three other reports, takes no account of treating with single fields per day, and does not account for the very large difference between the crude complication rate and the actuarial complication rate when the data come from a lung cancer cohort (Dische et al. 1981; Schultheiss et al. 1986). Nonetheless, 45 Gy became the standard dose limit for decades and is probably the most widely quoted spinal cord standard. The notion that 45 Gy produces a 5% incidence of RM, however, has been completely dispelled.

In a nasopharynx cancer clinical trial from 1971 (RTOG 71-02), the dose limit to the spinal cord was 50 Gy in 25 fractions. In 1973 (RTOG 73-02 lung cancer), the dose limit was 45 Gy in 25 fractions. Interestingly, Philip Rubin was the Protocol Design Chair for this trial. With very few exceptions, e.g., RTOG 93-09, the dose limits in RTOG trial remained at 45 Gy in 22–25 fractions throughout the rest of the century. Currently the dose limit is typically 48 Gy in 2 Gy fractions, but doses as high as 50 Gy are considered an acceptable variation. This transition occurred certainly by 2006, but not much earlier.

The final word, as of this writing, on the treatment planning dose limits to the spinal cord should be this. Do not irradiate the cord to a higher dose than necessary. This could easily be as low as 30 Gy. Furthermore, if a field reduction is planned, consideration should be given to implementing it as a concomitant (concurrent) boost. This technique minimizes the dose per fraction to all normal tissues, thereby further reducing the risk of injury. It is not the best practice to treat a region to what is considered the spinal cord limit and then reduce the fields to spare the cord. By treating the target volume with the boost included from the beginning of treatment, the spinal cord is irradiated at every fraction, ultimately to the same total dose, but with a lower dose per fraction. This technique will enhance the repair of occult damage in case there is a need to retreat at a future date. If a low dose to the cord is practicable, it would greatly facilitate retreatment if required for recurrent or new disease.

If a relatively high dose to the spinal cord is unavoidable if the tumor is to be treated definitively, the complication rate after 50 Gy in 2 Gy fractions is certainly less than 1%. Of course, the complication is devastating.

1.2 The Advent of New Technologies

It has been said that the advent of Co-60 units in the late 1950s also harbingered an increase in radiation myelopathy possibly resulting from an increase in the use of more intense treatments (fewer fractions). Perhaps that is true since large doses per fraction were not commonly employed with kilovoltage treatments and the disproportionate increase in late complications with high doses per fraction had not been established. However, it is also true that the technological improvement in the delivery of dose afforded by Co-60 machines resulted in (or at least coincided with) a dramatic increase in the number of patients receiving radiation therapy.

Another common belief among some practitioners of the mid-twentieth century was that the rapid deployment of linear accelerators once they became commercially available had the indirect effect of increasing radiation myelopathy occurrences. Large abutting fields were often treated with Co-60 without using a gap between field edges since the large source size and lower photon energy resulted in a broad penumbra that effectively smeared the dose at the field edges. However, the sharper penumbra of the linear accelerator would result in a significant hot spot if adjacent fields were not adequately gapped. The possibility of myelopathies due to field overlap was mentioned in a number of studies.

1.3 Hypofractionation

In the early-twentieth century, fractionation was commonly specified by the number of treatment days with the specific number of fractions often omitted as being understood or irrelevant. By the 1960s, the importance of the number of fractions in which a therapeutic dose was given was well appreciated. The NSD formula was proposed by Ellis as a mathematical model that demonstrated the interaction among total dose, number of fractions, and overall treatment time in dose regimens that produced equivalent outcomes, i.e., were isoeffective (Ellis 1969). This formula was a substantial improvement over the similar power law model of Strandquist that only included overall treatment time and not number of fractions. For nearly 20 years following its publication, the NSD formula and its much more accessible (but mathematically equivalent) successor, the TDF formula, informed clinical practice and guided protocol development (Orton and Ellis 1973). Unfortunately, the NSD was developed using data from acute normal skin reactions and from skin cancers only. In other words, it did not appropriately manage the fractionation effects seen in late normal tissue reactions. The disasters that subsequently resulted from the application of the NSD model represent a fundamental error of statistical analysis—that of applying the results of analysis to situations outside the sphere from which the data were taken.

For various reasons, including scarceness of resources, patient convenience, costs, etc., there was interest in the late 1960s and well into the 1970s in shortening courses of radiation, especially for sicker patients. To achieve equivalent results, the NSD formula was often deployed, at least as a check, to estimate hypofractionated doses that would produce approximately the same effect as the conventionally fractionated course. This exercise was reasonably effective in many cases, but not for late effects. The hypofractionated doses calculated by using the NSD produced more tissue damage in late responding tissues than would have been predicted if the dose regimens were isoeffective for both late effects and acute effects. As this lesson was learned, experimental studies confirmed that the exponents for late versus acute effects in an NSD-like formula would need to be different (Wara et al. 1973). The exponent for N increased for late effects and the exponent for T decreased, possibly to zero. The literature is unfortunately rich with reports of radiation myelopathy resulting from doses per fraction of 4–6 Gy that were expected to be safe according to the NSD. A very common example from this period was used in the treatment of lung cancer: 20 Gy in 5 fractions followed by a 2–4-week break and then another 20 Gy in 5.

1.4　The Spaghetti Plot: Not Culinary Intrigue, But Rather a Watershed in Radiobiology

As it became apparent that the exponent of N in a power law isoeffect formula like the NSD was dependent on the type of end point being observed (early versus late), a new model began to supplement, and then to supplant, these power law models. The model was the linear-quadratic (LQ) model of cell survival, which began being applied to clinical applications around 1980. For many years, cell survival experiments were used as a tool to elucidate the underlying mechanisms of radiation damage at the cellular level. A number of hypotheses regarding radiation action were put forward, and the different mechanisms generally resulted in different mathematical forms of the cell survival curve. In the end, the LQ model displaced all the others, probably because of its simplicity and its ability to be extended to clinical applications.

In the LQ model of cell survival, the fraction of surviving cells after a single dose d of radiation is given by

$$S(d) = \exp\left(-\alpha d - \beta d^2\right). \tag{1.1}$$

Very briefly, during the late 1970s and early 1980s, it was assumed that clinical radiation responses were the result of the death of specific target cells. For tumors, the target cells were clonogens; for normal tissues, they were often assumed to be specific parenchymal or stromal cells. Taking a clonogenic cell as the target cell for tumor control is essentially just restating the definition of clonogen. However, it is probably a gross oversimplification to assume that normal tissue responses, especially late effects, can be modeled as the depletion of a single population of cells to a given level. It is even more unlikely that numerical results from an *in vitro* cell survival model would be applicable to the clinical expression of a late injury months or years after treatment.

Nonetheless, the LQ model has proven to be widely versatile as an isoeffect equation. For N equal fractions of dose d, Eq. (1.1) can be manipulated to yield

$$\ln(S)/\beta = \alpha/\beta \cdot D + Dd \tag{1.2}$$

where D is the total dose ($d \times N$). In the clinical application of the LQ model, it is assumed that values of D and d for which the right-hand side of Eq. (1.1) is constant represent a constant level of injury for which the value of α/β has been determined. Thus if the value of α/β is known, then equivalent results for two different treatment regimens defined by (D_1, d_1) and (D_2, d_2) will be achieved if

$$\alpha/\beta \cdot D_1 + D_1 d_1 = \alpha/\beta \cdot D_2 + D_2 d_2 \tag{1.3}$$

In 1982, Thames et al. published the very important discovery that the values of α/β for tumors and acute normal tissue effects are almost invariably larger than those for late normal tissue effects (Thames et al. 1982). Figure 1 from that paper that illustrates this discovery is shown here in Fig. 1.1. This famous figure has been dubbed the "spaghetti plot" by its creators. The important consequence illustrated in this figure is that for late effects, a greater level of tissue damage results as one reduces the number of fractions in a treatment regimen and keeps the acute effects (or tumor response) constant. This explains why so many attempts at hypofractionation resulted in unexpectedly high incidences of RM.

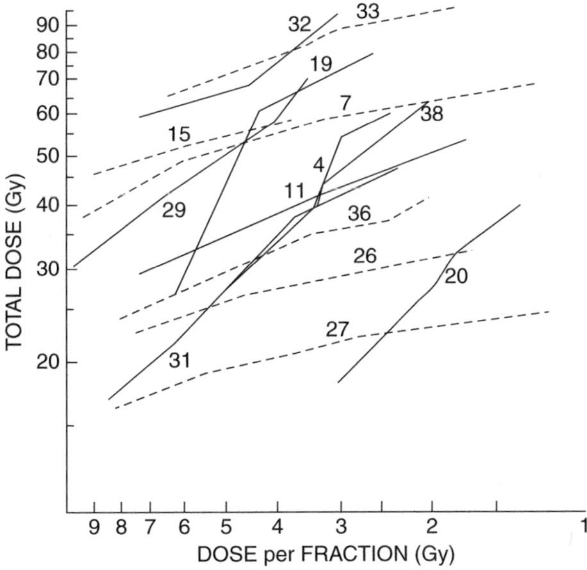

Fig. 1.1 "Total dose for isoeffect as a function of dose per fraction. Numbers refer to sources of data for early (dashed lines) and late (solid lines) effects (relevant tissues are identified with references in Table 3). The curves describing late isoeffects have steeper slopes than those for early isoeffects, with the implication for the dose-survival curves of the underlying target cells that the ratio β/α of parameters of the linear-quadratic survival model is greater for late responses." The Spaghetti plot from Thames et al. (1982). Reprinted from International Journal of Radiation Oncology*Biology*Physics, Vol 8/2, Howard D Thames, H Rodney Withers, Lester J Peters, Gilbert H Fletcher, Changes in early and late radiation responses with altered dose fractionation: Implications for dose-survival relationships, 219–226, Copyright (1982), with permission from Elsevier.

"Obviously" if hypofractionation decreased the (undefined) therapeutic ratio, then hyperfractionation should increase it. That is, by deploying hyperfractionation it was hoped to increase tumor control without increasing late normal tissue damage. To a large degree this strategy worked, but of course an increase in acute effects was seen and had to be managed. The logistical solution to using fraction sizes of about 1 Gy was to treat with two fractions per day. To reduce the treatment time, sometimes treatment was delivered in three daily fractions. As will be discussed later, this stratagem resulted in a surprising rate of RM.

1.5   Pathogenesis and Treatment

Radiation myelopathy was identified early as a demyelinating process (Stevenson and Eckhardt 1945), and it was understood that most demyelinating diseases were a result of or closely associated with inflammation. This led to the early and persistent use of the term "radiation myelitis," even though inflammation as a primary actor in the pathogenesis of RM was not established. Presciently, Boden suggests "that a sufficiently high dose could cause direct necrosis of nerve tissue, and a somewhat lower dose might produce slowly developing blood vessel changes, leading later to focal necrosis" (Boden 1948). In broad strokes, this hypothesis was validated during the second half of the twentieth century in both humans and experimental animals; however, necrosis of nerve tissue is not currently understood to occur in the absence of vascular injury. Nonetheless, lesions characterized by demyelination, gliosis, and white matter necrosis are reported at higher doses and with shorter latency than those whose primary or sole pathological features include vascular damage with parenchymal damage secondary to it (Schultheiss et al. 1984; van der Kogel 1979). Not only was this parenchymal-versus-vascular pathogenesis debated for RM, these alternative pathways became the subject of dispute for nearly all late effects throughout much of the history of clinical radiobiology. This is another example of how RM has served as the archetype for late radiation injury. Chapter 2 is devoted to the pathogenesis of RM. Interestingly, Boden goes on to say that it is of "academic interest only . . . to elucidate the precise intermediate pathological changes that connect the radiation and the damage to the nervous tissue."

This chapter ends with a mixed status report and outlook. Encouragingly, the technology of radiation treatment planning and delivery has advanced to the point that the risk of radiation myelopathy is avoidable in all but the rarest of situations. Even the potential for mistakes has been diminished as a result of improved safety programs (American Society for Radiation Oncology 2019). Of course avoiding a complication is preferable to treating it, even if the treatment is successful. Treatment for RM has been the subject of study and many editorials, but very limited success. Traditional methods of hyperbaric oxygen and corticosteroids have had both limited and short-lived success. Even when treatments are discovered that may have a benefit, "the route and timing of drug administration relative to irradiation appear to be critical in determining the type and magnitude of the combined effect" (Ang and Stephens 1994). A number of experimental studies are addressed in Sect. 5.3.

Recently attention has turned to growth factors and other cytokines (Nieder et al. 2002; Andratschke et al. 2004; Nieder et al. 2004; Andratschke et al. 2005; Nieder et al. 2005; Nieder et al. 2007). Potential subjects include IGF-1, bFGF, PDGF, VEGF and inhibitors, and the more traditional WR-2721 (amifostine). Standard dose modifying factors for amifostine are not impressive, but Spence et al. quote a value of 1.3 (Spence et al. 1986). This however was derived in an *ad hoc* manner and does not represent the standard understanding of what DMF means. More recently attention has focused on the role of bevacizumab in combating radiation-induced CNS necrosis. Some evidence suggests that bevacizumab can improve the radiological appearance of RM lesions, but it does not have a dramatic effect on clinical outcomes (Psimaras et al. 2016; Tye et al. 2014). In addition to Ang and Stephen's statement regarding timing of drug administration relative to irradiation, the same can be said of drug administration relative to the onset of symptoms, which is probably implicitly included in their statement. Ace inhibitors have been suggested as a possible CNS radioprotector, and two attempts have been made to demonstrate an effect in RM (Clausi et al. 2018; Saager et al. 2020). In Chap. 5 it will be shown that both of these studies failed to produce evidence of an effect on the dose-response curve. Rezvani et al. transplanted neural stem cells into the spinal cords of rats 90 days after irradiation with 22 Gy, a dose that caused RM in 12/12 animals in the control group. The RM-free survival was significantly improved for the transplanted animals. Histological evidence indicated that demyelination or white matter necrosis was not observed in 89% of the transplanted animals.

At this point there is no approved treatment for frank or impending RM. The best candidates for investigation appear to be VEGF inhibitors or other methods to reduce blood-brain barrier breakdown and bevacizumab or related compounds. Neural stem cell transplantation is intriguing, but the single study of Rezvani et al. must be repeated and many more studies conceived and performed.

References

Ahlbom HE. The results of radiotherapy of Hypopharyngeal cancer at the Radiumhemmet, Stockholm, 1930 to 1939. Acta Radiol. 1941;22(1–2):155–71. https://doi.org/10.3109/00016924109171582.

Alajouanine T, Lhermitte F, Cambier J, Gautier JC. Late post-radiotherapy lesions of the central nervous system (apropos of an anatomo-clinical observation of cervical myelopathy). Rev Obstet Ginecol Venez. 1961;105:9–21.

American Society for Radiation Oncology (2019) Safety is no accident: A framework for quality radiation oncology care. Available from https://www.astro.org/ASTRO/media/ASTRO/Patient%20Care%20and%20Research/PDFs/Safety_is_No_Accident.pdf.

Andratschke NH, Nieder C, Price RE, Rivera B, Ang KK. Potential role of growth factors in diminishing radiation therapy neural tissue injury. Semin Oncol. 2005;32(SUPPL. 3):67–70. https://doi.org/10.1053/j.seminoncol.2005.03.012.

Andratschke NH, Nieder C, Price RE, Rivera B, Tucker SL, Ang KK. Modulation of rodent spinal cord radiation tolerance by administration of platelet-derived growth factor. Int J Radiat Oncol Biol Phys. 2004;60(4):1257–63. https://doi.org/10.1016/j.ijrobp.2004.07.703.

Ang KK, Stephens LC. Prevention and management of radiation myelopathy. Oncology. 1994;8(11):71–6.

Atkins HL, Tretter P. Time-dose considerations in radiation myelopathy. Acta Radiol. 1966;5(1):79–94. https://doi.org/10.3109/02841856609139546.

Boden G. Radiation myelitis of the cervical spinal cord. Br J Radiol. 1948;21(249):464–9. https://doi.org/10.1259/0007-1285-21-249-464.

Buschke F. Predetermination of avoidable or unavoidable radiation hazards and complications-calculated risks. Am J Roentgenol Radium Ther Nucl Med. 1963;89(3):649–53.

Clausi MG, Stessin AM, Tsirka SE, Ryu S. Mitigation of radiation myelopathy and reduction of microglial infiltration by Ramipril. ACE Inhibitor Spinal Cord. 2018;56(8):733–40. https://doi.org/10.1038/s41393-018-0158-z.

Dische S, Martin WMC, Anderson P. Radiation myelopathy in patients treated for carcinoma of bronchus using a six fraction regime of radiotherapy. Br J Radiol. 1981;54(637):29–35. https://doi.org/10.1259/0007-1285-54-637-29.

Dynes JB. Radiation myelopathy. Trans Am Neurol Assoc. 1960;85:51–5.

Ellis F. Dose, time and fractionation: a clinical hypothesis. Clin Radiol. 1969;20(1):1–7.

Fletcher G, Million RR. Malignant tumors of the nasopharynx. Am J Roentgenol Radium Therapy Nucl Med. 1965:93:44–55.

Innes JRM, Carsten A. Demyelinating or Malacic myelopathy: a delayed effect of localized X-irradiation in experimental rats. Arch Neurol. 1961;4(2):190–9. https://doi.org/10.1001/archneur.1961.00450080072008.

Kajanti MJ. Magnus Strandqvist: 50th anniversary of his doctoral thesis. Acta Oncol. 1994;33(7):735–8.

Lampe I. Radiation tolerance of the central nervous system. Prog Radiat Ther. 1958;1:224–36.

Locksmith JP, Powers WE. Permanent radiation myelopathy. Am J Roentgenol Radium Therapy, Nucl Med. 1968;102(4):916–26.

Nieder C, Andratschke N, Astner ST. Experimental concepts for toxicity prevention and tissue restoration after central nervous system irradiation. Radiat Oncol. 2007;2(1) https://doi.org/10.1186/1748-717X-2-23.

Nieder C, Andratschke N, Price RE, Rivera B, Ang KK. Innovative prevention strategies for radiation necrosis of the central nervous system. Anticancer Res. 2002;22(2A):1017–23.

Nieder C, Andratschke NH, Wiedenmann N, Molls M. Prevention of radiation-induced central nervous system toxicity: a role for amifostine? Anticancer Res. 2004;24(6):3803–9.

Nieder C, Price RE, Rivera B, Andratschke N, Ang KK. Effects of insulin-like growth factor-1 (IGF-1) and amifostine in spinal cord reirradiation. Strahlenther Onkol. 2005;181(11):691–5. https://doi.org/10.1007/s00066-005-1464-x.

Orton CG, Ellis F. A simplification in the use of the NSD concept in practical radiotherapy. Br J Radiol. 1973;46(547):529–37.

Pallis CA, Louis S, Morgan RL. Radiation myelopathy. Brain. 1961;84(3):460–79. https://doi.org/10.1093/brain/84.3.460.

Phillips TL, Buschke F. Radiation tolerance of the thoracic spinal cord. Am J Roentgenol Radium Therapy, Nucl Med. 1969;105(3):659–64.

Psimaras D, Tafani C, Ducray F, Leclercq D, Feuvret L, Delattre JY, Ricard D. Bevacizumab in late-onset radiation-induced myelopathy. Neurology. 2016;86(5):454–7. https://doi.org/10.1212/WNL.0000000000002345.

Rubin P, Casarett G. Direction for clinical radiation pathology. The tolerance dose. Front Radiat Ther Oncol. 1972;6(1–16):6.

Rubin PJ, Coper R, Phillips TL. Radiation biology and radiation oncology syllabus (set RT 1: radiation oncology). Chicago, IL: American College of Radiology; 1975.

Saager M, Hahn EW, Peschke P, Brons S, Huber PE, Debus J, Karger CP. Ramipril reduces incidence and prolongates latency time of radiation-induced rat myelopathy after photon and carbon ion irradiation. J Radiat Res. 2020;61(5):791–8. https://doi.org/10.1093/jrr/rraa042.

Schultheiss TE, Higgins EM, El-Mahdi AM. The latent period in clinical radiation myelopathy. Int J Radiat Oncol Biol Phys. 1984;10(7):1109–15. https://doi.org/10.1016/0360-3016(84)90184-6.

Schultheiss TE, Thames HD, Peters LJ, Dixon DO. Effect of latency on calculated complication rates. Int J Radiat Oncol Biol Phys. 1986;12(10):1861–5. https://doi.org/10.1016/0360-3016(86)90331-7.

Spence AM, Krohn KA, Edmondson SW, Steele JE, Rasey JS. Radioprotection in rat spinal cord with WR-2721 following cerebral lateral intraventricular injection. Int J Radiat Oncol Biol Phys. 1986;12(8):1479–82. https://doi.org/10.1016/0360-3016(86)90198-7.

Stevenson LD, Eckhardt RE. Myelomalacia of the cervical portion of the spinal cord, probably the result of roentgen therapy. Arch Pathol. 1945;39(2):109–12.

Strandquist M. A study of the cumulative effects of fractionated X-ray treatment based on the experience gained at Radiumhemmet with the treatment of 280 cases of carcinoma of skin and lip. Acta Radiol. 1944;55:300.

Thames HD, Rodney Withers H, Peters LJ, Fletcher GH. Changes in early and late radiation responses with altered dose fractionation: implications for dose-survival relationships. Int J Radiat Oncol Biol Phys. 1982;8(2):219–26. https://doi.org/10.1016/0360-3016(82)90517-X.

Tye K, Engelhard HH, Slavin KV, Nicholas MK, Chmura SJ, Kwok Y, Ho DS, Weichselbaum RR, Koshy M. An analysis of radiation necrosis of the central nervous system treated with bevacizumab. J Neuro-Oncol. 2014;117(2):321–7. https://doi.org/10.1007/s11060-014-1391-8.

van der Kogel AJ. Late effects of radiation on the spinal cord; dose-effect relationships and pathogenesis. RBI; 1979.

Verjaal A. Radiation myelopathy. Ned Tijdschr Geneeskd. 1964;108:1123–7.

Wara WM, Phillips TL, Margolis LW, Smith V. Radiation pneumonitis: a new approach to the derivation of time-dose factors. Cancer. 1973;32(3):547–52.

Wara WM, Phillips TL, Sheline GE, Schwade JG. Radiation tolerance of the spinal cord. Cancer. 1975;35(6):1558–62. https://doi.org/10.1002/1097-0142(197506)35:6<1558::AID-CNCR2820350613>3.0.CO;2-7.

Pathology and Pathogenesis

2

2.1 Some Spinal Cord Anatomy

Figure 2.1 shows a basic illustration of the gross anatomy of the spinal cord (Cota 2019). The gray matter consists of neuron cell bodies and supporting vasculature as well as glial cells and unmyelinated axons, all in a butterfly or "*H*" pattern as seen in cross section. The gray matter is surrounded by white matter consisting of the axons, their myelin sheaths (cytoplasmic processes of oligodendrocytes), supporting vasculature, and other glial cells (predominantly astrocytes in the healthy cord). Despite the stellite morphology of astrocytes, the "stars" of the white matter are the nerve fibers (axons), surrounded by the myelin sheaths.

Nerve roots (bundles of axons) exit dorsally or ventrally at every vertebral level. The bulge seen on the dorsal root is the dorsal root ganglion. The oligodendrocytes exist only in the central nervous system, and as the nerve roots exit the spinal cord there is a transition from oligodendrocytes to Schwann cells as the myelinating cells

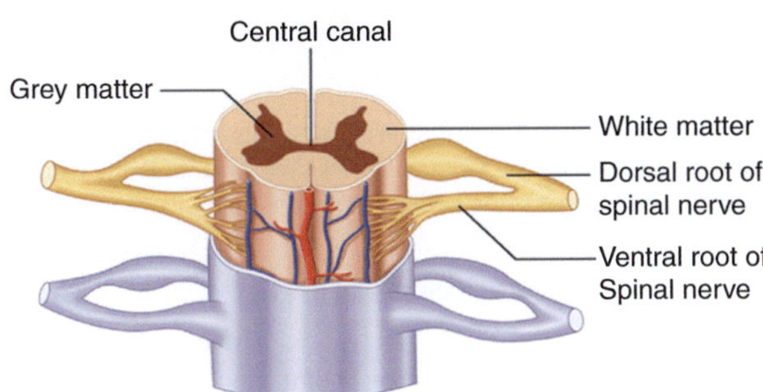

Fig. 2.1 Basic anatomy of the spinal cord. Reprinted with permission from Springer Nature: Springer Publishing Co., Spinal Cord Anatomy by Alan Gonzalez Cota in Deer's Treatment of Pain: Timothy R. Deer, Jason E. Pope, Tim J. Lamer, David Provenzano (Eds), Copyright 2019.

© Springer Nature Switzerland AG 2022
T. Schultheiss, *Radiation Myelopathy*,
https://doi.org/10.1007/978-3-030-94658-6_2

for the axons, which remains true for all peripheral axons. Also depicted in this picture is the anterior spinal artery, which traverses the length of the cord (red), and anterior spinal veins, which are part of the anterior coronal venous plexus (blue) that loosely surrounds the cord on its surface.

The dorsal root contains the afferent nerves sending sensory information to the brain. The dorsal root ganglion contains the cell bodies of those neurons. The ventral root contains the efferent nerve fibers that send motor information to the muscles.

Figure 2.2 shows some of the major spinal tracts in the white matter (Cota 2019). Table 2.1 gives the function of these tracts. Loss of myelin (demyelination) occurs in a very broad spectrum of injury and disease and is fundamentally uninformative with respect to etiology or pathogenesis. Of course it is a hallmark of RM as well. In addition to demyelination and primary white matter changes, various vasculopathies, which are seen in other conditions as well, also occur in the white matter in RM. Although changes are not strictly limited to the white matter in all cases,

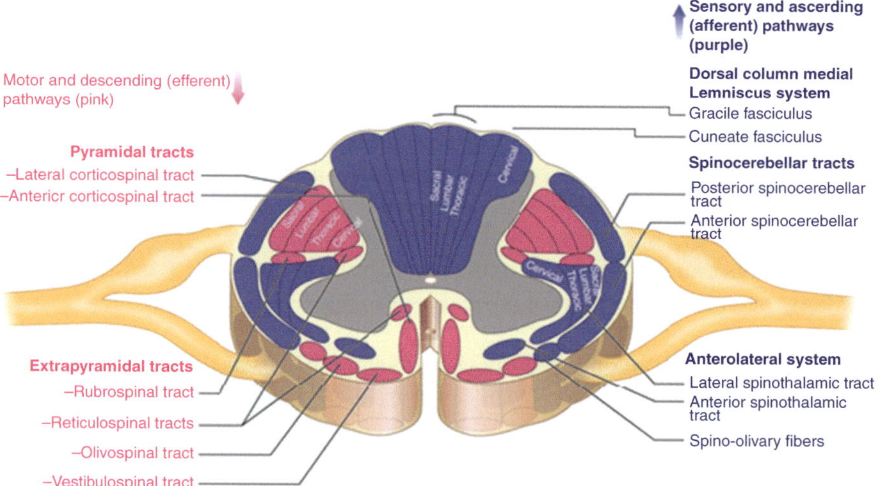

Fig. 2.2 Spinal tracts. Reprinted with permission from Springer Nature: Springer Publishing Co., Spinal Cord Anatomy by Alan Gonzalez Cota in Deer's Treatment of Pain: Timothy R. Deer, Jason E. Pope, Tim J. Lamer, David Provenzano (Eds), Copyright 2019.

Table 2.1 Spinal tracts and systems and their function

Tracts	Function
Pyramidal tracts	Voluntary control of the musculature of the body and face
Extrapyramidal tracts	Involuntary and automatic control of all muscles for muscle tone, balance, posture, and locomotion
Dorsal column medial lemniscus system	Fine touch (tactile sensation), vibration, and proprioception
Spinocerebellar tracts	Proprioception
Anterolateral system	Sensory loss for pain and temperature

incursion into the gray matter may result from the expansion of a lesion that originated in the white matter.

Because the inflammatory/immune response is important in RM, the lymphatic system of the spinal cord must be addressed, albeit in an abbreviated manner. Writing in 2013, Iliff and Nedergaard describe for the first time a system by which waste products are flushed from the CNS in a process similar to that in other organs with more obvious lymphatic channels (Iliff and Nedergaard 2013). They hypothesize that "CSF passes through the para-arterial space surrounding arteries; the space is bound by the albuminal surface of the blood vessel and the apical processes of astrocytes. Water channels called aquaporin 4 (AQP4) on the vascular endfeet of astrocytes (Nagelhus et al. 2004) facilitate convective flow out of the para-arterial space and into the interstitial space (see the Fig. [2.3]). As CSF exchanges with the ISF, vectorial convective fluxes drive waste products away from the arteries and toward the veins. ISF and its constituents then enter the paravenous space. As ISF [interstitial fluid] exits the brain through the paravenous route, it reaches lymphatic vessels in the neck, and eventually returns its contents to the systemic circulation (Nedergaard 2013)."

Some significant amendments to this process have been proposed, especially regarding the role of the AQP4 water channels (Abbott et al. 2018; Dupont et al. 2019). However, the component especially applicable to RM is the paravenous route for egress of ISF containing proteins, toxins, etc., and this has been largely unchallenged. (Many authors are using the term "peri-vascular" in preference to

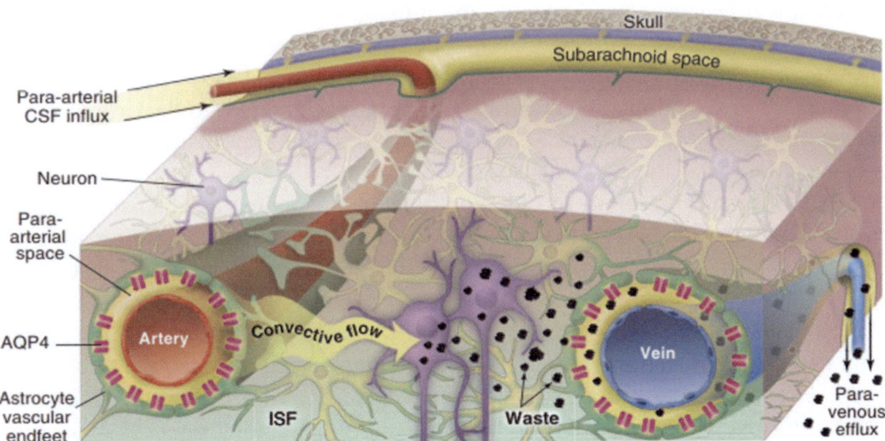

Go with the flow. Convective glymphatic fluxes of CSF and ISF propel the waste products of neuron metabolism into the paravenous space, from which they are directed into lymphatic vessels and ultimately return to the general circulation for clearance by the kidney and liver.

Fig. 2.3 An illustration of the glymphatic system, parts of which have been challenged conceptually. What has not been challenged and is of importance to radiation injury is the waste drainage to venules and veins, which can be impaired if the blood-brain barrier is disrupted leaving toxic metabolic byproducts unflushed in the white matter. From Nedergaard (2013). Reprinted with permission from AAAS.

"paravascular" to emphasize that this space surrounds the vasculature rather than lies beside it.) Although Nedergaard titled his article "Garbage truck of the brain," it is actually more like the sewage system.

2.2   Pathology

In symptomatic RM, the necropsied spinal cord from a human or experimental animal may be either swollen or atrophic. In dissected cross sections, the cord may have areas that are grossly discolored by necrosis and hemorrhage, indicating a more advanced stage of injury. In humans, pathological studies of the irradiated spinal cord are always limited to those cases that were symptomatic with neurological dysfunction and therefore with more advanced lesions. Animal studies have provided a clearer picture of the progression of lesions as well as an exploration of the relationship between pathological changes and the factors of dose and fractionation. It is through the detailed studies of several authors that we may extrapolate from the results in experimental animals to the findings in humans and ask whether it is appropriate to project the results from animal studies to humans. The answer is, "Not always."

Of course the earliest published material comes from human autopsies in the era when radiation myelopathy first became a clinical concern and subject of academic investigation. Several forms and causes of infectious myelitis were known at this time and thus the term "myelitis" was initially attached to radiation myelopathy. Indeed, it was because the response to the spinal cord to injury is limited in its variety that it can be said that no pathognomonic characteristics have been identified for the lesions of RM (or of radiation injury generally (Fajardo 1989)). Like many CNS diseases, injuries, and insults, there is a nearly strict limitation to the expression of injury within the white matter. Selective damage to the gray matter would in fact exclude radiation as the causative agent of injury.

Since Boden's first report of radiation myelopathy, a dual mechanism of injury has been hypothesized for RM (Boden 1948). In Boden's words, "a sufficiently high dose could cause direct necrosis of nerve tissue, and a somewhat lower dose might produce slowly developing blood vessel changes, leading later to focal necrosis." This explanation was at first merely hypothetical, but as material from autopsies and from experimental studies were published, this hypothesis was proven to be accurate, even if the pathogenesis remained elusive. van der Kogel and colleagues were the first to investigate systematically the relationships among latency, dose, fractionation, and type of lesion (van der Kogel 1977a, b, 1979; van der Kogel et al. 1982), but many others published closely related studies contemporaneously with van der Kogel's laboratory (Hubbard and Hopewell 1978; White and Hornsey 1978; Geraci 1979; Gibbs 1979; Leith et al. 1981).

In one rat strain, van der Kogel showed that white matter necrosis with little apparent vascular damage occurred sooner and at higher doses than vascular injuries of increased capillary density, telangiectasis, and hemorrhage (van der Kogel AJ 1980). Using published autopsy material combined with original studies on

Table 2.2 Spinal cord changes in radiation myelopathy (from Schultheiss et al. 1988)

A. White matter lesions	B. Vasculopathies[a]	C. Glial reactions

1. Demyelination: Isolated nerve fibers
2. Demyelination: Groups of nerve fibers (spongiosis)
3. "Inactive" malacia
  a. Spongiosis spheroids
  b. Scar
4. "Active" malacia
  a. Coagulative malacia
  b. Liquefactive malacia
    • Amorphous
    • Foam cell fields
    • Cystic
5. Hemorrhagic malacia
	1. None
	2. Increased vascularity
	3. Telangiectasis
	4. Hyaline degeneration and thickening
	5. Edema and fibrin exudation
	6. Peri-vascular fibrosis and inflammation
	7. Vasculitis
	8. Fibrinoid necrosis
	9. Thrombosis
	10. Hemorrhage
		1. Microglia/macrophages
		  a. Morphology
		    • Rod-shaped
		    • Foam cells
		    • Multinucleated
		  b. Patterns
		    • Diffuse
		    • Focal
		    • Peri-vascular
		2. Astrocytes
		  a. Morphology
		    • Inconspicuous
		    • Gemistocytic
		    • Fibrillary
		  b. Patterns
		    • Diffuse
		    • Focal

[a] With or without endothelial alterations of hyperchromasia, hypertrophy, anisokaryocytosis, and hyperplasia

rhesus monkeys, Schultheiss et al. found supporting results (Schultheiss et al. 1988; Schultheiss and Stephens 1992b). Schultheiss, Stephens, and Maor organized lesions by white matter lesions, vasculopathies, and glial reactions and then by severity as shown in Table 2.2 (Schultheiss et al. 1988). In this table, the sequence of elements does not represent an expected temporal sequence and the severity is related to the level of local changes, not to the level of consequent morbidity. The morbidity is mostly related to the size and location of the lesion (s). This categorization should be credited primarily to the renowned pathologist, L. C. Stephens, DVM, PhD.

White matter lesions listed in Table 2.2a refer primarily to changes of or that affect the axons and their myelin sheaths. Consequently, the mildest change and one that is likely to result in no symptoms is isolated demyelination of nerve fibers. However, demyelination of groups of nerve fibers can lead to symptoms. In coagulative necrosis, in contrast to liquefactive necrosis, the dead cells remain visible and structurally intact. Note that the term "malacia" and "necrosis" are often used loosely, but rather inaccurately as interchangeable. Malacia is a softening of tissue brought about because of the death and phagocytosis of cells, whereas necrosis refers to the actual death of those cells. Inactive malacia refers the regions where necrosis has stabilized and the remnants alone remain. Blood may fill a necrotic area and obscure the underlying white matter or occupy a cystic space.

In Table 2.2b, we see that there can be cases of RM where vasculopathies are absent. This can occur due to sampling error, timing of the necropsy relative to symptoms, or changes that are not apparent at the light microscopic level. All other vasculopathies listed are seen in radiation damage to essentially all normal tissues. Although their occurrence in RM is not unique, the consequent changes to the parenchyma and the accompanying symptoms are obviously tissue dependent. The vascular changes most often seen in human and experimental sections of irradiated spinal cord are endothelial alterations, telangiectasia, hyalinosis, and fibrinoid necrosis.

Finally in Table 2.2c, we see the glial reactions of the irradiated spinal cord. In this original version of this table, there are no changes listed for oligodendrocytes. This is because oligodendrocyte cell bodies are much less frequently seen than astrocytes and the major alteration involving oligos are observed in the myelin sheaths.

Microglia and astrocytes are components of the normal CNS and often evince vigorous reactions to radiation. Gliosis, proliferation or hypertrophy of glial cells in response to injury in the CNS, can be diffuse or focal. Focal gliosis may be concentrated in areas of white matter destruction, sometimes entirely replacing nerve fibers. One may also find a peri-vascular pattern of reactive glial cells.

Since the publication of these tables, the field of neuro-inflammation has seen tremendous growth, starting around 1990. It seems clear that a better understanding of the pathogenesis of RM is possible by adding the current understanding of the initial endothelial and peri-vascular targets (Coderre et al. 2006; Hopewell and van der Kogel 1999), the role of neuro-inflammation in other demyelinating lesions (Sosa and Smith 2017), and the potential contribution of various chemokines, cytokines, other pro-inflammatory molecules (Clausi et al. 2018) to the traditional understanding of the pathology of RM (Wong et al. 2015). A new interdisciplinary contribution on this subject would be very useful to the RM literature.

2.3   Pathogenesis

Pathogenesis is the route by which a disorder develops. Since the human spinal cord must be assessed almost exclusively by noninvasive or indirect methods, the pathogenesis of RM must be determined by animal models.

The pathogenesis of radiation injury is generally easier to study than its dose response. In experimental animals, the pathogenesis can be studied by irradiating animals to a dose known to elicit the response and then necropsying the animals at various times after treatment. Of course many variants may be added to this basic design. Obviously in humans, one is limited to confirming the animal findings by studying autopsy material. Naturally, it is generally assumed that the pathogenesis does not depend on fractionation and usually that it does not depend upon the amount by which the tolerance dose is exceeded. If these assumptions are true, single doses work as well as fractionated doses to study pathogenesis. However, single doses are not much help when studying clinically relevant dose response. In radiation myelopathy, these assumptions are almost true enough so that the study of the pathogenesis is not significantly complicated by whether or not one uses fractionated doses.

2.3.1 Cervical and Thoracic Levels

The majority of the early studies in rats (starting around 1960 and reaching an equilibrium of sorts before 1980) contained at least a component dedicated to the histopathology and pathogenesis of radiation injury. Several of these studies were performed by Albert van der Kogel and colleagues (van der Kogel and Barendsen 1974; van der Kogel 1977a; van der Kogel et al. 1982). He recaps extensively his findings and those of some of his contemporaries in his 1979 dissertation *Late effects of radiation on the spinal cord* (van der Kogel 1979). Much of his results are based on serial killing experiments where animals were killed prior to or very soon after expression of radiation injury, the timing of which had been established by previous experiments relating the latent period to the radiation dose. His findings are described in the following paragraphs, which borrow liberally from his Chap. 5 on pathogenesis.

The material consisted of mostly WAG/Rij (WR) rats, a descendent of the Wistar strain, supplemented in some cases with Brown Norway (BN) rats. Radiation results from spinal levels C5-T2 and L2-L5 were described in detail. The cervical spine was irradiated to 22 and 40 Gy for the serial killing experiments; these data were augmented with symptomatic rats from other experiments within this dose range. The dose range for the lumbar irradiations was 22–60 Gy.

Level C5-T2, 22 Gy At this dose, the mean latent period was 170 ± 7 days. Starting at 120 days, groups of four rats were killed every 10 days. No changes were seen in the spinal cord or nerve roots until 160 days. At 160 days, three rats showed neurological signs and two were normal. One normal rat showed extensive demyelination and necrosis in the lateral columns at C6 with modest gliosis at the edge of the lesions. The demyelinated region had many swollen axons. Vascular damage consisted of a few dilated capillaries with minute hemorrhages. No macrophages were observed. The other normal rat showed only a few swollen or demyelinated axons at C6.

The neurologically positive rats were killed at day 166. In all three, demyelinating lesions progressing into necrosis were irregularly distributed over the white matter, most commonly in the lateral and ventral columns. Again there was only a slight glial reaction noted, with hypertrophic astrocytes at the edges of the lesions. Gitter cells (fat laden microglia) were observed near the borders of the irradiated area. Vascular damage was as described above. Van der Kogel points out that the vascular damage appeared to be restricted to capillaries and venules, and that "fibrinoid degeneration of arteriolar walls . . . does not occur in the spinal cord," or at most that there were rare instances of hyaline degeneration or fibrinoid deposit in the arterioles. These findings indicate that the early vascular changes associated with demyelination and white matter necrosis are likely to be venous rather than arterial in nature. Damage to arterioles was seen in the lumbar region, which has a different pathogenesis (van der Kogel 1977a).

The animals irradiated to 40 Gy showed fundamentally similar changes to those at 22 Gy, but the changes occurred about 6 weeks earlier. The vascular changes were somewhat more severe with more fresh hemorrhages in both white and gray matter. However, 40 Gy is a very large dose, more than twice the threshold dose. It is unlikely that this dose is particularly relevant to the clinical material, but it is possible it could further elucidate the pathogenesis. What is more informative is the fact that there were no changes seen in the nerve roots or the ganglia despite the magnitude of the dose.

Progression from limited foci of demyelination to extensive necrosis occurred in about 10 days time.

No changes were seen in asymptomatic BN rats irradiated to 22 Gy because their latent period at this dose is longer than 160 days, which was the longest observation period for the serial killing experiment. The spinal cord changes in symptomatic BN rats and those irradiated to 40 Gy were the same as seen in the WR rats.

The changes described by van der Kogel and colleagues were generally supported in studies by Hopewell and Wright (Hopewell and Wright 1975) and Hubbard and Hopewell (1978). In these studies, Sprague Dawley rats were irradiated to 20 or 40 Gy × 1, 2, 4, 8, and 16 mm. The authors described extensive white matter necrosis after 40 Gy (single dose) with a field size of 8 or 16 mm on the cervical cord. They observed no marked glial reactions around the margins of necrosis, but saw some large mitotic figures (presumed to be astrocytes) around the margins of lesions but more commonly in irradiated but otherwise normal regions of the cord. There were no obvious vascular changes. At 20 Gy × 8 and 16 mm and smaller field sizes for 40 Gy, they found only vascular lesions consisting of fibrin deposits in vessel walls and lumens, a common reaction to vessel injury. Only 20% of the rats showed pathological changes at these regimens. A single animal showed white matter atrophy.

At field sizes of 1, 2, and 4 mm animals showed only focal necrosis, gitter cells, and generalized atrophy. There were occasional unspecified vascular lesions. Animals irradiated to 40 Gy had lesions more frequently, but the types of lesions were essentially the same at the two doses.

Oligodendrocyte and astrocyte cell populations were assayed at various times after 40 Gy by Hubbard and Hopewell (Hubbard and Hopewell 1979). After 8 weeks, both populations decreased significantly, but the oligos more than the astrocytes. They speculated that white matter necrosis could be due to mitotic cell death of oligos that had been stimulated back into the cell cycle as an attempt to repair early neuroglial loss (Hornsey et al. 1981b; Mastaglia et al. 1976; van der Kogel 1986). They discounted the possibility that white matter necrosis resulted from vascular injury.

In the cervical and lumbar regions of Wistar rats, Mastaglia et al. investigated early changes in individual fibers after 5–60 Gy (Mastaglia et al. 1976). Most of the animals were killed between 2 and 8 weeks (cervical) or 1–4 months (lumbar), with single animals in each dose group killed around 10 months. No necrotic lesions were observed. Single demyelinated fibers were found and remyelination was observed at later times. Unfortunately this experiment tells us little about late radiation necrosis. At these early times, no vascular or inflammatory response was seen and very little intercellular edema.

Hornsey et al. (1981b) showed early proliferation of neuroglia after 20 Gy. They related this to the early demyelination/remyelination as reported by Mastaglia but not to the late development of paresis and paralysis.

The general understanding of the pathogenesis of white matter necrosis in irradiated rats in the mid-1980s was that the vasculature was not involved and inflammation was not a significant part of the picture. Although this was not the uniform opinion of investigators who studied human material (see Chap. 1), there were two factors that obfuscated the potential role of the vasculature in the pathogenesis of white matter damage. First, an obvious vascular pathology occurs in rats at slightly lower doses and longer latent periods than the demyelinating lesions in white matter damage. The study by Schultheiss et al. suggested that these lesions, which are distinct in rats and are temporally separate, are likely to have latent periods in humans that have considerable overlap (Schultheiss et al. 1984). This was later verified by Schultheiss et al. (1988). Second, the human material was limited to autopsy cases and therefore presented a rather late picture of the expression of damage. Thus it would not be uncommon for lesions to be months old before being examined microscopically, whereas lesions in the rat were typically only days old. In humans, the greater age of the lesion could easily allow for damage to develop that would never be seen in experimental rodents. Even so, there were reports of white matter radionecrosis in humans with no apparent vascular injury.

Assuming that the white matter radionecrosis in rats was a result of an "avalanche" of mitotic cell death of oligodendrocytes (not ameliorated by glial precursors because of radiation damage to this cell compartment as well) was a logical position. Perhaps the most thorough exposition of this hypothesis, which has been somewhat simplified here, is given by van der Kogel (1986). However, the primary evidence against this proposed pathway is that the spinal cord tolerates denuded axons without necrosis being a result of the demyelination. This is true in multiple sclerosis and is mimicked by MS-like animal models (Blakemore and Patterson 1975). So radiation necrosis was not a necessary consequence of demyelination and white matter necrosis is not routinely seen in other demyelinating conditions.

In 1986, Myers et al. took a fresh look at "roles of glial and vascular elements" in early white matter necrosis in the rat (Myers et al. 1986). (The later vascular pathology alluded to above and discussed below was accepted as a separate pathway to RM.) They concluded that "there is no pathological pathway by which damage to oligodendrocytes alone could lead to massive tissue necrosis." They concluded that early vascular injury, including disruption of the blood-brain barrier (BBB), must be a component of the pathogenesis of white matter radionecrosis. (Although some readers may find it imprecise, "blood-brain barrier" will be used throughout this volume even when "blood-spinal cord barrier" is more specific.)

Writing in 1988, Delattre et al. went further in their assessment of the role of the microvasculature in necrotizing radiation injury to the white matter (Delattre et al. 1988). They state, "In contrast to previous observations employing similar models [observations showing little or no vasculopathy], we found histological evidence of vascular damage as well, including fibrinoid necrosis and thrombosis," and that "vascular damage is not only associated with, but precedes the development of radionecrosis and that partial correction of these vascular changes by dexamethasone is associated with transient symptomatic improvement." Calvo et al. (1988) made similar observations in the rat brain noting, "necrosis of the white matter was preceded by changes in the vasculature" of dilation of vessels and thickening of vessel walls, enlargement of endothelial cell nuclei, along with hypertrophy of astrocytes (Calvo et al. 1988). The authors conclude that in this model, white matter necrosis is not a result of primary cell killing of glial cells or glial progenitors, but is secondary to damage of astrocyte/endothelial cell nexus (Hopewell et al. 1989).

In 1990, Hornsey et al. tested their vascular hypothesis (Myers et al. 1986) by administering vasoactive drugs to rats receiving 25 and 27 Gy, with irradiated control animals receiving 23, 25, or 27 Gy (Hornsey et al. 1990). Drug treatments were dipyridamole (increases blood flow and reduces thrombosis), verapamil (a calcium channel blocker), or desferrioxamine (reduces reperfusion injury). Desferrioxamine animals were also given a low iron diet. Verapamil had no effect, but the dipyridamole and the desferrioxamine shifted the dose-response curve. Re-analysis shows that the two drug regimens had a dose modifying factor of 1.062 and 1.052, respectively. The DMF for dipyridamole is significant at $p = 0.05$, but not for desferrioxamine. In addition, the response rates for myelopathy are not significantly different at 25 and 27 Gy for either drug. Nonetheless, this study is cited as evidence for an important role of vascular injury in the pathology of white matter radionecrosis (Hopewell et al. 1989).

In a similar study, Gutin et al., noting the prevention of vasogenic edema in rat brain by the administration of difluoromethylornithine (DFMO) (Koenig et al. 1983, 1989), fed rats a diet that included long-term DFMO administration (Gutin et al. 1990). Polyamine (PA) levels were elevated at 20 weeks in irradiated rats, as was the spinal cord water content, indicating edema. DFMO diets reduced putrescine to undetectable levels after 10–12 weeks. The dose-response curve for paralysis was shifted slightly (DMF=1.03 by re-analysis), but not significantly, although DMFO-treated rats consistently had lower paralysis rates. These results are suggestive of a role for vascular damage in white matter necrosis, but they do not confirm that vascular damage is a necessary precursor to WMN.

So as of 1990, there was an open controversy regarding the pathogenesis of the so-called early delayed necrotizing white matter lesion of the cervical and thoracic levels of the rat (and by extension, human) spinal cord. The controversy did not extend to the lumbar cord and nerve roots. Nor was the pathogenesis of the late delayed RM lesion debated. Was this lesion, largely defined by white matter necrosis alone and malacia, a result of direct cell kill and sterilization of glial cells (oligodendroglia) and glial precursor cells or was there an essential component of vascular injury?

2.3.2 Lumbar Cord

The pathology and pathogenesis of the lumbar cord is surprisingly different from that of the higher levels of the spinal cord. Although the dose response and latency are generally similar, the progression of the lesions is somewhat slower, but more importantly different regions of the nervous parenchyma are involved (van der Kogel and Barendsen 1974; Bradley et al. 1977; White and Hornsey 1978). As far as the true lumbar cord extends in the caudal direction, no white matter necrosis is seen as a result of radiation in rats. The damage occurs in the nerve roots, where demyelination followed by necrosis is observed. The neurons of the irradiated nerve root ganglia remain intact, but axonal degeneration occurs outward from the irradiated region toward the neuron cell body to which the axon belongs. The work of Bradley in particular indicates that axonal degeneration seems to occur while the Schwann cells still myelinate the fibers. Inflammatory response seems limited to phagocytosis. The only nuclei observed of the degenerating Schwann cells were pyknotic, indicating that they were not stimulated into mitosis as a result of radiation.

This histology leads to questions that remain unanswered. Why do the axons seem to respond first? How is the lumbar true cord immune to damage? What is the role of the vasculature in the nerve root damage? And why is nerve root damage not seen at higher spinal cord levels? In the years following these early studies, all the efforts to elucidate the mechanism of radiation injury to the spinal cord were devoted to true myelopathy with little or no further effort given to investigate the pathogenesis of radiation radiculopathy.

Although the spinal cord in humans ends at the spinal level L1–2, Schultheiss et al. found one reported autopsy case of lumbar RM in their study of 53 cases from the literature (Palmer 1972; Schultheiss et al. 1988). The lesion extended from T11 to L2 with minimal necrosis but extensive demyelination. There was no examination of the nerve roots.

One might have thought that the profound difference in the response of the lumbar cord would have been a major area of research once it was recognized. In the rat, the lumbar cord is relatively longer than in the human. Thus it seems that discovering whatever prevents this cord level from exhibiting the same changes as the cervical and thoracic cord levels could have pointed the way to potential areas of prevention or treatment. Furthermore, any explanation of the pathogenesis of RM in the rat, which has been both a major effort and a major controversy, is incomplete as long as the lumbar response is not included.

2.3.3 Apoptosis

Brownson et al. (1963a, b) were probably the first investigators to investigate systematically the dose and time dependence of the apoptotic response of the CNS to radiation, although previous investigators had studied the acute response of the rat brain to radiation, noting isolated necrotic cells within 6 h of irradiation (Hicks and Montgomery 1952). Although these early investigators did not use the term "apoptosis," the pyknotic nuclei in their studies certainly represented apoptotic cells. (It seems that because the term "apoptosis" did not come into widespread use until the mid-1970s, investigators studying apoptosis after then simply assumed it had not previously been observed.) Brownson identified these cells as oligodendrocytes (in the Holtzman rat, descended from Sprague Dawleys). Brownson et al. divided the process of alteration into three stages. Although there is not complete overlap, there is similarity to the three stages of apoptosis as seen by Bursch et al. (1990) as referenced by Li et al. (1996). Our supplementary analysis of the Brownson data shows the number of altered (apoptotic) cells versus time can be approximated by a lognormal distribution with a mode of about 6.8 h, and their analysis put the mode at 6.2 h. Figure 2.4 reproduces their data. In a later series of papers investigating the apoptotic response of the spinal cord to irradiation, Li et al. found a similar result (Li et al. 1996; Li and Wong 1998, 2000). Two experiments showing the time distribution of apoptotic cells are depicted in Fig. 2.5. The peak of the distributions (mode) as given by the lognormal curves generated here is 7.8 h (Li et al. 1996), for the distribution not shown, it is 7.2 h (Li and Wong 1998). This seems to be good agreement with the Brownson data despite different techniques apply to brain versus spinal cord and the data being taken over 30 years apart. Using stains developed in the ensuing years, Li et al. also identified the cells as apoptotic oligodendrocytes. They found no changes in tissue architecture, no inflammation, all vessels grossly normal, and no obvious signs of demyelination except in a single animal at 20 weeks following 22-Gy single dose. They also found no evidence that neurons or vascular endothelial cells participated in apoptosis.

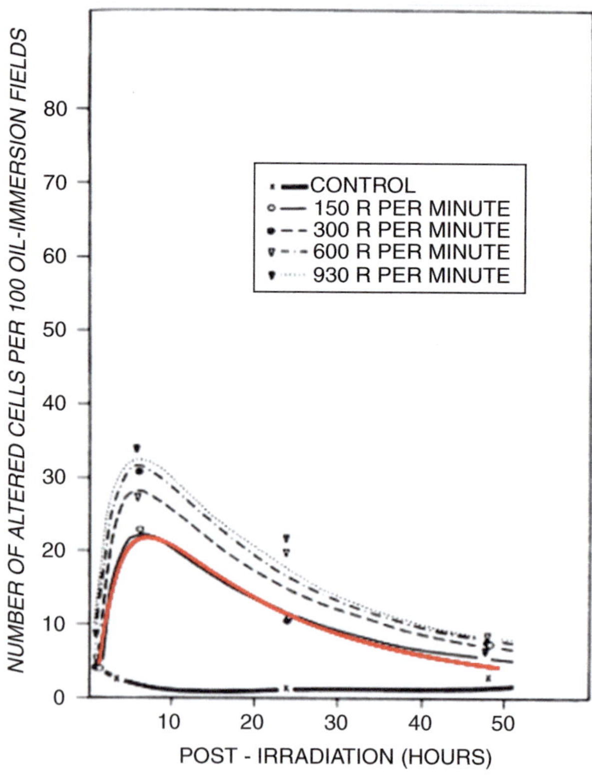

Fig. 2.4 Figure from Brownson et al. showing the time course of counts of pyknotic nuclei, now identified as apoptotic cells. Brownson et al. identified these cells as oligodendrocytes. Red line is a lognormal distribution referenced in the text. From Brownson et al. (1963b). Reprinted with permission from Wolters Kluwer.

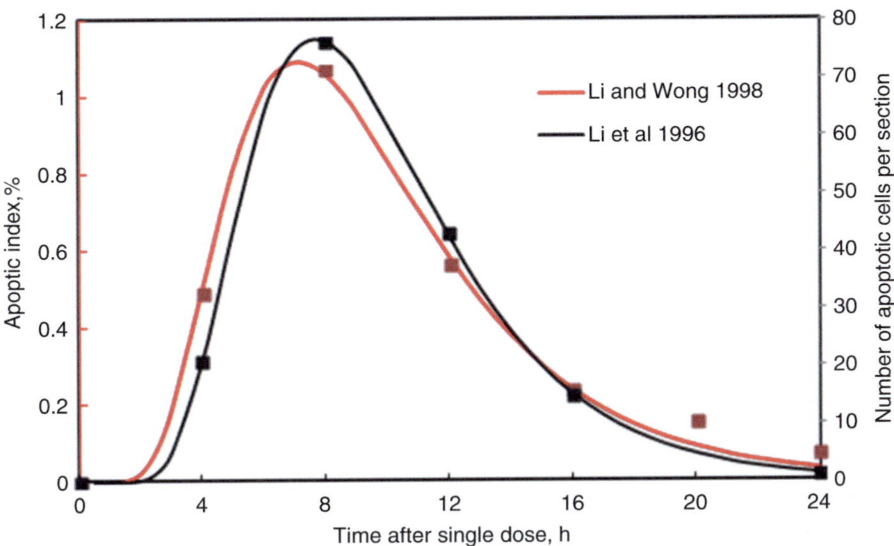

Fig. 2.5 The time course of development of apoptotic oligodendrocytes after single doses of radiation. The red curve and data are after 22 Gy and the black curve and data are after 30 Gy single dose. The right-hand ordinate is from Li et al. (1996) and the left-hand is from Li and Wong (1998).

Li et al. showed that the *kinetics* of the development of apoptosis was seemingly independent of dose up to 22 Gy. Brownson also found that doses up to 50 Gy all peaked around 6 h (with no data between 6 and 24 h); however, at 100 Gy the peak occurrence was delayed to 24 h. Li et al. also stated that "The present results suggest that there is a limited pool of apoptosis-sensitive cell in the rat spinal cord following irradiation. The dose response for apoptosis appeared to level off at a single dose of 8 Gy" (Li and Wong 1998). This apparently was based on a "total apoptotic yield" of 2.4 ± 0.2, 9.0 ± 0.3, and 10.9 ± 0.7% following 2, 8, and 22 Gy, respectively. However, they also state that using a different method, they found a total apoptotic yield of 15.4 ± 2.4% at 22 Gy. Furthermore, the data from Li et al. in 1996 show increasing total apoptotic response steadily and nearly linearly up to 30 Gy (Li et al. 1996). Brownson et al. show increased response up to approximately 100 Gy (Brownson et al. 1963a).

Using BrdU staining, Li et al. showed that glial cells proliferated just before WMN around 140 days. However, despite the fact that they stained for oligodendroglia and astrocytes differentially, they did not mention which cell population was responsible for the proliferation. Of course, this appears to be a moot point since they point out that mature oligodendroglia do not divide (Keirstead and Blakemore 1997). Of course this result was shown in 1981 by Hornsey et al. (1981a).

After about 2000, the research direction involving oligodendrocyte apoptosis after radiation seems not to have been pursued. The likely reason is that the BNCT research described above showed that the dose to the parenchyma of the spinal cord (axons and glial cells) was largely irrelevant to the development of RM. In 2015,

Wong et al. conclude "At present, there is no evidence of any causative association between acute oligodendroglial apoptosis after radiation and the late demyelinating events observed" (Wong et al. 2015). Although these authors were less definitive about the potential relationship between early apoptosis of endothelial cells and late WMN, Li et al. observed that 10 h post irradiation 84% of apoptotic cells were oligodendroglial while only 3% were endothelial cells (Li et al. 2004). In 2005 Nordal and Wong assume late disruption of the BBB is associated with mitotic death of endothelial cells (Nordal and Wong 2005).

2.3.4 BNCT Studies

A series of boron neutron capture therapy experiments was conceived and executed in the 1990s that was designed to irradiate the endothelium and the white matter parenchyma differentially, thereby being able to infer which cell compartment was responsible for the clinical response. In the first experiment (Morris et al. 1994b), rats were irradiated with thermal neutrons (TN) or thermal neutrons in the presence of borocaptate sodium (BSH). This capture agent does not cross the blood-brain barrier and the high LET particles emitted from the prompt fission of B-11 have a range of 5–9 μm. Thus a dose was delivered to the endothelium from the circulating BSH, but no more than 5.7 Gy physical dose was delivered to the white matter parenchyma, presumably from the TN. Nonetheless, radiation myelopathy was observed at typical times for white matter necrosis following irradiation (19 weeks). This was felt to be conclusive evidence that the endothelium is the initial radiation target for white matter necrosis since rats irradiated to 11.4 or 57 Gy with TN in the absence of BSH showed no evidence of parenchymal changes.

In two similar follow-up experiments, dose-response curves were obtained for RM in rats after irradiation with neutrons only, neutrons in the presence of BSH, and neutrons in the presence of p-boronophenylalanine (BPA), which is a capture agent that does cross the blood-brain barrier. There were major differences in the distribution of the radiation from the capture reactions. Despite this, the histology of the subsequent lesions was the same in using the two agents and also the same as seen with the neutron beam only. This lead to the conclusion that selective WMN was a vascular mediated process (Hopewell and van der Kogel 1999), presumably under the assumption that identical lesions implied identical targets and the only target in common was the endothelium (Morris et al. 1994a; Coderre et al. 2006).

These studies were inspired by the work of Slatkin et al. on BNCT induced CNS radiation syndrome in mice (Slatkin et al. 1988). By producing dose gradients from the blood to the brain parenchyma, the authors found support for the "endothelial pathogenesis" of acute brain injury. Although the end point in this study was acute death (< 4 days post irradiation), the hypothesis, the design of the experiment, and the result strongly parallel the RM studies using BNCT. We may infer that the definition of "endothelial pathogenesis" is "endothelial cell damage is a major determinant of acute lethality from the CNS radiation syndrome."

Studies were also undertaken to determine if isoeffective doses of x-rays, TN, TN + BPA, and TN + BSH were also isoeffective for glial progenitor cell survival (Coderre et al. 2006). The investigators used $\frac{1}{3}$ of the D_{50} for paralysis as the presumed isoeffect dose. A higher dose was not used because the assay would be unable to measure the very low cell survival of the glial progenitors at the higher dose.

There are multiple problems with the assumption that $\frac{1}{3}$ D_{50} is isoeffective across different types of radiation. (1) D_{50} itself is isoeffective *only* in the sense that it is the dose that induces 50% paralysis. It is acknowledged that many different cell types contribute to the pathogenesis of paralysis by WMN and there is no a priori reason to assume that for radiation treatments that involve different particles types and energies, these cells types are damaged in exactly the same proportion for every beam type. The response of various cell compartments would vary depending upon the blend of LET's and RBE's of the various particles and energies. Thus D_{50} is not necessarily biologically isoeffective, just numerically. (2) There is little justification to the assumption that $f D_{50}$ (where f is a constant presumably close to unity) is isoeffective for different beams. Even for this to be true for rate of paralysis from different beams, it would depend on the dose-response curves being parallel. (3) The problems with assuming parallelness of dose-response curves include (a) slopes of dose-response functions are notoriously hard to estimate accurately and (b) the slope depends on the exact mathematical model used to fit the data (probit, logistic, etc. against dose or log dose). Indeed, D_{50} is common to essentially all models of dose response, but in many ways it is the behavior of the slope that distinguishes the various models. (4) For steep dose-response functions as determined in these studies, the value of $\frac{1}{3}$ D_{50} is a very large extrapolation (measured in standard deviations of D_{50}) from any data for which the response was actually measured. Indeed there is truly no basis for making the assertion that $\frac{1}{3}$ D_{50} is isoeffective for "paralysis from white matter necrosis" when the paralysis rate at this dose is immeasurably small.

It is actually a straightforward exercise to demonstrate that $f \cdot D_{50}$ is *not* isoeffective, even for paralysis, much less for cell survival. If $f D_{50}$ is isoeffective then so would f times other doses, for example $f D_{80}$. Taking the published BSH dose-response curve for total dose, the value of D_{80} is 31.4 Gy and the value of D_{20} is 29.3. Taking the ratio of these two doses we get a value for f of 0.933 and $D_{20} = 0.933 \cdot D_{80}$. For the BPA dose-response curve, we have D_{80} is 15.2 Gy. Applying f to this value gives 14.2 Gy. This dose is associated with a response of about 50% and not 20%. Thus the assumption that $f D x$ is isoeffective fails even within a region very near the data and for exactly the same outcome. Extrapolating to doses far from the measured data to endpoints that were not measured to begin with is not warranted.

Despite the issues with the isoeffect-dose experiment, the BNCT studies by Morris et al. proved that the response of the endothelium to radiation is sufficient to trigger white matter necrosis, irrespective of the direct dose to the parenchyma. The exact details of the pathogenesis of white matter necrosis are not fully understood, but many subsequent studies have been done to refine further knowledge in this area.

2.3.5 Roles of Cytokines

Hopewell (1998) concluded that peri-vascular edema leads to ischemia and reperfusion injury. This idea was supported largely by the work of Hornsey et al. (1990). In Hornsey's study, it was stated unequivocally that after irradiation of the spinal cord, "the first pathological changes which are observed are extravasation of red blood cells and oedema shortly followed by compression of blood vessels, infarction, and ischemic necrosis." The three papers cited in support were: Myers et al. (1986), a thought paper without new data; Tamaki et al. (1984), a study in which no animals were irradiated; and Blakemore and Palmer (1982), a study in which the spinal cord of cats was irradiated from T11 to L4 (a regions that responds differently) to a (very high) single dose of 40 Gy. Both gray and white matter lesions were seen. Blakemore and Palmer also made a point of the similarity between the radiation lesions and those of decompression sickness in goats, where the infarct was conclusively attributable to nitrogen bubbles and led to rather than followed edema. Thus the Hornsey's statement regarding infarction and ischemic necrosis is largely unsupported, thereby reducing support for Hopewell's assumptions regarding ischemia and reperfusion injury. It is well known that brain ischemia can lead to edema, which in turns lead to a cycle of ischemia and edema. However, this cycle is started by the ischemic process.

In 1995, Stewart et al. from Wong's group found that endothelial cells of microvessels were reduced by more than 50% at 3 months following a relatively large dose of 25 Gy in Fischer 344 rats. This occurred only in white matter, not gray matter, and pericyte numbers were statistically unchanged (Stewart et al. 1995). At 93 days, BBB changes were just being seen as measured by extravasation of horseradish peroxidase (HRP), and "extravasation was widespread and much more pronounced at day 114." They postulated that the unrepaired BBB "may lead to edema, ischemia, and necrosis of the white matter."

Writing in 1999 Tsao et al. from the same group, postulate that late BBB breakdown leads to vasogenic edema leading to hypoxia leading to VEGF expression in astrocytes leading to further BBB breakdown (Tsao et al. 1999). This cycle repeats itself rather than being interrupted because "endothelial cells fail to proliferate due to radiation injury"; the repeated cycle results in white matter necrosis, but by an unstated mechanism.

In 2004, Wong and van der Kogel, the heads of the two most prolific research groups involved in exploring RM and it pathogenesis wrote "Mechanisms of radiation injury to the central nervous system: Implications for neuroprotection" in *Molecular Interventions* (Wong and Van der Kogel 2004). They explored extensively what was known regarding genetic alterations, activation of cytokines, adhesion molecules, growth factors, structural alterations of microvasculature, and reviewed the potential role of apoptosis, all with a view to identify areas of possible intervention. In a single figure they described a model of cellular responses in RM as

- Neuro-inflammation
- Hypoxia
- VEGF upregulation
- Late barrier breakdown

leading finally to necrosis.

Finally in 2015 Wong et al. reviewed the pathobiology of RM (Wong et al. 2015). This review offers no new hypotheses on the pathogenesis of RM or the specific mechanism that leads to white matter necrosis. The word "avalanche" often appears in the experimental literature to explain the conclusion of a process for which the details are somewhat vague but depend solely on mitotic cell death (Michalowski 1981).

It is clear or at least widely agreed upon, that the primary or initial target of radiation white matter necrosis is the endothelium. It is also broadly accepted that the first microscopic signs leading to paralysis include edema. A viable explanation for the cause of this edema is that endothelial cell loss results in BBB disruption that in turns promotes edema (Ljubimova et al. 1991; Lyubimova and Hopewell 2004; Stewart et al. 1995). At this point, it is useful to consider the role of the CNS lymphatic system, specifically the mechanism by which interstitial fluid and metabolites exit the CNS. The metabolic rate of the spinal cord is about one-third that of the brain, so its need to rid itself of cellular waste products is substantial. The evolving understanding of how this takes place critically depends upon the delivery of fluids (CSF and ISF) to venules along the peri-vascular spaces (PVS) (Abbott et al. 2018). The outer (CNS parenchymal) boundary of the PVS is provided by the astrocytic foot processes. In microvessels, the inner boundary is composed of the endothelial cells or pericytes. Thus the security of the space is intimately related to the integrity of the BBB.

It has been shown that reactive astrocytes cause BBB disruption and breakdown by the production of VEGF-A (Chapouly et al. 2015; Li et al. 2001; Nordal et al. 2004). Argwal et al. identify VEGF-mediated downregulation of CLN-5 (a transmembrane tight junction protein) as a significant mechanism in BBB breakdown (Argaw et al. 2009). Upregulation of VEGF is a mechanism that also contributes to the repair of BBB disruption. However, when disruption of the BBB occurs in the background of radiation damage that has exceeded the spinal cord tolerance, the vascular endothelium and reactive astrocytes that would otherwise be components of the repair process are incapable of performing their normal role. Lyubimova and Hopewell as well as Stewart et al. showed the number of endothelial cells are significantly decreased in number as BBB damage is just beginning to be expressed after radiation to slightly more than the tolerance dose (Lyubimova and Hopewell 2004; Stewart et al. 1995). Contemporaneously with VEGF positivity in astrocytes, hypoxia was observed in glial cells (at least 43% of which were astrocytes) near areas of necrosis, both focal and confluent, but also near disruption of the BBB (Nordal et al. 2004). Li et al. propose that "hypoxia induces VEGF expression in reactive astrocytes, which in turn leads to further increase in vascular permeability and BSCB disruption" (Li et al. 2001). In Clausi et al., treatment with the ACE inhibitor Ramipril reduced VEGF-A expression and demyelination by more than 50% in irradiated animals. The incidence of myelopathy was not affected, but the latency increased in the treated group (Clausi et al. 2018).

Haddadi et al. measured VEGF in spinal cords of irradiated rats at 20 and 22 weeks following 22 Gy single dose to the cervical spinal cord (Haddadi et al. 2013). An additional group was also treated with melatonin. "VEGF levels of the irradiated groups are significantly higher than the control groups (*$p < 0.05$ radiation vs. control groups). Melatonin significantly reduced VEGF levels in the spinal cords of rats subjected to irradiation (**$p < 0.05$ radiation + melatonin vs. radiation groups). Data are the mean ± SEM of six rats." They also observed that RM cases in the melatonin-treated group were delayed or reduced.

The failing, if it can be called that, of the theories of the pathogenesis of RM is that they are limited to the lesion of WMN. The pathogenesis of the well-established pathology of vascular injury leading to RM is not frequently addressed. We know that during the latent period the consequences of the initial radiation damage are gradually growing toward what is to become a locally catastrophic event in the white matter. It seems implausible for the lesion of WMN and the vascular lesion to have separate paths toward that event. We suggest that a necessary component of the WMN lesion is the late breakdown of the BBB, and that this may be averted or quickly repaired if the damage to the venous endothelium is not too great. Nevertheless, the general endothelial damage and glial responses would continue. This continuing expression of damage would then ultimately lead to the vascular lesions described in Table 2.2b. Of course breakdown of the BBB would ultimately occur, but not in the absence of obvious vascular damage.

By 2000, it was clear that the pathogenesis of RM involved an initial insult to the vascular endothelium, an initiation of neuro-inflammation involving cytokines, chemokines, and growth factors, culminating in proliferation of endothelial cells just prior to WMN (Sawaya et al. 1994). Blood-brain barrier disruption occurs at least twice—once immediately following radiation and again prior to expression of injury (Rubin et al. 1994; Stewart et al. 1995). Although cytokines were at first proposed in the protection from or treatment of radiation damage, it was soon recognized that they participate in or enhance radiation injury (McBride et al. 1989; Michalowski 1990; Chiang and McBride 1991; Schultheiss and Stephens 1992c; Hopewell 1994). In most hypotheses regarding WMN following spinal cord irradiation, after the onset of BBB breakdown, edema, and appearance of VEGF in activated astrocytes, it is postulated that secondary injury leads to necrosis. Typically, this secondary injury is somewhat nonspecific, but often ischemia and infarction and/or cytokine activity are invoked. Another possibility is that the disruption of the BBB necessarily disrupts the peri-venous space, resulting in impaired efflux of the continuously produced cytotoxic products of CNS metabolic activity (Verheggen et al. 2018; Wang et al. 2020). Buildup of these toxins is also a candidate for the ultimate cause of necrosis.

Even if this conjecture is ultimately validated or supplanted by a better explanation, a complete theory of the pathogenesis of RM must also explain other characteristics of the spinal response to radiation. These include: the general absence of a gray matter lesion, the relative absence of lesions in the lumbar cord, the white matter lesion of the lumbar nerve roots that is absent in higher nerve roots, and the nearly complete recovery of radiation damage at lower (but nontrivial) doses and long time intervals.

2.4   Traumatic Injury, Inflammation, and Covid-19

The pathophysiology of traumatic spinal cord injury comprises a two-part sequence—the initial mechanical trauma and the consequent secondary damage (Donnelly and Popovich 2008; Kwon et al. 2004; Rowland et al. 2008). The forces of the primary injury to the spinal cord disrupt axons and the supporting glial cells and vasculature (Profyris et al. 2004). The secondary injury is a result of various (primarily inflammatory) responses to the initial damage. These responses produce further cellular damage (Profyris et al. 2004; Rowland et al. 2008). In RM, there is no initial mechanical injury, but the secondary injury has many pathological features in common with RM. Since far more scientific study has been devoted to traumatic SC injury, it is likely that a better understanding of RM can be achieved by consulting this literature.

Possibly the most important element of the pathophysiology of traumatic SC injury is the inflammatory/immune response. This response plays an important role in both the repair of damage and the progression of injury (Chan 2008). However whether the immune response cascade is beneficial or deleterious is not fully resolved (Donnelly and Popovich 2008; Fleming et al. 2006; Rowland et al. 2008). Typically, the first cells to arrive are neutrophils, within hours of the insult, followed soon by exogenous macrophages and endogenous activated microglia (Fitch and Silver 2008). A neutrophilic response is not a major hallmark of RM. In general, inflammatory responses can have both neurotoxic and neuroprotective effects; which of these predominates depends upon the sequence and magnitude of expression, and the cells on which they act (Kwon et al. 2004). During the post-injury inflammatory cascade in the CNS, astrocytes and microglia are activated and neutrophils are recruited in the first 24 h. Blood monocytes migrate into the injury site over the course of several days (Dahlstrand et al. 1995). These can enter the CNS through the remote blood-CSF barrier at the ventricular choroid plexus of the brain (Damoiseaux et al. 1994) or through a disruption in the BBB resulting at the site of the injury. The neutrophils prepare the area for phagocytic repair; however, further damage can be caused when the neutrophilic response is especially brisk. Macrophages and microglia both produce factors that can promote axonal regeneration and growth, as well as neurotoxic cytokines and molecules (Fleming et al. 2006).

Neural progenitor cells are relatively quiescent in normal CNS tissue, but they become active and proliferate in response to trauma (Meletis et al. 2008). Most rodent studies have found neural progenitor cells to produce mostly astrocytes, but also some oligodendrocytes and their progenitors, which is similar to the results from cell transplantation (Barreiro-Iglesias 2010).

Because of many common histopathological traits between Covid-19 organ damage and radiation therapy complication (Rios et al. 2021), radiation oncologists should be alert to the possible increase in late effects from radiation in patients with or who have recovered from Covid-19. Although CNS damage from Covid-19 is an acute response, there is a significant overlap in the pathology (Nuovo et al. 2021), and numerous cases of transverse myelitis have been reported, including pediatric cases (Zachariadis et al. 2020; Munz et al. 2020; Kaur et al. 2020).

2.5   Conclusion

It may be that Boden was correct in his assertion that unraveling the pathogenesis is a feckless exercise, if interventions are not possible. This is perhaps more true for RM resulting from WMN than from massive vascular damage. If RM from gross vascular damage occurs later and at lower doses, then protecting against WMN would only delay the expression of RM until the expression of severe vascular damage results in irreversible neurological injury. Certainly the attention given to interventions using growth factors and other pharmacological interventions were directed at interfering with early WMN and not late vascular damage. Part of the reason for this is that WMN could be reliably induced and at well-established times in the rat model, whereas the late vascular injury occurs in a relatively narrow dose range, just above the lowest tolerance doses but below doses causing WMN. In rats, this is a very few Gy (single dose) at most.

References

Abbott NJ, Pizzo ME, Preston JE, Janigro D, Thorne RG. The role of brain barriers in fluid movement in the CNS: is there a 'glymphatic' system? Acta Neuropathol. 2018;135(3):387–407. https://doi.org/10.1007/s00401-018-1812-4.

Argaw AT, Gurfein BT, Zhang Y, Zameer A, John GR. VEGF-mediated disruption of endothelial CLN-5 promotes blood-brain barrier breakdown. Proc Natl Acad Sci. 2009;106(6):1977–82.

Barreiro-Iglesias A. Targeting ependymal stem cells in vivo as a non-invasive therapy for spinal cord injury. Dis Model Mech. 2010;3(11–12):667–8. https://doi.org/10.1242/dmm.006643.

Blakemore W, Palmer A. Delayed infarction of spinal cord white matter following X-irradiation. J Pathol. 1982;137(4):273–80.

Blakemore WF, Patterson RC. Observations on the interactions of Schwann cells and astrocytes following X-irradiation of neonatal rat spinal cord. J Neurocytol. 1975;4(5):573–85. https://doi.org/10.1007/BF01351538.

Boden G. Radiation myelitis of the cervical spinal cord. Br J Radiol. 1948;21(249):464–9. https://doi.org/10.1259/0007-1285-21-249-464.

Bradley WG, Fewings JD, Cumming WJK, Harrison RM. Delayed myeloradiculopathy produced by spinal X-irradiation in the rat. J Neurol Sci. 1977;31(1):63–82. https://doi.org/10.1016/0022-510X(77)90006-5.

Brownson RH, Suter DB, Diller DA. Acute brain damage induced by low dosage x-irradiation. Neurology. 1963a;13(3):181–91. https://doi.org/10.1212/wnl.13.3.181.

Brownson RH, Suter DB, Oliver JL, Diller DA. Acute brain damage induced by x-irradiation with special reference to rate and recovery factors. Neurology. 1963b;13(12):1011–20. https://doi.org/10.1212/wnl.13.12.1011.

Bursch W, Paffe S, Putz B, Barthel G, Schulte-hermann R. Determination of the length of the histological stages of apoptosis in normal liver and in altered hepatic foci of rats. Carcinogenesis. 1990;11(5):847–53. https://doi.org/10.1093/carcin/11.5.847.

Calvo W, Hopewell J, Reinhold H, Yeung T. Time-and dose-related changes in the white matter of the rat brain after single doses of X rays. Br J Radiol. 1988;61(731):1043–52.

Chan CC. Inflammation: beneficial or detrimental after spinal cord injury? Recent Pat CNS Drug Discov. 2008;3(3):189–99. https://doi.org/10.2174/157488908786242434.

Chapouly C, Tadesse Argaw A, Horng S, Castro K, Zhang J, Asp L, Loo H, Laitman BM, Mariani JN, Straus Farber R. Astrocytic TYMP and VEGFA drive blood–brain barrier opening in inflammatory central nervous system lesions. Brain. 2015;138(6):1548–67.

Chiang C-S, McBride WH. Radiation enhances tumor necrosis factor α production by murine brain cells. Brain Res. 1991;566(1–2):265–9.

Clausi MG, Stessin AM, Tsirka SE, Ryu S. Mitigation of radiation myelopathy and reduction of microglial infiltration by Ramipril, ACE inhibitor. Spinal Cord. 2018;56(8):733–40. https://doi.org/10.1038/s41393-018-0158-z.

Coderre JA, Morris GM, Micca PL, Hopewell JW, Verhagen I, Kleiboer BJ, van der Kogel AJ. Late effects of radiation on the central nervous system: role of vascular endothelial damage and glial stem cell survival. Radiat Res. 2006;166(3):495–503. https://doi.org/10.1667/RR3597.1.

Cota AG. Spinal cord anatomy. In: Deer TR, Pope JE, Lamer TJ, Provenzano D, editors. Deer's treatment of pain: an illustrated guide for practitioners. Cham: Springer International Publishing; 2019. p. 43–8. https://doi.org/10.1007/978-3-030-12281-2_6.

Dahlstrand J, Lardelli M, Lendahl U. Nestin mRNA expression correlates with the central nervous system progenitor cell state in many, but not all, regions of developing central nervous system. Brain Res Dev Brain Res. 1995;84(1):109–29. https://doi.org/10.1016/0165-3806(94)00162-s.

Damoiseaux JG, Döpp EA, Calame W, Chao D, MacPherson GG, Dijkstra CD. Rat macrophage lysosomal membrane antigen recognized by monoclonal antibody ED1. Immunology. 1994;83(1):140–7.

Delattre JY, Rosenblum MK, Thaler HT, Mandell L, Shapiro WR, Posner JB. A model of radiation myelopathy in the rat: pathology, regional capillary permeability changes and treatment with dexamethasone. Brain. 1988;111(6):1319–36. https://doi.org/10.1093/brain/111.6.1319.

Donnelly DJ, Popovich PG. Inflammation and its role in neuroprotection, axonal regeneration and functional recovery after spinal cord injury. Exp Neurol. 2008;209(2):378–88. https://doi.org/10.1016/j.expneurol.2007.06.009.

Dupont G, Schmidt C, Yilmaz E, Oskouian RJ, Macchi V, de Caro R, Tubbs RS. Our current understanding of the lymphatics of the brain and spinal cord. Clin Anat. 2019;32(1):117–21. https://doi.org/10.1002/ca.23308.

Fajardo LF. Morphologic patterns of radiation Injury1. Front Radiat Ther Oncol. 1989:75–84. https://doi.org/10.1159/000416572.

Fitch MT, Silver J. CNS injury, glial scars, and inflammation: inhibitory extracellular matrices and regeneration failure. Exp Neurol. 2008;209(2):294–301. https://doi.org/10.1016/j.expneurol.2007.05.014.

Fleming JC, Norenberg MD, Ramsay DA, Dekaban GA, Marcillo AE, Saenz AD, Pasquale-Styles M, Dietrich WD, Weaver LC. The cellular inflammatory response in human spinal cords after injury. Brain. 2006;129(Pt 12):3249–69. https://doi.org/10.1093/brain/awl296.

Geraci JP. Dose-latency period for radiation-induced myelitis in rodents. Radiology. 1979;132(1):238–9.

Gibbs SJ. Dose-latency period for radiation-induced myelitis in rodents. Radiology. 1979;132(1):239–40.

Gutin PH, McDermott MW, Ross G, Chan PH, Chen SF, Levin KJ, Babuna O, Marton LJ. Polyamine accumulation and vasogenic oedema in the genesis of late delayed radiation injury of the central nervous system (CNS). Acta Neurochir Suppl. 1990;51:372–4.

Haddadi G, Shirazi A, Sepehrizadeh Z, Mahdavi SR, Haddadi M. Radioprotective effect of melatonin on the cervical spinal cord in irradiated rats. Cell J. 2013;14(4):246–53.

Hicks SP, Montgomery POB. Effects of acute radiation on the adult mammalian central nervous system. Proc Soc Exp Biol Med. 1952;80(1):15–8. https://doi.org/10.3181/00379727-80-19505.

Hopewell JW. Models of CNS radiation damage during space flight. Adv Space Res. 1994;14(10):433–42. https://doi.org/10.1016/0273-1177(94)90497-9.

Hopewell JW. Radiation injury to the central nervous system. Med Pediatr Oncol. 1998;30:1–9.

Hopewell JW, Calvo W, Campling D, Reinhold HS, Rezvani M, Yeung TK. Effects of radiation on the microvasculature. Front Rad Ther Oncol. 1989:85–95. https://doi.org/10.1159/000416573.

Hopewell JW, van der Kogel AJ. Pathophysiological mechanisms leading to the development of late radiation-induced damage to the central nervous system. In: Wiegel T, Hinkelbein W, Brock M, Hoell T, editors. Controversies in neuro-oncology, Frontiers of radiation therapy and oncology, vol. 33. Basel: Karger; 1999. p. 265–75.

Hopewell JW, Wright EA. The effects of dose and field size on late radiation damage to the rat spinal cord. Int J Radiat Biol. 1975;28(4):325–33. https://doi.org/10.1080/09553007514551111.

Hornsey S, Myers R, Coultas P, Rogers M, White A. Turnover of proliferative cells in the spinal cord after X irradiation and its relation to time-dependent repair of radiation damage. Br J Radiol. 1981a;54(648):1081–5.

Hornsey S, Myers R, Coultas PG, Rogers MA, White A. Turnover of proliferative cells in the spinal cord after X irradiation and its relation to time-dependent repair of radiation damage. Br J Radiol. 1981b;54(648):1081–5. https://doi.org/10.1259/0007-1285-54-648-1081.

Hornsey S, Myers R, Jenkinson T. The reduction of radiation damage to the spinal cord by post-irradiation administration of vasoactive drugs. Int J Radiat Oncol Biol Phys. 1990;18(6):1437–42. https://doi.org/10.1016/0360-3016(90)90319-F.

Hubbard BM, Hopewell JW. The dose-latent period relationship in the irradiated cervical spinal cord of the rat. Radiology. 1978;128(3):779–81. https://doi.org/10.1148/128.3.779.

Hubbard BM, Hopewell JW. Changes in the neuroglial cell populations of the rat spinal cord after local X-irradiation. Br J Radiol. 1979;52(622):816–21. https://doi.org/10.1259/0007-1285-52-622-816.

Iliff JJ, Nedergaard M. Is there a cerebral lymphatic system? Stroke. 2013;44(SUPPL. 1):S93–5. https://doi.org/10.1161/STROKEAHA.112.678698.

Kaur H, Mason JA, Bajracharya M, McGee J, Gunderson MD, Hart BL, Dehority W, Link N, Moore B, Phillips JP. Transverse myelitis in a child with COVID-19. Pediatr Neurol. 2020;112:5.

Keirstead HS, Blakemore WF. Identification of post-mitotic oligodendrocytes incapable of remyelination within the demyelinated adult spinal cord. J Neuropathol Exp Neurol. 1997;56(11):1191–201.

Koenig H, Goldstone AD, Lu CY. Blood brain barrier breakdown in brain edema following cold injury is mediated by microvascular polyamines. Biochem Biophys Res Commun. 1983;116(3):1039–48.

Koenig H, Goldstone AD, Lu CY. Polyamines mediate the reversible opening of the blood-brain barrier by the intracarotid infusion of hyperosmolal mannitol. Brain Res. 1989;483(1):110–6.

van der Kogel AJ. Radiation-induced nerve root degeneration and hypertrophic neuropathy in the lumbosacral spinal cord of rats: the relation with changes in aging rats. Acta Neuropathol. 1977a;39(2):139–45. https://doi.org/10.1007/BF00703320.

van der Kogel AJ. Radiation tolerance of the rat spinal cord: time dose relationships. Radiology. 1977b;122(2):505–9. https://doi.org/10.1148/122.2.505.

van der Kogel AJ. Late effects of radiation on the spinal cord; dose-effect relationships and pathogenesis. RBI; 1979.

van der Kogel AJ (1980) Mechanisms of late radiation injury in the spinal cord. In Radiation Biology in Cancer Research, R.E. Meyn and H.R. Withers (Eds.) New York, Raven Press.

van der Kogel AJ. Radiation-induced damage in the central nervous system: an interpretation of target cell responses. Br J Cancer. 1986;53(SUPPL. 7):207–17.

van der Kogel AJ, Barendsen GW. Late effects of spinal cord irradiation with 300 kV X rays and 15 MeV neutrons. Br J Radiol. 1974;47(559):393–8. https://doi.org/10.1259/0007-1285-47-559-393.

van der Kogel AJ, Sissingh HA, Zoetelief J. Effect of X rays and neutrons on repair and regeneration in the rat spinal cord. Int J Radiat Oncol Biol Phys. 1982;8(12):2095–7. https://doi.org/10.1016/0360-3016(82)90551-X.

Kwon BK, Tetzlaff W, Grauer JN, Beiner J, Vaccaro AR. Pathophysiology and pharmacologic treatment of acute spinal cord injury. Spine J. 2004;4(4):451–64. https://doi.org/10.1016/j.spinee.2003.07.007.

Leith JT, Keith Dewyngaert J, Glicksman AS. Radiation myelopathy in the rat: an interpretation of dose effect relationships. Int J Radiat Oncol Biol Phys. 1981;7(12):1673–7. https://doi.org/10.1016/0360-3016(81)90191-7.

Li YQ, Ballinger JR, Nordal RA, Su ZF, Wong CS. Hypoxia in radiation-induced blood-spinal cord barrier breakdown. Cancer Res. 2001;61(8):3348–54.

Li Y-Q, Chen P, Jain V, Reilly RM, Wong CS. Early radiation-induced endothelial cell loss and blood–spinal cord barrier breakdown in the rat spinal cord. Radiat Res. 2004;161(2):143–52.

Li YQ, Jay V, Wong CS. Oligodendrocytes in the adult rat spinal cord undergo radiation-induced apoptosis. Cancer Res. 1996;56(23):5417–22.

Li YQ, Wong CS. Apoptosis and its relationship with cell proliferation in the irradiated rat spinal cord. Int J Radiat Biol. 1998;74(4):405–17.

Li YQ, Wong CS. Radiation-induced apoptosis in the neonatal and adult rat spinal cord. Radiat Res. 2000;154(3):268–76. https://doi.org/10.1667/0033-7587(2000)154[0268:RIAITN]2.0.CO;2.

Ljubimova NV, Levitman MK, Plotnikova ED, Eidus LK. Endothelial cell population dynamics in rat brain after local irradiation. Br J Radiol. 1991;64(766):934–40. https://doi.org/10.1259/0007-1285-64-766-934.

Lyubimova N, Hopewell J. Experimental evidence to support the hypothesis that damage to vascular endothelium plays the primary role in the development of late radiation-induced CNS injury. Br J Radiol. 2004;77(918):488–92.

Mastaglia FL, McDonald WI, Watson JV, Yogendran K. Effects of x-radiation on the spinal cord: an experimental study of the morphological changes in central nerve fibres. Brain. 1976;99(1):101–22. https://doi.org/10.1093/brain/99.1.101.

McBride WH, Mason K, Withers HR, Davis C. Effect of interleukin 1, inflammation, and surgery on the incidence of adhesion formation and death after abdominal irradiation in mice. Cancer Res. 1989;49(1):169–73.

Meletis K, Barnabé-Heider F, Carlén M, Evergren E, Tomilin N, Shupliakov O, Frisén J. Spinal cord injury reveals multilineage differentiation of ependymal cells. PLoS Biol. 2008;6(7):e182. https://doi.org/10.1371/journal.pbio.0060182.

Michalowski A. Effects of radiation on normal tissues: hypothetical mechanisms and limitations of in situ assays of clonogenicity. Radiat Environ Biophys. 1981;19(3):157–72. https://doi.org/10.1007/BF01324183.

Michalowski A. On radiation damage to normal tissues and its treatment: I. Growth factors. Acta Oncol. 1990;29(8):1017–23. https://doi.org/10.3109/02841869009091793.

Morris GM, Coderre JA, Hopewell JW, Micca PL, Nawrocky MM, Liu HB, Bywaters A. Response of the central nervous system to boron neutron capture irradiation: evaluation using rat spinal cord model. Radiother Oncol. 1994a;32(3):249–55. https://doi.org/10.1016/0167-8140(94)90024-8.

Morris GM, Coderre JA, Whitehouse EM, Micca P, Hopewell JW. Boron neutron capture therapy: a guide to the understanding of the pathogenesis of late radiation damage to the rat spinal cord. Int J Radiat Oncol Biol Phys. 1994b;28(5):1107–12. https://doi.org/10.1016/0360-3016(94)90484-7.

Munz M, Wessendorf S, Koretsis G, Tewald F, Baegi R, Krämer S, Geissler M, Reinhard M. Acute transverse myelitis after COVID-19 pneumonia. J Neurol. 2020;267:2196–7.

Myers R, Rogers MA, Hornsey S. A reappraisal of the roles of glial and vascular elements in the development of white matter necrosis in irradiated rat spinal cord. Br J Cancer. 1986;53(SUPPL. 7):221–3.

Nagelhus E, Mathiisen T, Ottersen O. Aquaporin-4 in the central nervous system: cellular and subcellular distribution and coexpression with KIR4. 1. Neuroscience. 2004;129(4):905–13.

Nedergaard M. Garbage truck of the brain. Science. 2013;340(6140):1529–30. https://doi.org/10.1126/science.1240514.

Nordal RA, Nagy A, Pintilie M, Wong CS. Hypoxia and hypoxia-inducible factor-1 target genes in central nervous system radiation injury: a role for vascular endothelial growth factor. Clin Cancer Res. 2004;10(10):3342–53. https://doi.org/10.1158/1078-0432.ccr-03-0426.

Nordal RA, Wong CS. Molecular targets in radiation-induced blood-brain barrier disruption. Int J Radiat Oncol Biol Phys. 2005;62(1):279–87. https://doi.org/10.1016/j.ijrobp.2005.01.039.

Nuovo GJ, Magro C, Shaffer T, Awad H, Suster D, Mikhail S, He B, Michaille J-J, Liechty B, Tili E. Endothelial cell damage is the central part of COVID-19 and a mouse model induced by injection of the S1 subunit of the spike protein. Ann Diagn Pathol. 2021;51:151682.

Palmer JJ. Radiation myelopathy. Brain. 1972;95(1):109–22. https://doi.org/10.1093/brain/95.1.109.

Profyris C, Cheema SS, Zang D, Azari MF, Boyle K, Petratos S. Degenerative and regenerative mechanisms governing spinal cord injury. Neurobiol Dis. 2004;15(3):415–36. https://doi.org/10.1016/j.nbd.2003.11.015.

Rios CI, Cassatt DR, Hollingsworth BA, Satyamitra MM, Tadesse YS, Taliaferro LP, Winters TA, DiCarlo AL. Commonalities between COVID-19 and radiation injury. Radiat Res. 2021;195(1):1–24.

Rowland JW, Hawryluk GW, Kwon B, Fehlings MG. Current status of acute spinal cord injury pathophysiology and emerging therapies: promise on the horizon. Neurosurg Focus. 2008;25(5):E2. https://doi.org/10.3171/foc.2008.25.11.e2.

Rubin P, Gash DM, Hansen JT, Nelson DF, Williams JP. Disruption of the blood-brain barrier as the primary effect of CNS irradiation. Radiother Oncol. 1994;31(1):51–60. https://doi.org/10.1016/0167-8140(94)90413-8.

Sawaya R, Rayford A, Kono S, Ang KK, Feng Y, Stephens LC, Rao JS. Plasminogen activator inhibitor-1 in the pathogenesis of delayed radiation damage in rat spinal cord in vivo. J Neurosurg. 1994;81(3):381–7. https://doi.org/10.3171/jns.1994.81.3.0381.

Schultheiss TE, Higgins EM, El-Mahdi AM. The latent period in clinical radiation myelopathy. Int J Radiat Oncol Biol Phys. 1984;10(7):1109–15. https://doi.org/10.1016/0360-3016(84)90184-6.

Schultheiss TE, Stephens LC. The pathogenesis of radiation myelopathy: widening the circle. Int J Radiat Oncol Biol Phys. 1992b;23(5):1089–91. https://doi.org/10.1016/0360-3016(92)90920-D.

Schultheiss TE, Stephens LC. Permanent radiation myelopathy. Br J Radiol. 1992c;65(777):737–53.

Schultheiss TE, Stephens LC, Maor MH. Analysis of the histopathology of radiation myelopathy. Int J Radiat Oncol Biol Phys. 1988;14(1):27–32. https://doi.org/10.1016/0360-3016(88)90046-6.

Slatkin D, Stoner R, Rosander K, Kalef-Ezra J, Laissue J. Central nervous system radiation syndrome in mice from preferential 10B (n, alpha) 7Li irradiation of brain vasculature. Proc Natl Acad Sci. 1988;85(11):4020–4.

Sosa SM, Smith KJ. Understanding a role for hypoxia in lesion formation and location in the deep and periventricular white matter in small vessel disease and multiple sclerosis. Clin Sci. 2017;131(20):2503–24.

Stewart PA, Vinters HV, Wong CS. Blood-spinal cord barrier function and morphometry after single doses of X-rays in rat spinal cord. Int J Radiat Oncol Biol Phys. 1995;32(3):703–11. https://doi.org/10.1016/0360-3016(94)00594-B.

Tamaki K, Sadoshima S, Baumbach GL, Iadecola C, Reis DJ, Heistad DD. Evidence that disruption of the blood-brain barrier precedes reduction in cerebral blood flow in hypertensive encephalopathy. Hypertension. 1984;6(2_pt_2):I75.

Tsao MN, Li YQ, Lu G, Xu Y, Wong CS. Upregulation of vascular endothelial growth factor is associated with radiation-induced blood-spinal cord barrier breakdown. J Neuropathol Exp Neurol. 1999;58(10):1051–60. https://doi.org/10.1097/00005072-199910000-00003.

Verheggen IC, Van Boxtel M, Verhey F, Jansen J, Backes W. Interaction between blood-brain barrier and glymphatic system in solute clearance. Neurosci Biobehav Rev. 2018;90:26–33.

Wang Z, Zhang Y, Hu F, Ding J, Wang X. Pathogenesis and pathophysiology of idiopathic normal pressure hydrocephalus. CNS Neurosci Ther. 2020;26(12):1230–40. https://doi.org/10.1111/cns.13526.

White A, Hornsey S. Radiation damage to the rat spinal cord: the effect of single and fractionated doses of X rays. Br J Radiol. 1978;51(607):515–23. https://doi.org/10.1259/0007-1285-51-607-515.

White A, Hornsey S. Time dependent repair of radiation damage in the rat spinal cord after X-rays and neutrons. Eur J Cancer (1965). 1980;16(7):957–62. https://doi.org/10.1016/0014-2964(80)90335-7.

Wong CS, Fehlings MG, Sahgal A. Pathobiology of radiation myelopathy and strategies to mitigate injury. Spinal Cord. 2015;53(8):574–80. https://doi.org/10.1038/sc.2015.43.

Wong CS, van der Kogel AJ. Mechanisms of radiation injury to the central nervous system: implications for neuroprotection. Mol Interv. 2004;4(5):273–84. https://doi.org/10.1124/mi.4.5.7.

Zachariadis A, Tulbu A, Strambo D, Dumoulin A, Di Virgilio G. Transverse myelitis related to COVID-19 infection. J Neurol. 2020;267(12):3459–61.

Statistics of Dose Response

3

Readers who (understandably) skip this chapter will probably return to it later. It is important for understanding the basis for some analyses of the literature that will appear in the subsequent chapters. The reader may wish to use this chapter as a reference to explore more deeply some of the issues related to statistical methods used (or skipped) in the literature discussed later.

Many scientists use statistical procedures commonly available in statistical software packages. There often seems to be the tacit assumption that if the data can be entered into the format required by the software, then the procedure can be legitimately deployed. For example, one could use linear regression to fit binomial data from a dose-response experiment, with the pairs of (dose, response rate) playing the roles of the independent and dependent variables. The results would probably not be useful. In other cases, the data may not meet the requirements of the procedure. For example in linear regression, all observed values of the independent variable, y, are assumed to be normally distributed with the same variance; and, very importantly, the regressor variable x is assumed to be (and must be) independent of y. Meeting these requirements, among others, is necessary for the regression to be valid, or at least for the results to mean what they are understood to mean in statistical texts. Using the wrong statistical method, one cannot test the validity of the model, and the biology that one was attempting to elucidate is not clarified.

Another serious error made by investigators is to apply a reasonable or appropriate model, such as a linear logistic model, but subsequently ignore the fact that the model does not fit the data by normative model assessment criteria. Most software packages will provide estimates of the parameters in the model independent of whether the model fits the data, and often this fact is not obvious to the investigator (or their assistant who is actually performing the analysis). *Parameter estimates of a model that does not fit the data are irrelevant.* The situation may be salvageable, but being unaware of whether the model fits is scientifically unsound.

The purpose of this chapter is not to present a comprehensive exposition of statistical procedures in statistical analysis of dose response, but rather to direct the reader to those procedures to use for specific and rather limited types of data

© Springer Nature Switzerland AG 2022

T. Schultheiss, *Radiation Myelopathy*,

https://doi.org/10.1007/978-3-030-94658-6_3

analysis. The focus will be on the analysis of binomial dose-response data with, of course, special emphasis on radiation myelopathy. Additionally, other metrics in the statistical analysis of dose-response data in radiation myelopathy will be mentioned.

3.1 Experimental Design

Experimental design is the aspect of statistical analysis that is generally given the least attention. For most of us, it is the part about which we know the least. In the analysis of retrospective clinical data, the role of experimental design is played by the selection of the subjects and variates to be included in the analysis. For prospective animal experiments in RM, the independent variables are selectable, and it is the organization of these variables into covariate patterns, along with the number of animals per covariate pattern that comprise the fundamental components of the experimental design. A *covariate pattern* is simply a particular set of independent covariates that define the extent of the experiment in covariate space. For example in a group of rats being irradiated to the same field length, total dose, and dose per fraction, the specific set of those values, e.g., 20 mm, 35 Gy, 5 Gy, define one covariate pattern. In single-fraction dose-response experiments, each dose comprises a covariate pattern.

Statistical power is the probability that a hypothesis test will indicate an effect when such effect exists, i.e., it correctly rejects the null hypothesis when it is false. The estimated power depends on the experimental design and assumptions about the magnitude of the effect. Although formulas exist for calculating the power of an experimental design under specific conditions, the most versatile method is to generate random outcomes for the experiment based on an estimate of the dose-response function, i.e., a Monte Carlo technique. Each generated outcome is subjected to statistical analysis and the power is estimated by the fraction of outcomes where the generating model fits the data.

The fundamental objective in experimental design is for a design to be powerful enough to detect an effect if one exists or otherwise to elucidate at a statistically significant level the role of covariates (biological factors) in producing the effect being studied. It is a tragic waste of time and animals' lives to design an experiment incapable of producing meaningful results, i.e., one that is underpowered.

The fundamental independent variables in the dose-response assessment of RM are dose per fraction and number of fractions. It is often stated that overall treatment time does not have much influence on the response rate when the treatment is given within typical clinical scenarios. However, this assertion has been challenged (Herbert 1989). By varying dose per fraction (or total dose) and the number of fractions, fractionation effect models such as the LQ model or the power-law model can be explored. Additional independent variables come into play depending upon the effect being investigated. Interfraction interval (when less than 24 h) is used to explore the effect of incomplete repair. Field length is varied when exploring the volume effect, and various dichotomous variables are used to study the effects of differing treatment conditions such as the effects of radiation protectors or sensitizers, treatment in hyperbaric oxygen, etc. How these variables appear in the equation defining the dose response impacts the type of analysis that can be performed or that is appropriate.

Inbred animals have been almost exclusively used in rodent experiments. Random bred animals have been used in large animal experiments. Inbred animals are genetically stable, but that does not mean that they are stable with regard to slight changes in experimental conditions. Thus animals from the same source used in experiments separated in time or location may exhibit different responses as a result of different experimental conditions despite the investigators' effort to keep conditions constant. It is feckless to hypothesize what the conditions may be that result in an altered response since these would depend upon the strain, the laboratory, the delivery equipment, etc.

3.2   Experimental Data

Typical dose-response experiments in radiation myelopathy are, like all dose-response experiments, composed of the stimulus variables and the response variables. The response variable in RM is (nearly always) the presence/absence (values of 1 or 0, respectively) of evidence of neurological injury to the spinal cord that occurs within a certain period following treatment. How this neurological deficit is defined depends upon the investigators. (In unusual cases, the response variable is the time to expression of radiation myelopathy, the latent period.) The stimulus variables are the independent variates *aka* predictor variables, explanatory variables, or regressor variables. They obviously comprise the "dose" part of "dose-response." In addition, they may also include any biological variables, conditions, or other treatments that may affect the expression of radiation myelopathy.

Throughout this book, radiation dose-response experiments will use the following conventions: r is the number of responders for a specific dose schedule, m is the number of subjects at that dose level, the number of dose schedules in a given experiment will be n, i will be the index between 1 and n that identifies a dose level (or covariate pattern), P is the response at a given dose level. Thus $P_i = r_i/m_i$. $\hat{P}_i$ is the model-estimated response probability. Total doses are designated by D, the dose per fraction by d, and the number of fractions by N. We shall assume that the covariate pattern consists of g elements including the constant.

By far, the major problem with experimental data in clinical radiobiology is power, i.e., the ability to measure the response meaningfully and quantitatively and possibly to compare it to some alternative. "Clinical radiobiology," in this context, means the *in vivo* response of organs, tissues, or organisms to radiation rather than the response of cells or tissue cultures *in vitro*. For RM, and for many other endpoints, the response is quantal, i.e., present or absent. So radiobiological experiments in RM may be designed to determine one of the following:

The dose response for single or multiple fractions.
The dose response at various doses per fraction.

The effect of adding of a putative radiation protector or sensitizer.
The effect of changing the LET of the radiation.
The effect of treating under non-standard conditions, e.g., in hyperbaric oxygen, shortened interfraction intervals, etc.
The effects of other altered fractionation such as top-up doses.
The effect of treating different volumes or regions of the spinal cord.
The retreatment response.

All of these experiments require groups of animals being treated with specific covariate patterns as defined by the dose, dose per fraction, field length, etc. The problem of an underpowered study arises when too few animals are allocated to each covariate pattern or dose group. The certain sign of this problem is achieving too many groups with either 0 or 100% responses. Acknowledging that the dose-response function is simply a cumulative probability distribution function (pdf), then one can see in principle that the power to distinguish between alternative treatment patterns amounts to being able to distinguish between pdf's. Experiments that are composed of mostly extreme responses and groups with 5–10 subjects will predictably have little chance to distinguish one distribution from another.

In regression of binomial data, there is a general rule of thumb that there should be a minimum of ten events per predictor variable (Vittinghoff and McCulloch 2006). For logistic regression applied to data with extreme values of 100% response, these events are not counted in that number or should be counted as a single event at most. Vittinghof and McCulloch show that in favorable cases, 5–9 events per variable may suffice, but fewer than this generally results in poor estimates of confidence intervals, biased parameter estimates, and excess type I errors, i.e., excessive rejection of the null hypothesis.

3.2.1 Separation

For a dose-response experiment with a single variate (dose), *separation* exists when all 0% responses can be separated from all 100% responses by a single dose point (quasi-separation) or a gap in dose with the response at all lower doses of 0% and the response at all higher doses of 100%. See Fig. 3.1. Multivariate models can have separation defined by a plane in the variate space. Of course the single variate situation is much easier to visualize. It is easy to see that if one is estimating a median tolerance parameter and a slope parameter, that essentially all fitting methods will fail, and any parameter estimates will be specious with separated data. Specifically, *MLE's of model parameters do not exist in the case of separation* and no model assessment is possible. Some have recommended solving the problem by specifying a maximum value for the slope of the dose-response function, which would otherwise have been infinite, or taking a Bayesian approach, which results in the final estimates being contingent on prior knowledge (Williams 1986). A better solution would be to repeat the experiment armed with knowledge of where the D_{50} is bracketed and use an appropriate number of subjects per dose group.

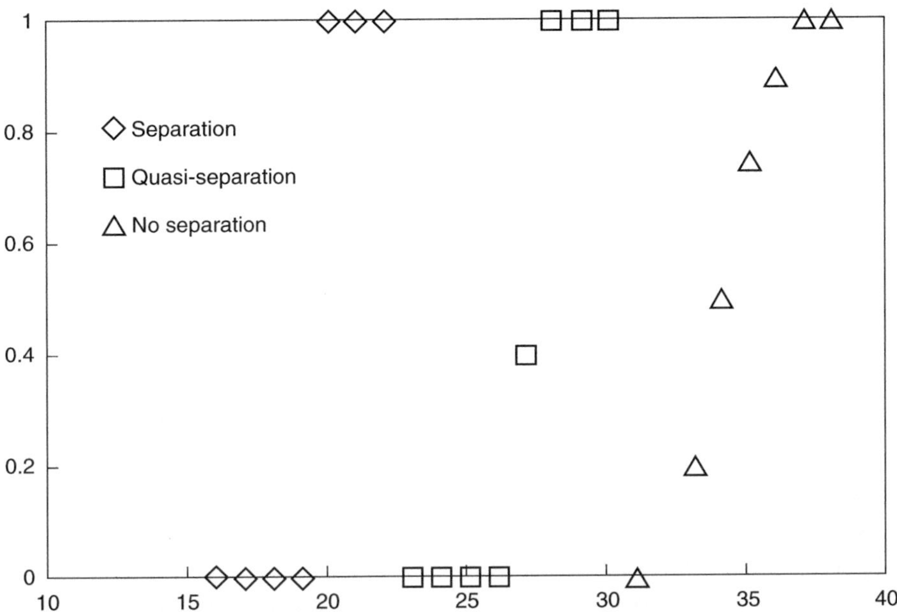

Fig. 3.1 Showing separation where all 0% x values are less than all 100% x values; quasi-separation where one point separates 0% values from 100% values; and no separation.

One macro, i.e., subroutine, for goodness-of-fit test written for SAS (goflogfit) in 2005 states: https://github.com/friendly/SAS-macros/blob/master/goflogit.sas

```
* * * * * * * * * * * * * * * * * * * * * * * * * * * * * * * * * * * * * * * * *
**                          WARNING!!!!                          **
**      THE MACRO FOUND A SEPARATION IN YOUR DATASET.            **
**        ML-ESTIMATORS DO NOT EXIST IN THIS CASE AND           **
**          AND ALL REPORTED RESULTS ARE NOT VALID !!!          **
**              PLEASE CHECK YOUR DATA!!!                        **
* * * * * * * * * * * * * * * * * * * * * * * * * * * * * * * * * * * * * * * * *
```

For the most part, the problem of separation was largely ignored in radiobiology studies until the mid-1990s.

3.2.2 More on Extreme Values

Even if an underpowered experiment does not result in separation of the dose-response points, an excess of extreme values is likely to result, especially for steep dose-response functions. Take for example the component of a fractionation experiment given in Table 3.1. The original doses were fractionated, but for simplicity only their single dose equivalent of the fractionated schedule is given in the table. The doses have been rounded to the nearest 0.1 Gy.

Table 3.1 Actual dose-response data (originally reported as fractionated data, reproduced here with dose expressed as single dose equivalents), showing an extreme case of quasi-separation

r_i	m_i	D_i
0	9	17.4
0	9	18.2
8	9	19.0
9	9	19.5
9	9	20.2
9	9	21.0
9	9	21.7
9	9	23.0
9	9	24.2
9	9	25.0
6	6	25.5

Of the 11 points, using 96 animals, only one of the points has a non-extreme response, neither 0 nor 100%. There are many reasons to avoid extreme values, but one can consider this issue as a reflection of how resources are allocated in an experimental design. In a dose-response experiment, each dose or covariate pattern can be viewed as constituting a pair of cells in a contingency table. Guidelines exist regarding the minimum number of events (or non-events) per cell that is necessary in order to satisfy certain assumptions regarding the goodness-of-fit statistics. Generally it is necessary that both r_i and m_i-r_i are always greater than one and should be greater than 5 in 80% of cases (Cochran 1954; Garson 2016). It should be obvious that the latter will not hold if $m_i \leq 10$.

If $r_i = 0$ then that cell contains 0 events and likewise if $r_i = m_i$, then the complementary cell contains 0 non-events. The resources allocated to obtaining either of these extreme values will have been underutilized at best and squandered at worst.

In fact, a point that is nearly uniformly overlooked in radiation oncology literature is that the norm is to ignore extreme values. In 1938 Bliss wrote "However, when some of the dosages show 0 or 100% mortality and the corresponding corrected probits are determined primarily from the evidence of the remaining observations, their range of variation is restricted and the conventional rule for the number of degrees of freedom is not strictly applicable. An empirical scheme which has proved convenient is to count as one observation the lots at the upper (or lower) end of the curve, enough of which are combined until the sum of the expected number of survivors (or fatalities) exceeds one individual" (Bliss 1938; Cochran 1954). In 1953 Berkson wrote "To consider all observations of 100% as random samples from true P's which are never 100%, as statistical theory of quantal response does, is quite in contradiction of the known biologic facts. Where theory and fact contrast so violently, it is foolhardy to press the theory very hard. It is the considered opinion of the present writer that observations with 0 or 100% should not be used at all—that when they occur, another experiment should be performed with different dosages at values where observations of 0 or 100% are very unlikely, but this is probably an extreme position that will not be generally acceptable. I may point out, however, that in the widely advocated Karber [18, 30] method of estimate of L.D. 50, if at two successive doses on observation of zero per cent mortality is made, only the larger of the two doses is used, and similarly only the smallest of several consecutive doses which show 100% mortality is considered, in making the calculation. There seems to be something unreasonable in never using certain observations in one good method of estimation and always using them in another. My suggestion, then, is to use at most one such observation at each extremity" (Berkson 1953).

Some standard methods for assessing the contribution of each point to the fit of the model to the data will be discussed below and will illustrate this point. By deploying these methods, one can get a better understanding of the overall fit and also identify both the points that are essentially superfluous as well as those that are outliers, or both.

3.3   Analyzing Dose-Response Data

3.3.1   The Dose-Response Function

We relate the stimulus variables to the outcome variable via a mathematical model. In the physical sciences or any situation where the independent variable is controlled and the dependent variable is measured, the most common model used to describe the relationship is the linear model. In such a case we might say "y depends linearly on x." Whereas this may be true, that is not what is meant by a linear model in statistics.

In linear models in statistics, "linear" does *not* refer to the independent variates, i.e., the x's, but rather their unknown coefficients, β, as in the equation

$$z = G(y) = \beta_0 + \beta_1 f(x) \tag{3.1}$$

where z is a function of the measured dependent variable, y; G is the function used to make z depend linearly on the values of β; and $f(x)$ either is x or is a function of the independent variate, x. So z is the G transform of y, and it transforms the model from nonlinear to a linear one. Typically the transform $G(\cdot)$ maps the value of y, whose range is typically (0, 1), into the range $(-\infty, \infty)$ or possibly $(0, \infty)$. Generally it is assumed that $f(x) = x$, and Eq. (3.1) is called a simple linear equation. In the multivariate case

$$\mathbf{z} = G(y) = \beta \cdot \mathbf{X} = \beta_0 + \beta_1 \cdot x_1 + \beta_2 \cdot x_2 \ldots \tag{3.2}$$

where $\mathbf{z}$ is the column vector of responses, β is a vector of unknown coefficients, and $\mathbf{X}$ is a matrix of values of the independent variables. Thus

$$\mathbf{X} = \begin{bmatrix} 1 & x_{11} & \cdots & x_{(g-1)1} \\ \vdots & & \ddots & \vdots \\ 1 & x_{1n} & \cdots & x_{(g-1)n} \end{bmatrix} \tag{3.3}$$

and is known as the design matrix, where g is the number of fitted parameters and n is the number of observations. Using the nomenclature of generalized linear models (GLM), the right-hand side of Eq. (3.2) is called the *systematic component* and sometimes the *linear predictor*. *If* z itself is the measured response (typically of a continuous variable), then Eq. (3.2) represents linear regression and the transform G $(\cdot)$ is the identity function. However, more generally z is a transform of the measured response, y. This transform is called the *link function*. In the case of radiation myelopathy, it takes the value of the observed response, which may vary from 0 to 1, and maps it into the range $(-\infty, \infty)$.

For binary data, $m = 1$, and the response variable r has values of 0 or 1. For binomial data, the response variable is r/m. Obviously P is bounded by 0 and 1, and there are a number of transforms that can map the range $(0,1)$ to $(-\infty,\infty)$, the latter being the range of most distribution functions. Probably the most common transform in GLM's and certainly for biomedical application is the logistic function with

$$G(P) = \ln\big(P/(1-P)\big) \tag{3.4}$$

from which

$$P = \frac{e^{\beta \cdot \mathbf{x}}}{1 + e^{\beta \cdot \mathbf{x}}} \tag{3.5}$$

is easily obtained. Other link functions are given in Table 3.2. Equations 3.4 and 3.5 are used for logistic regression.

For many decades, the equation

$$P(D) = \frac{1}{1 + \left(\dfrac{D_{50}}{D}\right)^k} \tag{3.6}$$

was a common form in which the logistic dose-response function appeared in the RM literature (Leith et al. 1982). From Eqs. 3.5 and 3.6 we have, $\beta_0 = k{\cdot}\ln(D_{50})$,

Table 3.2 Common link functions for binomial data

Link function name	Link function $z=G(P)$	Inverse link function $P = G^{-1}(z)$
logit	$\ln[P/(1-P)]$	$\exp(z)/[1 + \exp(z)]$
probit	$\Phi^{-1}(P)$	$\Phi(z)$
complementary log-log	$-\ln[-\ln(1-P)]$	$1-\exp[-\exp(-z)]$
log-log	$-\ln[-\ln(P)]$	$\exp[-\exp(-z)]$

$\beta_1 = -k$, $x_0 = 1$, and $x_1 = \ln(D)$, which results in a linear logistic function. Suppose there is a factor, x_2, in addition to dose that influences the dose response, and the presence or absence of this factor is denoted by values of 0 and 1. This factor then could represent a radiation protector or sensitizer, the use of a different beam quality other than x-rays, or some other response modifying treatment. Then

$$z = \beta_0 + \beta_1 \cdot \ln(D) + \beta_2 \cdot x_2. \tag{3.7}$$

Noting that $P/(1-P)$ in Eq. (3.4) above defines the *odds*, then it is straightforward to show that for treatments with $x_2 = 1$ compared to treatment with $x_2 = 0$, the odds ratio is e^{β_2}. [If x_2 is not a dichotomous variable but a continuous one, then e^{β_2} is the odds ratio for a one unit increase in x_2.]

Using a different formulation for z, that is

$$z = \beta_0 + \beta_1 D + \beta_2 \cdot Dd, \tag{3.8}$$

where d is the dose per fraction in a fractionated scheme, the α/β ratio in the LQ model can be estimated by β_1/β_2 (Taylor 1990). However, if we are assuming that this α/β ratio represents the curvature of the survival function of the putative targets, then the Poisson transform might be more appropriate to use than the logistic transform, and the Poisson error structure would apply. Because the pathogenesis of late effects is so complex, the proposition that isoeffects for RM are equivalent to isosurvival of a target cell compartment (or group) is difficult to justify and probably wrong. Thus a number of link functions other than the Poisson function would be appropriate, and the precision of the data is rarely sufficient for the error structure to be apparent.

Using

$$z = \beta_0 + \beta_1 \cdot \ln(D) + \beta_2 \cdot \ln(N) \tag{3.9}$$

gives us the systematic component for the power-law formulation of dose response where the effective dose DN^{ε} is a variation of the NSD or TDF. The Lyman and Wolbarst (*aka* LBK model) model uses the normal transform for $G(\cdot)$ rather than the logistic (Lyman and Wolbarst 1987, 1989), but is not linear since

$$z = \left(D - \beta_1 v^{-\beta_2}\right) / \left(\beta_0 \beta_1 v^{-\beta_2}\right) \tag{3.10}$$

where in their original nomenclature $\beta_0 = m$, $\beta_1 = D_{50}$, and $\beta_2 = n$.

Many dose-response models for RM have the LQ model for fractionation effects embedded within them. The systematic component would be something like Eq. (5.6) where D becomes the effective dose D_{eff}, and for the LQ model D_{eff} is computed using

$$D_{\text{eff}} = D_i \frac{(d_i + \alpha/\beta)}{(d + \alpha/\beta)} \tag{3.11}$$

where D_{eff} is the dose given in dose per fraction d that is equivalent to D_i given in dose per fraction d_i. This generally causes the dose-response model to be nonlinear. Nonlinearity is not really a drawback for parameter estimation, but it does eliminate some convenient features of GLM's, especially the regression diagnostics to be discussed below.

In this volume, a mildly nonconventional application of the GLM methodology will be used. Rather strictly following the structure of Eq. (3.2), the signs of the $\beta_i x_i$ elements may be negative where it is desirable to express the value of β_i as positive.

3.3.2   Collinearity

For fractionation studies in RM, the experimental design generally consists of various doses and doses per fraction or number of fractions. Because the total dose and the number of fractions are often correlated in experimental data, collinearity between variates can easily, but not necessarily exist. Figure 3.2 shows a fractionation experiment by Wong and colleagues. Figure 3.2 shows dose-response design with single doses and five different doses per fraction (Wong et al. 1992). There is some collinearity built into the design (all data points). The points in red highlight the non-extreme responses (the ones that count), making the effective collinearity clearly evident. As Herbert (1989) points out, significant collinearity results in inflated estimates of parameters and their variances (and covariances), and can even result in parameters having the wrong sign. Collinearity can be hidden within what seems to be a design without collinearity as a result of allocating too few subjects per covariate pattern, resulting in excessive numbers of extreme values of responses whose covariate patterns then basically drop out of the design matrix.

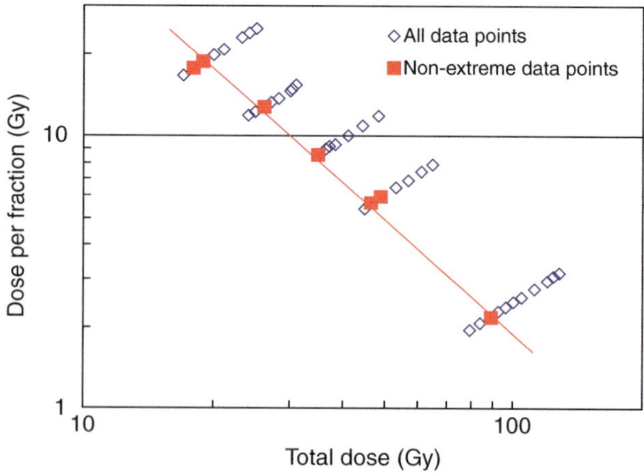

Fig. 3.2 Experimental design showing all data points (empty diamonds) and non-extreme data points (solid squares) plotted on a log-log scale. In simple dose-response experiments collinearity is difficult to avoid in steep dose-response functions, but much more information can be elicited from the data if care is taken to avoid extreme responses.

3.4   Fitting Models

3.4.1   Maximum Likelihood Estimation

The most useful method of fitting a model to binomial dose-response data is nearly always maximum likelihood estimation. Its statistical properties (consistency, efficiency, asymptotically unbiased, asymptotically normal, etc.) are compelling and the optimization process is easy to set up. Note that minimum least squares fit used in linear regression is a maximum likelihood procedure for simple linear models.

Maximum likelihood estimation (MLE) has been used in radiation response experiments since at least 1959 (Best 1959). Thames et al. (Thames, 86) preferred to use the informal term "direct method" despite the fact that "maximum likelihood estimation" and its uses had been well described in the radiation oncology literature (Herring 1980; Metz et al. 1982). For example predating Thames et al. by 6 years, Herring published not only a development of MLE, but also gave a methodology for confidence region estimation and model assessment, and used standard statistical nomenclature throughout rather than adding confusion by adding to the lexicon. He deployed these methods against the LQ model using data from Douglas and Fowler. Unfortunately, *ad hoc* isoeffect analyses, such as the Fe-plot, gained favor in radiobiological applications over this standard statistical technique. MLE gradually supplanted isoeffect analyses after the 1986 Thames publication. At least part of the problem with the relatively slow adoption of MLE techniques was a result of the tedious methods by which literature could be searched during this era; it was especially difficult if a useful paper was never or infrequently cited.

The likelihood function for a dose-response study is, apart from the combinatorial factor $m!/r!\,(m-r)!$,

$$L = \prod_{i=1}^{n} \left(P_i\right)^{r_i} \left(1 - P_i\right)^{m_i - r_i} \tag{3.12}$$

and

$$LL = \sum_{i=1}^{n} r_i \ln\left(P_i\right) + \left(m_i - r_i\right) \ln\left(1 - P_i\right) \tag{3.13}$$

where there are r_i responses in m_i subjects with $P_i = r_i/m_i$. P_i is the dose-response function given, for example, by Eq. (3.5). Maximizing LL as a function of β_0, β_1, and β_2 give their maximum likelihood estimates. The fact that they are estimates rather than "true" values is designated by a caret, e.g., $\hat{\beta}$. The value of P for a given covariate pattern is then also designated by a caret, $\hat{P}$. Note that although LL is in fact the log of the likelihood, it is often referred to as the likelihood since the likelihood itself is never used in practice, it is LL whose features are consistent with the normal regression model.

Once the maximum likelihood estimates of the parameters of the model are in hand, the confidence limits on those estimates may be obtained. There are numerous ways to obtain these confidence limits, but discussion of confidence limits will follow after a few paragraphs on a more important topic, model assessment.

3.4.2 Model Assessment

Strictly speaking, an assessment of the adequacy of the fitted model should precede any attempt at interpreting it. (Hosmer et al. 2013).

It should be emphasized at the outset that I do not necessarily recommend either (1) [Likelihood ratio] or (2) [Pearson χ^2] as useful statistics for detecting lack of fit when the data are sparse. (McCullagh 1986), parenthetic terms added.

The relationship between the fit of the model and the interpretation of the data has an interesting history in RM experiments. Until the 1990s, investigators rarely assessed the model fit, so the interpretation and discussion of biological parameters in dose-response models were unencumbered by the issue of whether the data analysis could support the model-based conclusions. Even now in the clinical literature of radiation oncology, logistic regression and proportional hazards models are commonly deployed without stating whether the model fits or whether the data meet the underlying assumptions of the model. Yet hazard ratios or odds ratios are taken at face value. In experimental literature of RM, there are examples of publications in which the values of biological parameter and their confidence limits are discussed at length even though the model is a poor fit to the data (Kim et al. 1997). But even when the proffered models (typically the LQ model) do not fit the data, one still has the data from which qualitative conclusions might be drawn, but conclusions *based* on a model are not possible without some evidence that the model adequately describes the data.

It is important to understand what is meant by "the model fits the data." It means that the actual results and the expected results based on the model are in agreement to a meaningful and measurable degree. Among the things that it does not mean is that the model assessment confirms specific hypotheses regarding a biological process. It may be that phenomenological models can sometimes speak to the underlying biology, but too often researchers overinterpret their modeling results in that direction. Basing a mathematical model on biological processes does not preclude the possibility that other processes could produce similar or superior results.

Obviously the mathematical model is designed to determine the value of the dependent variable based on the values of the independent variables, i.e., the specific covariate patterns. The difference between the observed and the model-estimated value of the dependent variable is the residual. Ultimately, the model assessment is based on the values of the residuals. Taken together as a set, the values of the residuals follow a "pattern" that is determined by the nature of the data. For binomial data, there is more than one definition of residual (Duffy 1990), but generally the residuals as a group are distributed according to a specific statistical probability distribution function. In general most common distributions for residuals are the normal distribution, the χ^2 distribution, and the t distribution, but these do not comprise an exhaustive list.

The mathematical tools deployed in model assessment will depend on the nature of the experimental data. The dependent variable may be counts, a measured continuous variable, or binary or binomial data. Nearly all the data discussed or re-analyzed in this book will be binomial data. The term binary data and binomial data are frequently used interchangeably, but strictly speaking, the term *binary data* refers to data where dependent variable comes from a single subject and takes on the values of 0 and 1 (or can be interpreted as such), whereas binomial data refers to a group of identical binary elements.

The discussion of model assessment will be limited to models with binomial responses. However, binary data are generated when one or more independent variables are continuous and the data are not grouped. In that case, the responses for all covariate patterns are either 0 or 100%. Thus, clinical response data in radiation oncology, which likely include dose, dose per fraction, and volume as independent variates, are natively binary, but might be "groupable."

Once the model has been fitted to the data, most investigators start their model assessment by simply graphing the data and the model together to get a visual impression of how well the model and the data seem to agree.

The first step in quantitative model assessment is the calculation of the appropriate residuals. A residual is a measure for a given covariate pattern of the agreement between model prediction and the data. The *Pearson residual* for binomial data is

$$r_{Pi} = \frac{\left(r_i - m_i \hat{P}_i\right)}{\sqrt{m_i \hat{P}_i \left(1 - \hat{P}_i\right)}} \tag{3.14}$$

where the response for the ith covariate pattern is r_i/m_i. From these residuals, a summary or global statistic can be calculated and then a determination made whether its value implies a divergence or congruence of the model and the data. The global statistic is often a sum of squared residuals. As stated above, these residuals (and their squares) have a certain *pdf* (or they approximate one) and the cumulative value of this distribution is compared to the sum of residuals. One estimates whether the deviations of the model and the data are likely based on random chance.

Examination of the individual residuals is often the second phase of model assessment. Common summary statistics are often based on the *squares* of the residual values summed over all data points. The summary statistic based on the Pearson residuals is the Pearson χ^2 statistic

$$\chi^2 = \sum_{i=1}^{n} r_{Pi}^{\,2}. \tag{3.15}$$

This statistic is distributed, not surprisingly, as χ^2 with n–g degrees of freedom where n is the number of unique covariate patterns and g is the number of parameters to be estimated. The χ^2 statistic does come with a caution. The values of r_i and m_i-r_i should not be too small. This is typically taken to mean not less than 5. In dose response for radiation, this constraint is rarely met for an entire data set, so the researcher must be prepared to justify the use of the χ^2 statistic when the data are sparse because the squared residual cannot be relied upon to approximate a χ^2 distribution (McCullagh 1986; Collett 2002).

The likelihood ratio, another common goodness-of-fit statistics, is also asymptotically distributed as χ^2. The likelihood ratio is also known as the residual deviance (or total deviance but often shortened to just the deviance), D. That statistic is given by

$$\mathrm{LR} = -2\sum_{i=1}^{n} r_i \ln\left(\hat{P}_i\right) + \left(m_i - r_i\right)\ln\left(1 - \hat{P}_i\right)$$
$$+ 2\sum_{i=1}^{n} r_i \ln\left(\frac{r_i}{m_i}\right) + \left(m_i - r_i\right)\ln\left(1 - \frac{r_i}{m_i}\right) \tag{3.16}$$

where the first term is the maximized log likelihood and the second term is the likelihood with $P_i = r_i/m_i$, corresponding to the log likelihood for the saturated model. In practice the values for the likelihood ratio and the χ^2 value are usually very close, but they become increasingly different as the data become sparser. Because the maximum likelihood method for fitting the data will also minimize the LR, Collett suggests that it is more appropriate to use the LR than χ^2 when linear logistic model estimates are obtained by maximizing the likelihood function (Collett 2002). However, this seems to be a specious argument since the more appropriate statistic would be the one that produces the more accurate assessment of the agreement between the model and the data, which is not reflected by the value of the likelihood (Heinrich 2003). (Using Collett's logic it would appear that the χ^2 statistic would be appropriate when parameters are determined by minimizing the value χ^2.) In multiple simulations, Kuss reports that the deviance is anticonservative in sparse data and is generally outperformed by the Pearson χ^2 (Kuss 2002). However the modeling in his study was limited and general conclusions about the relative strengths of the Pearson χ^2 versus the LR may be limited as well.

Unfortunately, the nomenclature can become confusing. Above, the likelihood ratio for the fitted model is compared to the saturated model wherein $P_i = r_i/m_i$. If the fitted model is compared to the null model, i.e., a model with only a single constant such that $P_i = P$ for all i, then the resulting deviance (differences in log likelihoods) is the null deviance, which Kuss calls the deviance statistic. It is the metric for assessing whether the addition of covariates describing dose, etc. results in a significantly improved model. However, if the LR indicates a poor overall fit to the data, it would seem that the null deviance is irrelevant.

Kuss points out that neither the LR nor the Pearson χ^2 tests may be valid "in the case of sparse data, if the p-values are calculated from the χ^2 distribution, with the residual deviance D [*aka* LR] showing by far the more erratic behaviour" (Kuss 2002). In many of statistical manuscripts on this subject, "sparse data" refers to truly binary data, i.e., data where there is a single subject per covariate pattern. [This occurs generally when one or more of the covariates is a continuous variable.] In the Kuss paper, several situations were simulated, namely 1, 1 or 2, 2, 5, 10, and 1 to 10 subjects per covariate pattern. Kuss goes on to say that the "Pearson test χ^2 might be used and will give reliable results if a significant portion of covariate patterns has more than five individual observations." However, it is important to remember that the simulations performed in that paper had a minimum of 100 subjects. Thus with $m_i = 5$, there were 20 covariate patterns, and in the simulations 80% of these had non-extreme responses. In the case of a radiation dose-response analysis, this would mean 16 dose points that were not extreme responses. However, it is common with rodent experiments for the dose-response functions to be quite steep so 16 non-extreme dose-response points would need to be very closely spaced. In this field, sparse data can be taken to mean relatively few covariate patterns (often fewer than 4 per variable with non-separated data) and fewer than 8 subject animals per pattern. Experiments with 5–6 animals per dose group are incompatible with the use of either the Pearson χ^2 or the likelihood ratio.

Thus we may conclude that having eight animals per dose group *can* yield reliable results but only when the data are not separated and when the data points are spaced closely enough to have few if any extreme responses. Indeed, an extreme response of 0 or 100% for $m_i = 8$ contains the same information as the same point with $m_i = 1$ (Herbert 1993).

The Akaike information criterion (AIC) estimates a model's prediction error against new data (Akaike 1977). It is given by AIC $= -2 \cdot$ Log likelihood $+2 \cdot g$, where g is the number of estimated parameters. Therefore it is useful in comparing candidate models to each other with the higher AIC being preferred. However, this should not be considered as an absolute criterion. Other factors, such as parsimony, can play a role in model assessment.

3.4.3 Degrees of Freedom

For a given value of the χ^2 statistic, the cumulative distribution function decreases with increasing degrees of freedom. This means that the goodness-of-fit statistic improves with increasing degrees of freedom when the χ^2 is constant. So to achieve a better fit, all a researcher has to do is to add additional points to their dose-response experiments at the 0 or 100% levels, well away from the D_{50}, which will not materially affect the χ^2 statistic but by does increase the number of degrees of freedom. Of course if the researcher actually knew the value of D_{50}, they would crowd in more points near that dose, which would be more informative. But even if they do not know the value of D_{50}, they can achieve a good fit by additional high- and low-dose points, as long and they manage to get a couple of points without these extreme responses, i.e., between 0 and 100%.

This of course is specious logic. As stated above, these extreme values should not be included in the analysis *at all*, with the possible exception of the lowest dose 100% value and the highest dose 0% value (Berkson 1953). The number of degrees of freedom is intended to represent the net number of independent elements from which a parameter is calculated or the number that contribute to its calculation minus the number of constraints placed on that calculation. Therefore as a matter of principle, one should not include in the determination of the Pearson χ^2, extreme values whose contribution will necessarily be 0 because the expected and predicted values must be virtually identical.

In further support of limiting the number of degrees of freedom, although Cochran suggests that limiting the minimum expected value of r_i or $m_i - r_i$ to 5 is too conservative, he is somewhat adamant about keeping this minimum at 1 or larger (Cochran 1954). He is thereby advocating that extreme responses should not be used because of the failure of the Pearson residuals to be asymptotically distributed as χ^2.

Example Suppose we wish perform an experiment to determine the single dose D_{50} for RM in some strain of rats. Leith et al. published multiple sets of experiments on the spinal cord tolerance of the (Fischer) CDF™ rat to cyclotron produced He ion radiation. To analyze the data we use the LL from Eq. (3.11) and the dose-response function as expressed in Eq. (3.2) with $f_1(x) = \ln(D)$ as was done in the original work.

The data are given in Table 3.3. Some of these data were inferred from a graph since data were published in table form for only part of the experiments (Leith et al. 1975, 1982). Nonetheless, the data are useful for illustrative purposes.

The data represent several experiments performed at different times. The individual data points at the same dose have not been combined in Table 3.3, but it is fair to ask how the analysis might be different when these points are combined. The answer depends on the method used to choose the best estimates of the $\hat{\beta}$.

Table 3.3 Dose-response data from Leith et al. (1975) for He ion irradiation to the spinal cord in rats

Dose (Gy)	r	m	P_{obs}	$\hat{P}$	$\ln(L)$	χ^2
5	0	14	0	9.56E-11		
5	0	13	0	1.42E-09		
10	0	14	0	5.79E-07		
10	0	11	0	1.13E-05		
15	0	12	0	0.000677		
15	0	12	0	0.00182	−0.02182	0.021842
20	1	18	0.0588	0.07361	−3.83232	0.054493
20	1	12	0.0833	0.07361	−3.45003	0.016644
20	1	13	0.0769	0.14067	−3.78059	0.437062
25	9	13	0.6923	0.68425	−8.02610	0.003904
25	9	12	0.75	0.68425	−6.87329	0.240092
25	12	14	0.8571	0.68425	−3.42938	0.968459
30	11	12	0.9167	0.96995	−3.84060	1.169228
30	14	15	0.9333	0.96995	−3.93212	0.690265
35	15	15	1	0.99685	−0.04727	0.047347
35	15	15	1			

Since the maximum likelihood method is linear in r_i and m_i, then the $\hat{\beta}$ will be invariant under the combination of the data points with the same dose. However, if one were to use something like a minimum χ^2 method, slight differences in the output would be seen. For the same reasons, the χ^2 goodness of fit values will be somewhat different for different groupings.

This data set has a number of attributes that make it very useful for demonstrating the points in this chapter. First, note that there is no separation, i.e., a fair number of responses (8 groups over 3 unique doses) are non-extreme. Also there are 6 leading values of 0% and two trailing values of 100%. Thus only the 0% point with the highest dose will be included in the analysis (Berkson 1953). Finally, all points but one have at least 12 animals. Maximizing the log likelihood with respect to D_{50} and k, we obtain $\hat{D}_{50} = 23.5$ Gy and $\hat{k}=14.6$.

If we used Eq. (3.5), which is merely a change in the form, the same result is obtained. The p-value of the χ^2 using the individual data points is 0.94, a very high value. Combining the data points having the same doses gives $p(\chi^2) = 0.43$, indicating a good fit of the model to the data. Again, the number of degrees of freedom was determined by including only the data from 15.5 to 30 Gy because the points below and above these doses were superfluous extreme values.

This process is adequate for determining the value of D_{50}. If we were more concerned with whether the functional form of the dose-response function is appropriate, an examination of residuals would be necessary. This could be true, for example, if we wanted to assess the validity of the LQ model for an experiment with multiple fractionation schemes. Assessing whether the functional form is misspecified would require enough (high precision) data points to detect a pattern in the residuals.

Although these data were used for illustration purposes, it should be noted that this experiment is one of the most extensive single fraction experiments in the RM literature.

3.4.4 Confidence Intervals and Joint Confidence Regions

Although the absolute value of the likelihood function cannot be used to assess the goodness of fit, its shape near its maximum can be used to assess the confidence intervals or regions for the parameters in the model.

Cox (70) and Cox and Snell (85) suggest at least 3 methods. Strictly speaking, one of the methods applies directly only to the *linear* logistic model and will not be discussed here.

Method 3.1 The Fisher information is the negative of the Hessian matrix (matrix of second derivatives with respect to the model parameters, β_i) of the log likelihood function. The inverse of the information matrix, I, is the asymptotic covariance matrix. Following the Wald test, the 1-α confidence limits of β's may be approximated by $\hat{\beta}_i \pm k_\alpha \sqrt{I_{ii}^{-1}\left(\hat{\beta}_i\right)}$ where k_α is the $\alpha/2$ tail of the standard normal distribution, and

I_{ii}^{-1} is the ith diagonal of the inverse of the information matrix. (Note, for the 95% CI, $k_\alpha = 1.96$.) The math for this technique is considerably simplified for the case of the linear logistic model.

Method 3.2 Similarly to Eq. (3.12) defining the Pearson residual, the deviance residual is

$$d\left(r_i,\hat{P}_i\right) = \pm\left\{2\left[r_i\ln\left(\frac{r_i}{m_i\hat{P}_i}\right)+\left(m_i-r_i\right)\ln\left(\frac{m_i-r_i}{m_i\left(1-\hat{P}_i\right)}\right)\right]\right\}^{1/2} \tag{3.17}$$

where the sign is sgn $(r-m_i\hat{P}_i)$. To test the hypothesis that β_i (or some set of multiple β_i's) lies in a region around $\hat{\beta}$ we use.

$$2\times\left[L\left(\hat{\beta}\right)-L\left(\hat{\beta}_i^*\right)\right]\leq\chi^2_{df,\alpha} \tag{3.18}$$

where $\hat{\beta}_{\{i\}}^*$ is maximized with respect to the components of $\boldsymbol{\beta}$ that do *not* include the components in $\{i\}$ being tested for the confidence region, for the components of $\boldsymbol{\beta}$ not in $\{i\}$ are treated as nuisance parameters. For example we might wish to test that all $\hat{\beta}_{\{i\}} = 0$. The likelihood ratio of expression (3.18) is distributed as χ^2 with degrees of freedom equal to the dimensionality of $\hat{\beta}_{\{i\}}$. This statistic is also commonly used to test whether the addition of another term in z of Eq. (3.2) is significant. In this case, it applies only to nested models. For non-nested models the reader is referred to *Applied Linear Regression*, third edition (Hosmer et al. 2013).

Example from Leith Returning to the above example from the Leith et al. He ion irradiation, we will produce the 1-α joint confidence region of D_{50} and k. We use Method 3.2 above and search for the values of D_{50} and k that satisfy (3.18). Since there are no nuisance parameters, i.e., additional parameters in the model, we only need to search the region around $\hat{D}_{50}$, $\hat{k}$ for the contour that satisfies

$$2\times\left(LL\left(\hat{D}_{50},\hat{k}\right)-LL\left(\hat{D}_{50}^*,\hat{k}^*\right)\right)=\chi^2_{2,\alpha} \tag{3.19}$$

where $LL\left(\hat{D}_{50},\hat{k}\right)$ is now a constant and the values of $\left(\hat{D}_{50}^*,\hat{k}^*\right)$ that satisfy 3.19 are sought. The contours formed by the $\left(\hat{D}_{50}^*,\hat{k}^*\right)$ pairs are shown in Fig. 3.3 for two values of α. As a check, the 10,000 most probable combinations of the response data, based on the values of $\hat{D}_{50}$ and $\hat{k}$, were calculated. The cumulative mass density function for these 10,000 values was 99.96%, thus verifying that nearly all possibilities were covered. Calculating the optimum values of D_{50} and k for these response combinations, we find that those that lie within the 95% contour in Fig. 3.3 account for 91.3% of the cumulative distribution function for all the response data combined. Within the 50% contour, 50.3% of the mass density was accounted for. This exercise simultaneously demonstrates the general validity of the method while also showing that it is not exact. Indeed, there will be some inconsistencies among any methods used for determining confidence intervals or regions. Also shown in this figure is the confidence ellipse determined from the variance-covariance values.

Finally, it should be pointed out that the contours of Fig. 3.3 are well behaved as assessed by their visual symmetry. In other models of dose response with more complicated processes being modeled and with correlated covariates, the contours can take on what seem to be nearly arbitrary shapes, may be discontinuous surfaces, or may not even be closed curves. A simple example of a joint confidence region for two highly correlated variates is depicted in Fig. 3.4 (Rooney and Biegler 1999). Its features include the following. The individual confidence intervals form a rectangular region that does not reflect the correlation of the parameters. Second the range of

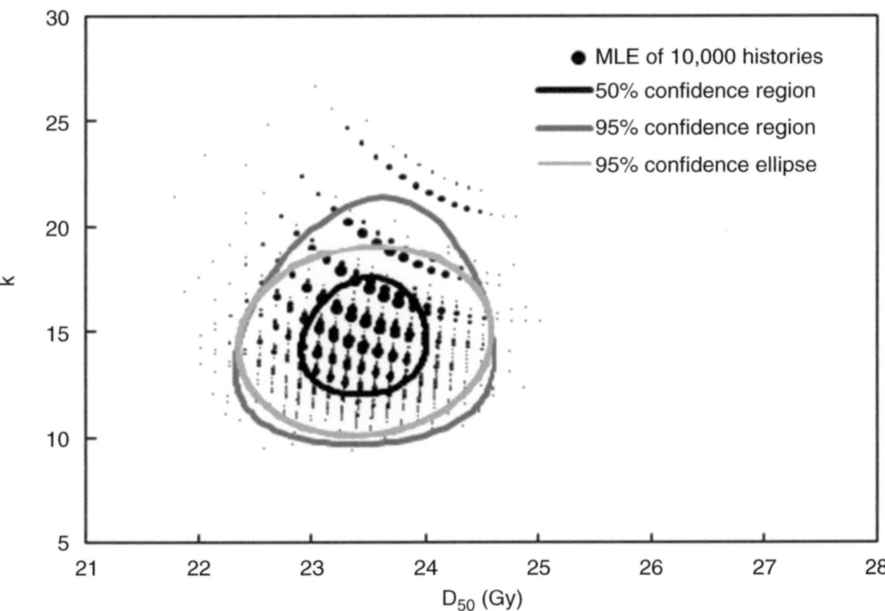

Fig. 3.3 Results of Monte Carlo simulation of data from Leith based on ML estimates of D_{50} and k of original data. The data points comprise a bubble plot of the 10,000 histories with the bubble size proportional to the mass density for each history and therefore obscure many points. The linear pattern in the data points reflects the quantal rather than continuous nature of the outcomes. The 95 and 50% confidence regions were determined by Method 3.2. The standard 95% confidence ellipse is shown in light gray. Note that the eccentricity of the ellipse and of all the data depends on the scale of the axes. The axes need not have the same scale since D_{50} and k do not share a common metric.

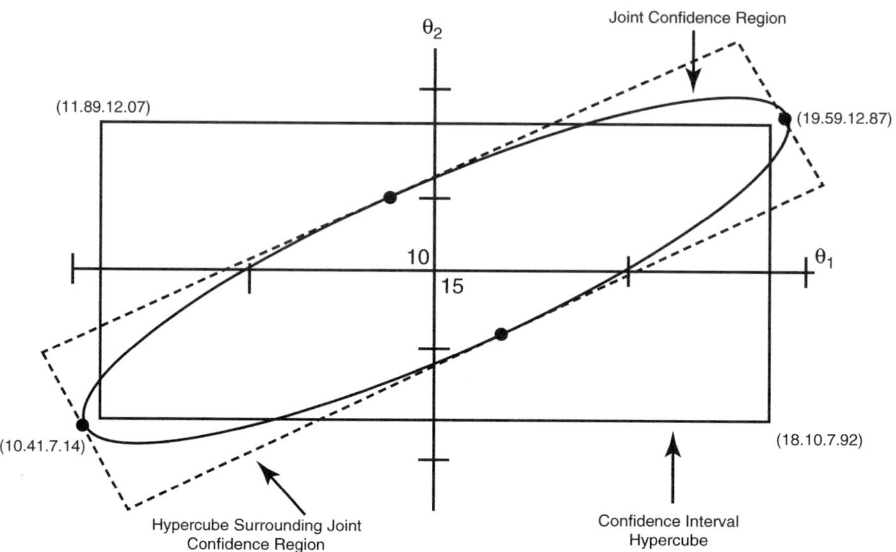

Fig. 3.4 Joint confidence region and individual confidence intervals. The point (15,10) is the maximum likelihood estimate of the parameters and the points defining the rectangle are the limits of the individual confidence intervals. Reprinted from Computer & Chemical Engineering, 23 (10), William C.Rooney and Lorenz T.Biegler, Incorporating joint confidence regions into design under uncertainty, page 1567,1999, with permission from Elsevier.

the individual confidence interval is always less than the range of their joint confidence region. Third, the largest and smallest values included in the joint confidence region do not necessarily correspond to the extreme points of the ellipsoid and in fact may include zero indicating a possible lack of effect.

3.4.5 Examination of Residuals

The model assessment should include more than the global goodness-of-fit assessment. At a minimum, it should include an examination of residuals, the differences between the fitted and the observed values of the response. The residual for the Pearson χ^2 is given in Eq. (3.13), and for the likelihood ratio, the residuals are the summands in (3.16). The residuals may be plotted against $\hat{P}$ or the independent variables or variables not in the model. Generally residuals are approximately normally distributed. Residual analysis is performed to reveal outliers, model misspecification, and possible missing terms.

Another regression diagnostic tool is the hat matrix. Although this name is used more commonly than others, it is more precisely known as the projection matrix. It is easiest to understand using linear regression. It takes the vector of observed responses and maps (projects) them into the vector of fitted responses, i.e., the "hatted" responses, hence the name. The hat matrix is furthermore known as the leverage matrix because its diagonal elements determine the influence each response value has on each corresponding fitted value, i.e., its leverage. In linear logistic regression, the hat matrix is given by

$$\mathbf{H} = \mathbf{V}^{1/2}\mathbf{X}\left(\mathbf{X'VX}\right)^{-1}\mathbf{X'V}^{1/2}. \tag{3.20}$$

where $\mathbf{V}$ is a diagonal matrix with elements $m_i\hat{P}_i\left(1-\hat{P}_i\right)$. In linear regression, $\mathbf{V}$ is replaced by the identity matrix, and $\hat{\mathbf{y}} = \mathbf{Hy}$ is exact. Because $\hat{P}$ must be determined iteratively rather than algebraically in a single step in logistic regression, it can be appreciated that no exact version of $\mathbf{H}$ exists. In *nonlinear* regression, the elements of $\mathbf{X}$ can be replaced by $\partial y/\partial x$ (or $\partial z/\partial x$ in *nonlinear logistic* regression).

The hat matrix diagonals play a role in residual analysis that is not essential to understanding the statistical analysis presented in this volume. The reader is referred to the section "Logistic Regression Diagnostics" in *Applied Linear Regression*, third edition, Chap. 5 (Hosmer et al. 2013).

3.5 Factors Other than Dose

3.5.1 Odds Ratio and Dose Modifying Factor

It is common in experimental radiation oncology to investigate factors that affect the dose response. These may include: other treatments including chemotherapeutics, radiation sensitizers, surgical intervention; altered treatment conditions including treatment under high pressure oxygen, treatment with other types of radiation; or treatment under conditions of comorbid diseases or conditions such as hypertension, diabetes, ataxia telangiectasia, autoimmune diseases; or other categorical variables of interest such as sex, age range, etc. Most of these situations can be handled statistically by treating the factor as a dose modifying factor or as a factor altering the odds ratio. Frequently these two are essentially equivalent.

Let P_N be the dose response under normal conditions and P_E be the dose response under the experimental conditions that differ from the normal conditions only by a single factor. We assume that the parameters of the dose-response functions for the normal and experiment conditions are the same, except the probability of injury differed by a constant odds ratio, OR. Thus

$$\text{OR} = \frac{\dfrac{P_N}{1-P_N}}{\dfrac{P_E}{1-P_E}}. \tag{3.21}$$

Rearranging, we get

$$P_E = \frac{P_N}{OR\left(1 - P_N\right) + P_N}$$

(3.22)

Substituting P_N from Eq. (3.5) into Eq. (3.18) we get

$$P_E = \frac{1}{1 + OR\left(\dfrac{D_{50}}{D}\right)^k}.$$

(3.23)

It is easily seen from Eq. (3.23) that in this model, the odds ratio is equivalent to a dose modifying factor (DMF) with DMF=OR$^{-1/k}$. Recall that this formulation of the logistic model is linear in ln (dose). A similar approach can be used in a GLM that is linear in dose, but in that case, the OR is dose dependent, which in turn means that the odds ratio is not constant.

3.5.2   Exact Interval for Proportion: Graphical Presentation of the Data

It is common but far from universal for the dose-response data points to be presented with error bars. Generally when they are present, the error bars represent ±one standard deviation. (A 95% confidence interval would be so large as to possibly cast doubt on the validity of the experiment.) The standard deviation of a proportion, p, is commonly based on the normal approximation, $[p\,(1-p)/n]^{1/2}$, where n is the number of subjects. Of course this approximation fails at small values of n and for values of p close to 0 or 1 (Morisette and Khorram 1998). An error of ±one standard deviation is generally associated with a central 68.3% confidence region, but for $p < 0.5$ this underestimates the size of the upper limit of this region and overestimates the size of the lower limit. This is illustrated in Fig. 3.5, where $p = 1/n$ is plotted versus n, along with the 68.3% exact confidence limits and $p \pm$ one standard deviation.

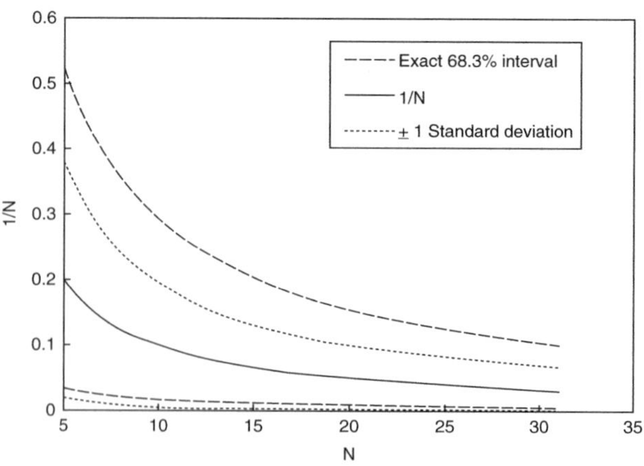

Fig. 3.5 Exact confidence interval for the ratio 1/N. The dotted line shows ±1 standard deviation, which is very often what authors chose to use. The dashed lines show the much larger exact confidence interval for 1/N.

References

Akaike H. On entropy maximization principle. In: Krishaiah PR, editor. Applications of Statistics. New York: North Holland; 1977. p. 27–41.

Berkson J. A statistically precise and relatively simple method of estimating the bioassay with quantal response, based on the logistic function. J Am Stat Assoc. 1953;48(263):565–99. https://doi.org/10.2307/2281010.

Best JB (1959) Maximum likelihood analysis of ionizing radiation induced mortality in whole body fractionated dose experiments. Health Physics 2 (2):139-156. https://doi.org/10.1097/00004032-195904000-00003

Bliss CI (1938) The determination of the dosage-mortality curve from small numbers. Quart. J. Pharm. Pharmacol. 11(2):192–216

Cochran WG. Some methods for strengthening the common χ 2 tests. Biometrics. 1954;10(4): 417–51.

Collett D. Modelling binary data. Boca Raton: CRC Press; 2002.

Duffy DE. On continuity-corrected residuals in logistic regression. Biometrika. 1990;77(2):287–93. https://doi.org/10.1093/biomet/77.2.287.

Garson GD. Logistic Regression: Binomial and Multinomial. 2016th ed. Asheboro: Statistical Associates Publishers; 2016.

Heinrich J (2003) Pitfalls of goodness-of-fit from likelihood. arXiv preprint physics/0310167.

Herbert D. Reflections on the LQ model. Does it 'fit'? (Does it matter?). In: Paliwal B, Fowler J, Herbert D, Kinsella T, Orton C, editors. Prediction of response in radiation therapy: Analytical models and modelling. New York: American Institute of Physics Inc; 1989. p. 400–516.

Herbert DE. Quality assessment and improvement of dose response models: Some effects of study weaknesses on study findings "Cest magnifique?". Madison: Medical Physics Publishing Co.; 1993.

Herring DF. Methods for extracting dose response curves from radiation therapy data, I: a unified approach. Int J Radiat Oncol Biol Phys. 1980;6(2):225–32. https://doi.org/10.1016/0360-3016(80)90042-5.

Hosmer DW, Lemeshow S, Sturdivant RX. Applied logistic regression. Wiley series in probability and statistics. 3rd ed. Hoboken: Wiley; 2013.

Kim JJ, Hao Y, Jang D, Wong CS. Lack of influence of sequence of top-up doses on repair kinetics in rat spinal cord. Radiother Oncol. 1997;43(2):211–7. https://doi.org/10.1016/s0167-8140(97)01928-2.

Kuss O. Global goodness-of-fit tests in logistic regression with sparse data. Stat Med. 2002;21(24):3789–801.

Leith JT, Lewinsky BS, Woodruff KH, Schilling WA, Lyman JT. Tolerance of the spinal cord of rats to irradiation with cyclotron-accelerated helium ions. Cancer. 1975;35(6):1692–700. https://doi.org/10.1002/1097-0142(197506)35:6<1692::AID-CNCR2820350631>3.0.CO;2-M.

Leith JT, McDonald M, Powers-Risius P, Bliven SF, Howard J. Response of rat spinal cord to single and fractionated doses of accelerated heavy ions. Radiat Res. 1982;89(1):176–93. https://doi.org/10.2307/3575694.

Lyman JT, Wolbarst AB. Optimization of radiation therapy III: a method of assessing complication probabilities from dose-volume histograms. Int J Radiat Oncol Biol Phys. 1987;13:103–9.

Lyman JT, Wolbarst AB. Optimization of radiation therapy, IV: A dose-volume histogram reduction algorithm. Int J Radiat Oncol Biol Phys. 1989;17(2):433–6. https://doi.org/10.1016/0360-3016(89)90462-8.

McCullagh P. The Conditional Distribution of Goodness-of-Fit Statistics for Discrete Data. J Am Stat Assoc. 1986;81(393):104–7. https://doi.org/10.1080/01621459.1986.10478244.

Metz CE, Tokars RP, Kronman HB, Griem ML. Maximum likelihood estimation of dose-response parameters for therapeutic operating characteristic (TOC) analysis of carcinoma of the nasopharynx. Int J Radiat Oncol Biol Phys. 1982;8(7):1185–92. https://doi.org/10.1016/0360-3016(82)90066-9.

Morisette JT, Khorram S. Exact binomial confidence interval for proportions. Photogramm Eng Remote Sens. 1998;64(4):281–2.

Rooney WC, Biegler LT. Incorporating joint confidence regions into design under uncertainty. Comput Chem Eng. 1999;23(10):1563–75.

Taylor JM. The design of in vivo multifraction experiments to estimate the alpha-beta ratio. Radiat Res. 1990;121(1):91–7.

Vittinghoff E, McCulloch CE. Relaxing the rule of ten events per variable in logistic and cox regression. Am J Epidemiol. 2006;165(6):710–8. https://doi.org/10.1093/aje/kwk052.

Williams DA. Interval estimation of the median lethal dose. Biometrics. 1986;42:641–5.

Wong CS, Minkin S, Hill RP. Linear-quadratic model underestimates sparing effect of small doses per fraction in rat spinal cord. Radiother Oncol. 1992;23(3):176–84. https://doi.org/10.1016/0167-8140(92)90328-R.

Radiation Myelopathy in Conventional Treatments

4

Although most clinical reports of RM distinguish between the levels of the spinal cord, it is common practice to treat the different levels as if they all have the same response to radiation. This assumption is generally made at the individual patient treatment planning level as well as in the design of dose limits in clinical trials. Certainly the anatomy of the spinal cord varies by level and more recent studies have shown that the radiation response is different. However, these differences are not yet well understood.

4.1 Relevant Aspects of Anatomy of the Spinal Cord by Level

The spinal cord is composed of two types of regions, the gray matter and the white matter. The gray matter is centrally located, roughly forming an H or butterfly shape. The white matter surrounds the gray matter and the cord is approximately bilaterally symmetric morphologically. The gray matter contains the cell bodies of the neurons. Axons, comprising the white matter (along with glial cells mentioned below), are the long threadlike appendage from the neurons that extend upward to the brain or downward to the muscle. These axons carry the electrical impulses to and from the brain. They are insulated with myelin, a fatty protuberance from oligodendrocytes (or oligodendroglial cells, occasionally simply "oligos") that wraps itself around a short segment of the axon. While the myelin is an insulator, it is essential to the secure conduction of electrical signals along the axon. In the central nervous system, axons are myelinated by the processes from the oligos, but in the peripheral nervous system, the axons are myelinated by Schwann cells that wrap themselves around the axons. Peripheral nerves are less sensitive to radiation than spinal cord white matter but probably not because they are myelinated differently. In animal models, nerve roots appear to have a sensitivity to radiation that is similar to the spinal cord. Astrocytes serve many functions in the brain and spinal cord, but one of their mechanical functions is to extend processes (endfeet) to support the

© Springer Nature Switzerland AG 2022
T. Schultheiss, *Radiation Myelopathy*,
https://doi.org/10.1007/978-3-030-94658-6_4

endothelial cells of capillaries and microvessels that form the blood-brain barrier. These endfeet are also involved in signaling, transport, and other maintenance and regulatory functions.

The intrinsic vascular supply of the spinal cord arises from the anterior spinal artery (aka anterior median spinal artery) and two posterior spinal arteries. The anterior spinal artery is in reality a series of longitudinal anastomoses that runs along the anterior midline at the median fissure. Branching from it at right angles and into the fissure are the central or sulcal arteries. After penetrating into the fissure, they then curve alternatively to the right or left into the cord. They branch into precapillaries and capillaries as they enter the white and gray matter. These arteries supply vertically overlapping regions of the central ventral portion of the spinal cord. The two posterior spinal arteries branch into an arterial plexus on the surface of the cord from which branch penetrating arteries into the spinal cord parenchyma. The region that this pial plexus supplies and the region supplied by the central arteries overlap in the ventral portion of the cord. The venous drainage in general closely follows the same architecture. As is usually the case, the veins are fewer in number but generally larger in caliber, and they have more anastomotic channels. Of potential importance to the radiation response, the endothelial cells of the venous capillaries have a different structure and function from those of the arterial capillaries (dela Paz and D'Amore 2009).

The cross-sectional area of the spinal cord varies in size depending primarily upon the number of axons that traverse that level. The number of vessels and glial cells vary according to the need for them created by the number of axons. Consequently, the number of radiation targets and the stochastic opportunity to create a necrotic region vary as well. We note that the spinal cord is neither homogeneous nor isotropic.

Although the size of the spinal cord varies significantly from one level to another, depending upon the number of axons traversing that level, the relative abundance of its vascular supply is sufficient to the needs of that level. Thus it cannot be said, from the perspective of radiation injury, that the thoracic level is less well vascularized than the cervical cord at the cellular level. The surgeon's perspective is quite different and to it we owe this somewhat confusing concept. The anterior spinal artery at the level of the thoracic cord has fewer anastomoses and a less redundant vascular supply so that cutting a thoracic ventral/radicular artery is more likely to result in infarction than in the cervical level.

4.2   Appearance of Symptoms

There have been many suggestions for categorizing RM into different syndromes. However, the various signs and symptoms can appear in almost any combination, depending upon the size and location of the lesion in the white matter. These attempts have primarily tried to classify types of RM based on the latency, severity of the neurological deficit, and whether the course of the condition stabilized. We now know, from numerous experimental studies and autopsies, that there is relatively little variety in the types of white matter lesions produced by radiation. Once the first symptoms appear, there is currently no way to predict the course or degree of the injury. The progression of the symptoms is fairly predictable, but they may halt in their course at any point, they sometimes improve, and the progression sometimes resumes after a relatively long interval of stability.

The first signs often go unnoticed by the patient until they are associated with a progression of symptoms. Initially, there are unilateral sensory deficits, decreased sensitivity to temperature, weakness, clumsiness, and a disruption in proprioception. By the time that the symptoms are severe enough to be reported by the patient, they will have intensified and are likely to include changes in gait or difficulty walking and often some degree of hemiparesis. Spasticity, pain, weakness, hyperreflexia, and incontinence are common. Brown-Sequard syndrome may also be present, but its etiology is probably not the typical hemitransection of the cord. It is classically the result of simultaneous injury to the corticospinal and spinothalamic tracts and the dorsal columns. However with RM, two or all of these regions may contain lesions that are independent of each other. A positive Babinski sign is common. Although Babinski sign is generally associated with upper motor neuron disease, in the case of RM it is the lesion in the white matter motor tracts that is responsible for the pathological reflex. Patients sometimes report a fall or injury as a precipitating event.

L'hermitte's sign (or phenomenon, syndrome, or most correctly "symptom") is the sensation of shock or tingling in the spine and limbs upon neck flexion. It can occur a few weeks after low-dose irradiation, with initial reports coming from results of radiation for Hodgkin disease or other treatments involving long radiation fields. It is understood to result from demyelination, principally in the cervical region. It is usually transient and relatively mild. However, it can accompany or presage permanent RM.

The progression of signs and symptoms of RM almost uniformly start distally and progress caudally up to the level of the spinal irradiation. Sensory deficits precede motor deficits. The severity of the symptoms also progresses, albeit nonmonotonically, but can include periods of stabilization and sometimes some level of improvement. Any improvement usually proves temporary. Symptoms are not associated with spinal cord levels higher than that which was irradiated.

A particularly succinct and yet complete description of the symptoms associated with RM comes from the 1966 paper by Atkins and Tretter (1966b). *"The most common complaint was weakness of one or both lower extremities. One woman noted a change in sensory perception when she could not feel warmth after placing her foot in a hot bath. Leg cramping or jerking was another early complaint. Others noted numbness as well as diminished or absent temperature sensation. Patients frequently had a sensory level with a decrease or loss of pain perception. Pain, sometimes of a burning nature, was early complaint, after which other symptoms and signs developed including loss of vibration sense and position, a Babinski reflex, ankle clonus, spasticity and incontinence or inability to initiate urination or defecation. Most patients developed a Brown-Sequard syndrome characterized by motor weakness and pyramidal tract signs in one lower extremity with alteration of sensory perception (particularly pain and temperature) in the other. Paraplegia developed as a late and ominous occurrence.*" Unfortunately in cervical RM, symptoms can progress to quadriplegia, leaving the patient highly susceptible to death from comorbid disease (less common now than decades ago) that include pneumonia and pulmonary embolism, renal failure, other vascular conditions, cachexia, and suicide.

The latent period for RM is that time between the end of treatment and the first symptom of the injury. For adults, this is never less than 4 months unless an exceptionally high dose was given or there were extrinsic factors that reduced the patient's radiation tolerance. The distribution of latent periods seems to follow a lognormal distribution (Schultheiss et al. 1984). The distribution also appears to be the same for the thoracic and cervical levels. One may infer from this that the radiation lesion develops at the same rate in the cervical and thoracic cord levels.

Although there are many more patients treated with thoracic radiation than cervical spine radiation, there are roughly equal numbers of reports of radiation myelopathy in the cervical region as in the thoracic region (Schultheiss et al. 1984). The reasons are probably that the consequences of cervical RM are generally more severe and patients treated in the cervical region are historically more likely to survive long enough to express a late radiation injury. One might arguably add to this that the doses used to treat head and neck cancers are often higher than those used for thoracic malignancies (or were during the late-twentieth century.) To this can be added the fact that there is usually more opportunity to spare the spinal cord during treatment planning of thoracic irradiation than cervical irradiation.

The cervical spinal cord is most commonly in the treatment fields for head and neck cancer. Using modern technology, the dose can be delivered with intensity modulation over large volumes. Prior to this achievement, complex field arrangements were the hallmark of head and neck radiation treatments. Mixed electron and photon fields were used, and anterior supraclavicular fields abutted lateral neck fields to treat nodal involvement. These complex field arrangements increase the likelihood that fields might overlap, and overlapping fields were often blamed for the occurrence of RM. This happened even when there was no other evidence of overlap other than the occurrence of RM at a dose that was unexpectedly low.

Because head and neck cancers were difficult to control with radiation and there are fewer vital normal structures to avoid, relatively high radiation doses were used. This in turn increased the risk of RM as a result of two factors. First, occasional geographic misses resulted in the spinal cord receiving an unintendedly high dose and second, attempts at dose escalation started from a higher base compared to thoracic treatments risking higher doses to the spinal cord. Such studies were often carried out with higher doses per fraction.

## 4.3	Dose Response by Level of Cord

One can find conflicting assertions in the early literature regarding the relative radiosensitivity of the cervical versus the thoracic levels of the spinal cord. Certainly by the mid-1980s, the dogma was that the thoracic cord has a lower tolerance than the cervical cord. The idea was so entrenched, it barely escaped becoming a question on the specialty boards for radiation oncology (circa 1988). The most probable reason for the perception of increased radiosensitivity was related to the common practice in the 1960s and 1970s of treating with parallel opposed fields but using only one field per day. (This technique survived into the 1980s at some centers.) Because this technique resulted in alternating high and low doses per fraction to the spinal cord, the biological spinal cord dose using this technique for AP/PA treatments of the lung was much higher than the midline dose. Therefore with the same physical dose, a higher biological dose was given to the thoracic cord because of a high dose per fraction in half the treatments. For head and neck treatments with opposed *lateral* fields, the spinal cord itself was at midline and the effective spinal cord dose was not significantly higher than the prescribed dose. Additionally, it was reasoned (wrongly) that since the thoracic cord had a less redundant blood supply, it was therefore more sensitive to vascular injury.

It was not until 2008 that the relative tolerance of the cervical and thoracic levels was assessed using dose-response data (Schultheiss 2008). Unfortunately, a good fit to a dose-response model was achievable only for the *cervical* cord. An update to that study, presented here, has produced dose-response curves for both the cervical and thoracic levels. Although the 2008 data for the thoracic cord had too much dispersion to produce an analyzable dose-response function, a good fit was ultimately achieved here by using only reports that were devoted specifically to radiation myelopathy (as opposed to reports of retrospective analyses of treatment regimes that happened to contain cases of RM) or reports from prospective clinical trials that include details of complications.

Logistic regression was used to model the dose response. Because many patients whose tolerance was exceeded would not survive their disease long enough to express a late effect like RM, the crude incidence found in literature reports cannot be used without modification. Following Schultheiss et al. (1986), if m_i patients were treated with covariate pattern i (total dose, dose per fraction, etc.), then number of patients expressing the radiation injury between time t and $t + dt$ would be given by the expression $m_i \cdot P_i \cdot S(t) \cdot f(t)dt$, where P_i is the probability of exceeding tolerance for the ith covariate pattern, $f(t)$ is the distribution of latent periods, and $S(t)$ is the probability of surviving to time t for the disease being treated. In words, the number of patients expressing the injury between t and $t + dt$ is:

[the number of patients irradiated]

    ×

[the probability that tolerance was exceeded]

    ×

[the probability that the injury was expressed between t and $t + dt$ (given that tolerance was exceeded)]

    ×

[the probability that patients survived long enough to express the injury, i.e., to time t].

Thus total number of expected myelopathy cases, r_i, for covariate pattern i is given by

$$r_i = m_i P_i \int_0^\infty S(t) f(t)\, dt \qquad (4.1)$$
$$= m_i P_i g_i$$

where the integral, g_i, is the probability that a patient whose tolerance was exceeded survives to express the injury.

Table 4.1 gives the references and the data included in the regression analysis. The publications from which the data were abstracted had to satisfy one of the following criteria: the paper was specifically dedicated to radiation myelopathy (or RM plus other late effects), or the paper was presenting results from a prospective clinical trial complete with complication information. In either case, the dose regimen had to be the same for all patients constituting the m_i patients at risk. In other

Table 4.1 Data from reports of radiation myelopathy or clinical trials with RM used in the logistic regression

Reference	Cervical (C), Thoracic (T)	Total dose (Gy)	Dose per fraction (Gy)	RM cases (r_i)	Total cases (m_i)	g[a]
McCunniff and Liang (1989)	C	60	2	1	17	0.929
	C	65	1.63	0	24[b]	0.929
Abbatucci et al. (1978)	C	54	3	7	21	0.836
Atkins and Tretter (1966a)	C	19	9.5	4	13	0.704
Marcus Jr. and Million (1990)	C	47.5	1.85	0	211	0.738
	C	52.5	1.85	0	22[b]	0.738
Jeremic et al. (1991)	C	60	2	2	19	0.891
Jeremic et al. (1991)	C	65	1.63	0	19[b]	0.891
Kim and Fayos (1981)	C	62.8	1.93	3	109	0.531
(Macbeth et al. 1996)	T	37.8	3.15	0	86	0.200
(Dische et al. 1981), Dische et al. (1986), and Dische et al. (1988)	T	34.1	5.68	8	87	0.260
	T	34.9	5.82	4	31	0.368
Hatlevoll et al. (1983)	T	18	6	8	157	0.217
	T	18	6	9	230	0.217
Macbeth et al. (1996)	T	17.9	8.93	3	524	0.174
	T	40.9	3.15	2	153	0.211
	T	31.5	5.25	0	36[b]	0.200
Simpson et al. (1982)	T	36	6	1	5	0.302
Fitzgerald Jr. et al. (1982)	T	41.2	4.12	6	200	0.205

[a]This factor represents the probability that a patient whose tolerance was exceeded will survive to express the injury

[b]As recommended by Berkson (1953), only a single data point was included that contained a 0% response; therefore these groups were included only in the fitting process but not the goodness-of-fit statistics

words, papers that did not meet the inclusion criteria were reports of retrospective studies that might have included RM data or papers where only ranges of doses and doses per fraction were given. By using this constraint, it was assumed that follow-up would be sufficiently rigorous to find all the cases of RM that occurred prior to the report.

The values of g in the last column of Table 4.1 were determined by numerical integration. The function $f(t)$ was taken from Schultheiss (2008), based on data from Schultheiss et al. (1984), where $f(t)$ was reported to be a mixture of two log-normal distributions with

$$f(t) = 0.414 \cdot \mathcal{N}\{\ln(10.9\,\mathrm{mo}), 0.405\} + 0.586 \cdot \mathcal{N}\{\ln(29.6\,\mathrm{mo}), 0.334\}.$$

$\mathcal{N}\{\mu, \sigma\}$ is the normal distribution with mean μ and standard deviation σ. The survival function, $S(t)$, for each covariate pattern was either determined by fitting an exponential survival function to the published survival data or by integrating from one time point to the next if individual survival times were given, again assuming exponential survival between data points. To illustrate the validity of the estimate of g, values of the actuarial incidence of RM were compared to values of $P(D)$ determined from Eq. (4.1). Note $P(D)$ determined in this manner is nonparametric in that it does not depend on a specific functional form of the dose-response function.

The results of the comparison between the incidence estimated by the actuarial method (using the original authors' estimates where available) and the estimate from Eq. (4.1) are shown in Fig. 4.1. No statistical analysis was performed, but the solid line goes through the origin with unit slope, showing where the two estimates would be equal. The three large outliers above the line result from the outsized effect that a late event has on the actuarial estimate when there are relative few subjects remaining in the cohort. These points come from reports of the incidence in

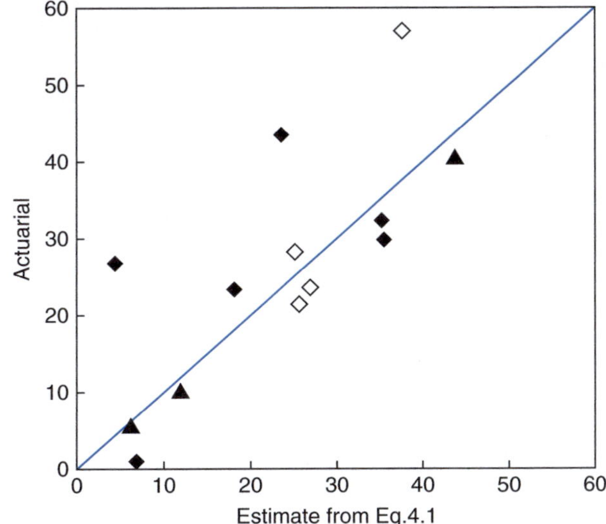

Fig. 4.1 Estimate of the rate of RM from Eq. (4.1a, b) compared to stated actuarial values from various reports. Solid points are those used in the dose-response analysis and open points were from papers not meeting the inclusion criteria. Triangles are cervical and diamonds are thoracic RM. The three outliers result from the outsized influence that late occurring complications have on the actuarial calculation.

Table 4.2 Excluded studies potentially containing RM data

Reference	Reason for exclusion
Abramson and Cavanaugh (1973)	Denominator unclear; cord blocking nonuniform
Choi et al. (1980)	Survival is NED; case of RM did not have full dose
Coy and Dolman (1971)	Tables and text were inconsistent; fractionation may not have been constant; denominator unclear
Eichhorn et al. (1972)	Tables were inconsistent; multiple techniques were used; question regarding 250 kV dosimetry
Guthrie et al. (1973)	No mention of RM
Hazra et al. (1974)	No mention of RM
Madden et al. (1979)	Retrospective; inconsistent numbers
Miller et al. (1977)	Abstract only
Scruggs et al. (1974)	Retrospective

patients treated for lung cancer with the last myelopathy case occurring at 39, 42, and 48 months when, as one would expect, there were very few survivors. Overall the agreement between the actuarial estimates and the estimates from Eq. (4.1) is encouraging. The data come from papers listed in Tables 4.1 and 4.2.

For the initial analysis, the cervical and thoracic data were analyzed separately. The logistic model used for the dose response is given in Eq. (4.2a) below:

$$\ln\left(P / \left(1 - P\right)\right) = \beta_0 + \beta_1 \cdot D + \beta_2 \cdot d \cdot D \tag{4.2a}$$

$$\ln\left(P / \left(1 - P\right)\right) = \beta_0 + \beta_1 \cdot D + \beta_2 \cdot d \cdot D + \beta_3 \delta_T \tag{4.2b}$$

where P is the probability of myelopathy. The ratio of β_2/β_1 can be interpreted as the α/β ratio of the LQ model (Taylor 1990) although in this formulation the α/β ratio is merely a fractionation parameter and not associated with a specific cell survival concept. The term δ_T accounts for the relative odds when it is the thoracic cord being irradiated ($\delta_T=1$ for the thoracic cord).

The log likelihood function, L, which was maximized with respect to the vector of unknown parameters, β, is

$$\ln\left(L\right) = \sum_{i=1}^{n} r_i \ln\left(g_i P_i\right) + \left(m_i - r_i\right) \ln\left(1 - g_i P_i\right) \tag{4.3}$$

[The first term in the sum can be expanded as $r_i \ln (g_i) + r_i \ln (P_i)$; $r_i \ln (g_i)$ does not contribute to the optimization since it does not depend on the unknown variables.]

In the following analyses, the recommendation of Berkson was followed. It states that if there are multiple consecutive values of extreme responses (0 or 100%), only the ones adjacent to non-extreme responses should be used (Berkson 1953). This applies only to consecutive extreme responses at the low-dose (0%) or high-dose (100%) ends of the design, not to extreme responses interspersed among non-extreme response values. In this analysis, only a single extreme response out of four was used for the cervical cord, but the one with the largest number of patients in the denominator (47.5 Gy, 0/211) was selected rather than the one for which the dose was highest. There were two extreme responses in the thoracic data, but they were not consecutive, therefore both were used.

In fact, no theoretical problem exists for using the complete set of data for determining the maximum likelihood estimates of β. However, in assessing the goodness-of-fit, the presence of an excessive number of uninformative extreme values, which do not truly affect the fit, inflate the number of degrees of freedom and consequently artificially "improve" the goodness-of-fit. Therefore, all the values were used in the maximum likelihood estimates but only one of the consecutive extreme values was used to assess the χ^2 value.

The data in Table 4.1 comprise the totality of published clinical data meeting the criteria stated above that were identified through literature searches, reviews, and decades of investigation, which in all but three cases involved direct contact with the authors. There are additional studies that seem at first to meet the inclusion criteria, but these have issues that confuse the determination of the observed incidence and were therefore excluded. These and other additional studies are included in Table 4.2. Because the data contained in these studies are exhaustive (as much as practicable) and irreproducible, a number of dose-response model are investigated below in the attempt to produce the most comprehensive exposition possible of the dose response of the spinal cord.

First we should compare the current results to the 2008 analysis. The cervical data in Table 4.1 were fitted to the regression model of Eq. (4.1a). The current data include the data from Kim that were not in the 2008 analysis. Two other data points were changed after a reassessment. It was judged on this re-analysis that for the McCunniff et al. data, a denominator of 17 was a better representation of the original data than value of 12 used in the 2008 paper, and for Abbatucci et al., the raw incidence of 7/21 was used in preference to the 7/15 in the 2008 study, which is supported by other references (Combes et al. 1975). The increase in both denominators is a result of included patients treated with smaller field sizes that were not included in the original analysis. In addition, the value of g_i for the Abbatucci paper was recalculated using data directly from a table in the original publication. The smaller value of α/β found here also results in a shift of the dose-response curve. All other data from the 2008 paper remained unchanged after a complete review.

Figure 4.2 shows the original (4.2a) result (2008) along with the current analysis (4.2b). Also shown here is the shift of the two data points whose y values changed as a result of the reassessment. The higher part of the dose-response curve is clearly shifted to the right, but the lower part (the clinical region below 60 Gy) exhibits a negligible change. The data point whose $D_{2\,Gy}$ shifted the most was from Atkins and Tretter with a very high dose per fraction (9.5 Gy), making it more susceptible to changes in the α/β ratio. In this figure, the D_{50} for the cervical cord is 76.1 Gy compared to 69.4 Gy in the 2008 analysis (Schultheiss 2008).

Figure 4.2b shows the current cervical-only fit (above) with the data points from the thoracic paper plotted on the same graph, but using the α/β ratio found for the cervical data analysis to determine $D_{2\,Gy}$. The procedure was then repeated, fitting

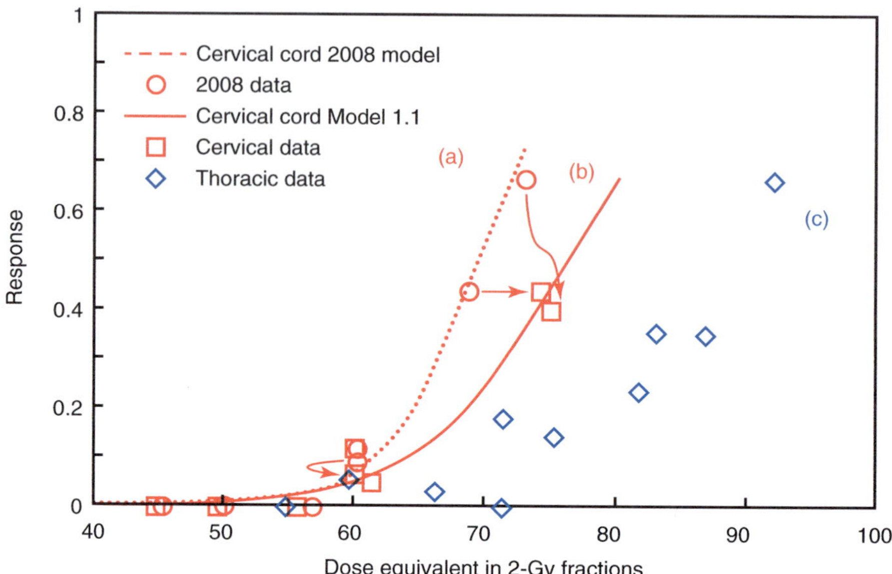

Fig. 4.2 Comparison of the 2008 dose-response model for cervical RM and the current Model 1.1. Also shown are the two data points that were revised between the 2008 publication and this volume; a curved arrow is drawn from the 2008 value to the current value. (**c**) shows the thoracic data point locations assuming the cervical α/β ratio from Model 1.1. The straight arrow represents the shift in the $D_{2\,Gy}$ calculated for the 2×9.5 Gy regimen from Atkins and Tretter resulting from different α/β ratios estimated for the curve (**a**) versus curve (**b**).

only the thoracic data and plotting the cervical data using α/β ratio found for the thoracic fit (Fig. 4.3). Of course, the y values for the data points do not depend on the fitted parameters since they are the observed response (in the absence of censoring), but the values of $D_{2\,Gy}$ depend on the fit through the estimated value of α/β. Therefore with different sets of the β parameter values found for the different regressions, the relative positions of the clinical data points shift.

The result was a good fit in each dataset as measured by the Pearson χ^2 statistic, with examination of residuals showing no concern regarding outliers and the Hat matrix diagonals not showing concern regarding points that seemed excessively influential in the fit (Hosmer et al. 2013). The maximum likelihood estimates of the α/β values were 0.57 and 1.41 Gy for the cervical and thoracic levels, respectively. The parameter values are given in Table 4.3.

In Fig. 4.2, the thoracic and cervical data points are well separated. However in Fig. 4.3, they all seem to follow the thoracic dose-response curve, even though *only* the thoracic data contribute to the fit in this case. The data point that does not follow the dose-response curve is the Atkins and Tretter 2×9.5 Gy.

The congruence between the cervical and thoracic data motivated pooling the data for regression. In the analysis of the pooled data, the χ^2 value indicated a poor fit ($p\,(\chi^2) \sim 0.1$), but this was based solely on the contribution of the Atkins and Tretter data point. Treating that data point as an outlier and re-analyzing, a good fit was obtained with an α/β ratio of 1.14 Gy.

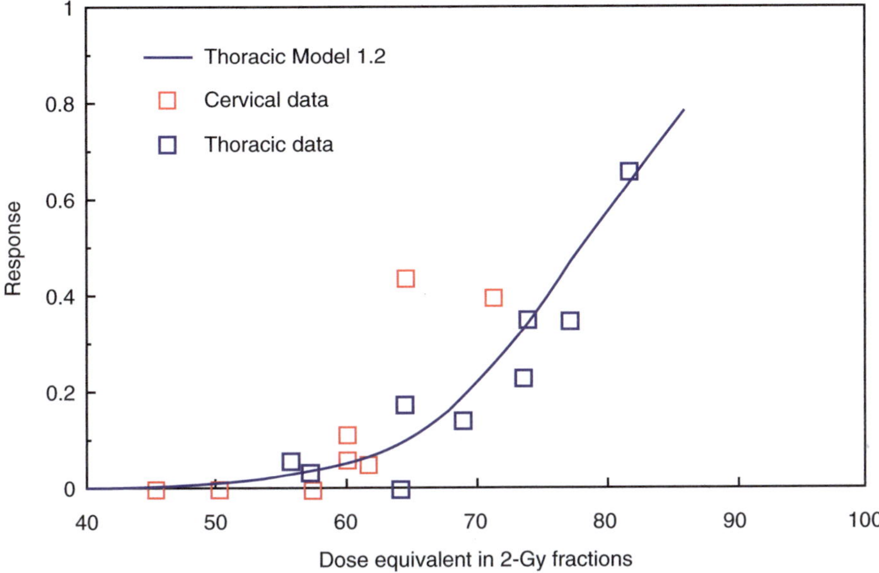

Fig. 4.3 Thoracic-only data fitted with Model 1.2. Also shown are the cervical data points using the α/β ratio from Model 1.2. The data overlap except for a single outlier from Atkins and Tretter.

This situation prompts a serious examination of whether to treat Atkins and Tretter's 2×9.5 Gy regimen as an outlier, especially considering that a good fit was obtained when it was included in the cervical-only analysis. The publication indicates that the treatment field was a single anterior *en face* 22.5 MV beam. So not only was the dosimetry quite simple, having no posterior field means that there was not possibility of a problem due to treating alternate fields on alternate days. Four of 13 patients who were treated with this technique developed RM, and survival times were given for all patients. There is very little chance that anything other than stochastic variation would cause the observed response to be in error. Finally, given that the estimated response rate was 0.437, there is only an 11% chance that the observed response would be less than 4 out of 13 patients. Therefore rejecting this data point out of hand is unwise, and a thorough examination of the possibility that the cervical and thoracic dose responses differ is necessary.

The simplest technique to assess the potential difference in the difference in responses for the cervical and thoracic levels of the spinal cord is to include an odds ratio for one of the levels. The model for this is given in Eq. (4.2b) where the term, $\beta_3 \cdot \delta_T$ was added to Eq. (4.2a) to account for the level of the cord. The variable, δ_T, has values of 0 for the cervical and 1 for the thoracic level. Thus, the coefficient of δ_T is the log odds and e^{β_3} is the odds ratio for RM of the thoracic spinal cord relative to the cervical cord at the same dose. The pooled analysis is shown in Fig. 4.4, and the parameter estimates are given in Table 4.3.

Modeling the difference in the responses of the cervical and thoracic cord as an odds ratio indicates only that data do not reject the hypothesis that there is a difference in the dose response in these two levels of the spinal cord, but it indicates

Table 4.3 Models for clinical dose response of the spinal cord

	Cervical only Model 1.1	Thoracic only Model 1.2	Pooled data Model 1.3[a]	Pooled + OR Model 2	Pooled + volume Model 3.1	Pooled + volume log-log link Model 3.2	Pooled Model 4	Pooled LBK Model 5
β_0	−14.92	−10.51	−12.59	−11.45	−12.87	−6.028	–	–
D_{50}	76.1	77.7	78.1	80.2	77.7	77.4	77.5	77.5
β_1 CI	0.044 −0.012,0.103	0.056 0.005,0.115	0.059 0.027,0.054	0.028 0.001,0.050	0.03 0.003,0.066	0.020 0.004,0.036	$k = 11.3$	$m = 0.16$
β_2	0.076	0.040	0.051	0.057	0.066	0.031	–	–
α/β	0.57	1.41	1.14	0.50	0.52	0.64	0.56 0.09,1.01	0.51 −0.02,1.04
OR (CI)	–	–	–	0.31 (0.11,0.82)	–	–	–	–
v CI	–	–	–	–	0.31 0.15,0.70	0.38 0.18,0.82	0.33	$v^{-n} = 1.13$
χ^2 (df)	3.2 (3)	5.2 (7)	12.8 (13)	13.1 (11)	12.2 (11)	10.4 (11)	10.9 (11)	11.5 (11)
p (χ^2)	0.36	0.64	0.46	0.28	0.35	0.50	0.45	0.40

CI=confidence interval, df degrees of freedom, OR odds ratio, v = relative volume
Model 1 = Eq. (4.2a); Model 2 = Eq. (4.2b); Model 3 = Eqs. (4.2a) and (4.4); Model 4 = Eqs. (4.4)–(4.6); Model 5 = Eq. (4.8)
[a]Atkins and Tretter study deleted as an outlier

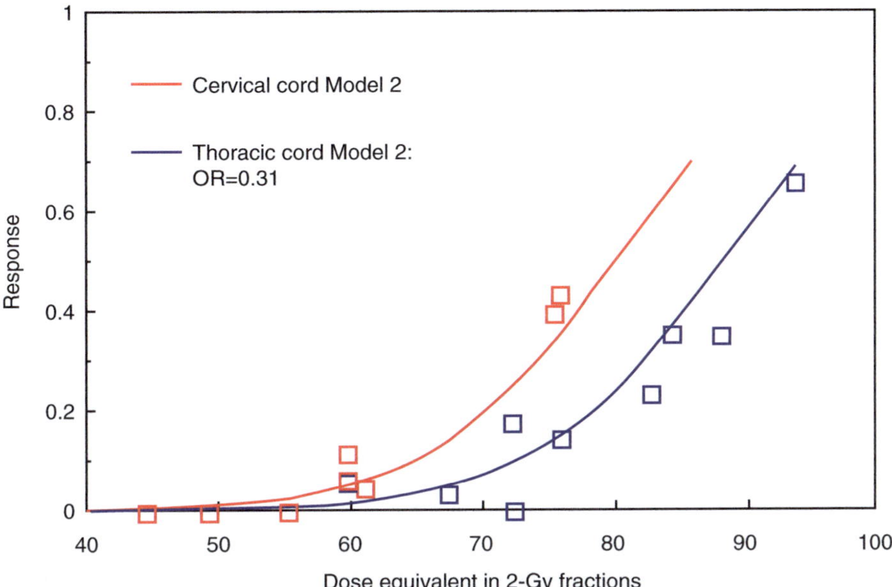

Fig. 4.4 Model 2 fitted to pooled cervical and thoracic data. The odds ratio applies to the thoracic response.

nothing about the underlying biology. The idea that there is an intrinsic difference in the cellular responses for the two levels is troublesome. This could result in different α/β values for the two cord levels. We can however pursue mechanistic models to account for the difference. One can hypothesize that the difference between the dose response of the two levels is due to a difference in the target volume (or the volume of targets) in the thoracic versus the cervical level. The simplest and yet a robust model for the volume effect is generally, but inexactly called the critical element model, but it does not have any fundamental dependence on elements. In this case, it is given by

$$P_T = 1 - \left[1 - P_C\right]^{v} \tag{4.4}$$

where P_C is the probability of injury for the cervical cord by Eq. (4.2a), P_T is the probability for the thoracic cord, and v is the volume of the thoracic cord (or the targets therein) relative to the cervical cord. Treating the relative volume, v, as a variate, its MLE estimate was 0.38 with a 95% confidence interval of 0.18–0.82. See Table 4.3.

Equation (4.2a) is a linear logistic function because it is linear in the elements of β. A similar form of the logistic dose-response function is

$$P = \cfrac{1}{1 + \left(\cfrac{D_{50}}{D}\right)^{k}} \tag{4.5}$$

which is also linear if we write $\beta_0 = k \cdot \ln(D_{50})$ and $\beta_1 = -k$. However, this expression is not linear in D but rather in $\ln(D)$. Furthermore, its linearity is maintained only for constant dose per fraction or constant number of fractions. For changing fractionation, D must be converted to an equivalent dose. Using $D_{2\,Gy}$, the dose delivered in 2-Gy fractions, as that equivalent, we have

$$D_{2\,Gy} = D\frac{\alpha/\beta+d}{\alpha/\beta+2} \tag{4.6}$$

where dose D is delivered in fractions of d. Substituting $D_{2\,Gy}$ for D in Eq. (4.5) maintains the logistic link function (see Chap. 3), but loses the linearity in the fitted parameter α/β. However, using this form of the dose-response function, α/β can be estimated directly. The parameters estimates are given in Table 4.3, along with the confidence limits on α/β. A slightly better fit is obtained, but the value of α/β is very similar.

The dose-response function found by modeling the volume effect seems to have tails that are too large, so a modification to the model was tested. Rather than using the logistic link, the log-log link was used. For this link P is determined by

$$\ln\left(-\ln(P)\right) = \beta_0 + \beta_1 \cdot D + \beta_2 \cdot d \cdot D, \tag{4.7}$$

and for this last model, the volume effect was modeled using Eq. (4.4). A slightly better fit is again achieved as seen in Table 4.3 and Fig. 4.5. Using this link for the odds ratio model (results not shown here) yields a slightly poorer but still adequate fit.

The final model tested on the data is the LBK model by Lyman and Wolbarst (1989). In this model, the link function is the inverse normal, Φ^{-1}, given by

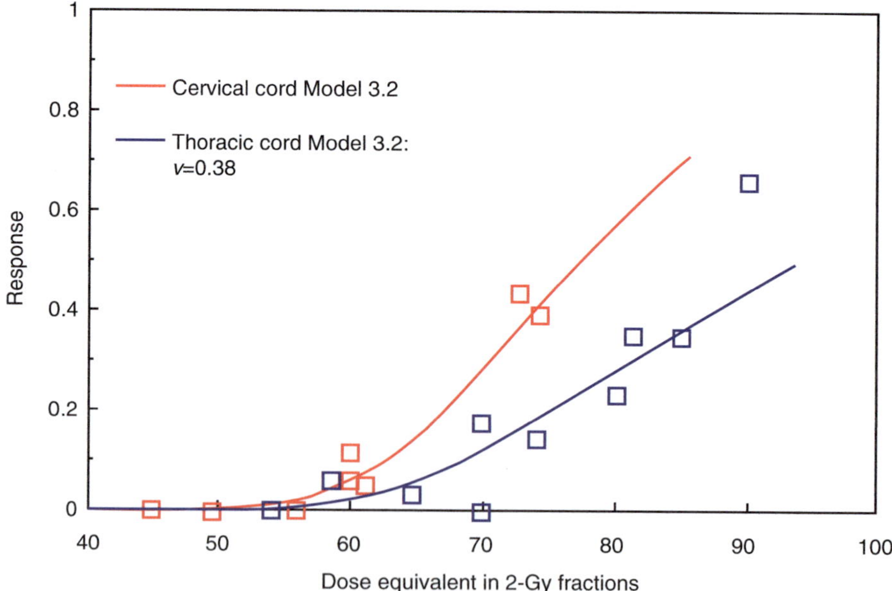

Fig. 4.5 Model 3.2 fitted to pooled cervical and thoracic data. The relative volume, v, applies to the thoracic response.

$$\Phi^{-1}\left(P\right) = \frac{\left(D - D_{50}v^{-n}\right)}{mD_{50}v^{-n}} \tag{4.8}$$

where D and D_{50} are given in 2-Gy fractions, m is the steepness parameter, v is the volume relative to some standard volume, and n is the volume sensitivity parameter. Using the current data, the relative volumes of the treated spinal cord are not available, and we have assumed a fixed target volume of the thoracic cord relative to the cervical cord, for which $v = 1$. For fitting purposes, this means that we cannot estimate n in Eq. (4.8) but only v^{-n}. Of course v cannot be estimated independently in this case either.

The parameter estimates for all models are given in Table 4.3. The fact that the 95% confidence interval for β_1 goes negative in two cases indicates that the confidence interval for α/β includes 0. In the original 2008 analysis of the cervical cord data, the α/β was 0.87 Gy with 95% confidence limits given as 0.54–1.19 Gy. This confidence interval was incorrect because of an error made in the calculation of the confidence limits using the likelihood profile. Upon recalculation, the lower confidence interval was found to be 0.06–1.54 Gy (Bender 2016). Recalculation of the parameter values resulted in no change and the χ^2 values were likewise unchanged. However, because of the recommendation to include only a single low-dose point with 0% response (Berkson 1953), the number of degrees of freedom drops from 5 to 2. The fit remains good. For the analysis presented here, the lower bound for the α/β ratio is very close to or less than 0 in all cases. The interpretation of this is that the dose term in Eq. (4.2) is less important than the dose-times-dose-per-fraction term, or equivalently, that the dose term barely reaches statistical significance at 95% confidence level.

Choosing which model to use, based on the limited data at hand, is almost a matter of taste. Of the models fitted to all the data (with no outliers), the volume model with the log-log link (Model 3.2) seems the best choice. However, we must decide whether fitting all the cervical and thoracic data with the same parameters (Model 1.3) is preferable to allowing for at least one parameter (volume) to distinguish the responses of the two spinal levels. If we treat the Atkins and Tretter data as an outlier and apply the probability model for volume using the log-log link, we can compare the results to the more parsimonious model 1 using the same data. In this case, the application of the Akaike Information Criterion indicates that the volume effects model significantly outperforms Model 1 and is to be preferred. This result strongly implies that the cervical cord is more sensitive than the thoracic cord, but this is clinically true only at doses that are higher than are clinically acceptable. Only when applying unusually high spinal cord doses should the difference in the responses become a consideration. Nonetheless, when making estimates of complication probabilities or converting doses from one fractionation schedule to another throughout the rest of this book, Model 4 result will be applied.

It is not certain that the volume effect accounts for the entire difference between the radiation tolerance of the cervical versus the thoracic cord. There are a number of additional factors that could play a role. In general, the cervical cord seems more prone to be the site of injury and disease. Traumatic spinal cord injury occurs more frequently in the cervical cord than in the thoracic cord, mostly for mechanical reasons (Ahuja et al. 2017; Mathison et al. 2008; Ulndreaj et al. 2017). However, mechanical stress on the spinal cord may also contribute to the initiation of RM (Schultheiss and Stephens 1992). The cervical cord is more likely to bear MS plaques than the thoracic cord (Hua et al. 2015; Vollmer 2007; Eden et al. 2019). In MS, the population distribution of plaques increases from C1 to C3 and then falls off distally (Hua et al. 2015; Fog 1950), although the cross-sectional distribution of MS plaques differs from the distribution of lesions in RM.

In humans, the mid thoracic cord is about one half the physical size of the cord at C4 (Frostell et al. 2016). In one study, the number of myelinated axons was measured to be about 1.1×10^6 at C3 down to 3.5×10^5 at L5 (Duval et al. 2019). In rodents, inflammatory cytokine and chemokine expression is level dependent after SCI (Hong et al. 2018). Pericyte number and coverage, and therefore the permeability of the BBB are different in the cervical versus thoracic cord (Winkler et al. 2012; Winkler et al. 2013).

In short, there are a number of anatomical, morphological, and neuro-inflammatory differences in specific aspects of the thoracic and cervical levels of the spinal cord related to response to injury and disease that *theoretically* could contribute to the difference in their responses to radiation injury (Ulndreaj et al. 2017). Which among these, if any, are relevant to the radiation response is speculative at this point.

The data from two studies support the idea that the thoracic cord is less sensitive to radiation than the cervical cord in the hyperbaric oxygen literature as well. Coy and Dolman observed three cases of RM in 17 patients who survived 9 months after treatment for lung cancer with eight fractions of approximately 4.8 Gy in three atmospheres of oxygen (Coy and Dolman 1971). This is a crude incidence of 18%. The original cohort was 20 patients, giving 85% survival in 9 months. Van den Brenk et al. observed seven cases in 38 head-and-neck cancer cases also surviving 9 months after HBO treatment at four atmospheres (van den Brenk et al. 1968). This is also a crude incidence of 18%. However, in this case the patients received only six fractions of 4.5 Gy ± 0.5 Gy to the cervical cord. The survival of the entire group was 67% at 9 months, but it was very diverse with respect to diagnosis. No matter what may account for the difference between the sensitivity of the two cord levels, it does not seem to be neutralized by treatment in HBO.

To conclude, it appears that the thoracic spinal cord is more radioresistant than the cervical cord at high doses, but the two levels have essentially the same observed responses at clinically relevant doses below 60 Gy. If there is indeed a difference in the dose response at higher doses, the cause remains elusive. In clinical practice, it would be unwise to rely on a high radiation tolerance for thoracic spinal cord. In the clinically relevant region (doses of 60 Gy or less), the dose-response data are statistically indistinguishable.

4.4   Dose Limits

Since the evolution of the first widely accepted dose limit of 45 Gy to the spinal cord, the field has not stopped searching for safe dose limits. Two widely promoted efforts to establish dose limits and to regularize grading of normal tissue complications were the Late Effects of Normal Tissues (LENT) and the Quantitative Analysis of Normal Tissue Effects in the Clinic (QUANTEC). Both of these resulted in special issues of the International Journal of Radiation Oncology Biology Physics (LENT volume 31, number 5 and QUANTEC volume 76, number 3, Supplement). Table 4.3 summarizes the dose limits established by these two efforts. Note that LENT and QUANTEC did not distinguish between the cervical and thoracic levels of the lung, and the same data (presented in Table 4.1) informed both efforts. The dose guidelines from LENT and QUANTEC are in basic agreement with each other and fundamentally agree with the results in Table 4.4.

Some may doubt whether the last entry of Table 4.4 is valid, i.e., that the $D_{50,2\,Gy}$ for the thoracic cord is nearly 100 Gy. Consider that Simpson observed one myelopathy case in five patients treated with 36 Gy in 6-Gy fractions (Simpson et al. 1982). Using $\alpha/\beta = 0.5$, this dose is equivalent to about 94 Gy. Thus, this entry seems plausible at least. In addition, it may seem like the complementary log-log link results in responses that are too low in the lower-dose region. In fact, the data are so sparse in the region below $D_{2\,Gy} \leq 50$ Gy that reliable estimates of response rates are not possible or have confidence intervals that are unserviceably large.

A Comment Regarding Complication Rate Versus Cumulative Incidence "Cumulative incidence is an unbiased estimate of cause-specific failure," i.e., the probability over time of observing a specific event in the patient population. In dose-response analysis, this is not what we are interested in. Dose-response analysis is about the probability of exceeding tolerance. Generally in radiation oncology, we assume that this probability is not disease dependent, but the cumulative incidence is very much disease and even stage dependent. Thus when Caplan et al. state that "the KM (Kaplan-Meier)

Table 4.4 Comparison of LENT (Schultheiss et al. 1995) and QUANTEC (Kirkpatrick et al. 2010) dose limits with current analysis (Model 3.2)

Estimated response	Dose	Source LENT/ QUANTEC	Estimates from Model 3.2	
			Cervical	Thoracic
<0.2%	45	L	<0.01%	<0.01%
~0.2%	50	Q	0.1%	<0.1%
<1%	54	Q	0.8%	0.3%
5%	57–61	L	3.7–6.5%	1.1–1.9%
6%	60	Q	6.8%	2.6%
<10%	61	Q	5.4%	2.1%
50%	68–73	L	22.1%–36.8%	9.0%–15.9%
50%	69.4	Q	77.7	94.2

curve greatly overestimates the probability of late complications," the accuracy of the statement depends on what is meant by the probability of complication (Caplan et al. 1995). In this case, they mean that the actuarial complication rate (KM) is significantly higher than the crude complication rate because patients die before they can express the complication. So when radiation oncologists advise a patient on the possibility of late complications, they must distinguish between the probability that the patient's tolerance will be exceeded versus the probability that the patient will experience the complication. One gets from the former to the latter mathematically by multiplying by the convolution of the survival function applicable to the patient with the distribution of latent periods (Eq. (4.1)). In practice this is not done of course.

Oxygen Effect While it is well known that malignant tumors are typically hypoxic, in healthy normal tissues, oxygen levels are sufficient to the needs of the tissue. Moreover, the CNS in particular is highly vascularized and therefore presumed to be well oxygenated. However, early published reports of treatment in hyperbaric oxygen, for the purpose of oxygenating putatively poorly perfused tumors, show an increase in the incidence of radiation myelopathy over that which is seen for treatment in air.

Coy et al. (1969) and Coy and Dolman (1971) published their experience with twice weekly fractionation schedules deployed in treating thoracic tumors. The typical tumor dose was 40 Gy given in eight twice weekly fractions using 2 AP/PA opposed fields daily. The spinal cord received doses ranging from 35 to 45 Gy, but the majority received about 42 Gy, which is common for this field arrangement. In the second of two reports, they described radiation myelopathy in three patients receiving averages doses of 38.7 Gy in eight fractions in HBO out of a cohort of 17 patients (18%) who survived a minimum of 9 months. In the previous report of patients treated in air with this fractionation, they reported three cases of radiation myelopathy in patients receiving a cord dose of 42 Gy out of 75 cases (4%) who received 40–45 Gy with twice weekly fractions and surviving 9 months also. Thus in HBO the incidence of myelopathy was about four times that observed in air despite the dose being somewhat lower.

van den Brenk et al. reported on a series of 357 H&N patients treated in HBO with high doses per fraction and treatment schedules ranging from 2 to 6 fractions (van den Brenk et al. 1968). In 2008, Schultheiss published a metaregression of data on the dose response of the cervical cord (Schultheiss 2008). Figure 3 of that publication shows the dose-response function for irradiation in air and the data points for the HBO patients for various treatment schedules. Virtually all of the HBO data points lie to the low-dose side of the in-air dose-response curve, indicating lower radiation tolerance in HBO.

In two somewhat related observations, Dische et al. observed a lower hemoglobin level in patients with myelopathy than in similarly treated patients without myelopathy, and Reinhold et al. observed a lower average blood pressure in myelopathy cases compared to nonmyelopathy cases in that study (Dische et al. 1986; Reinhold et al. 1976). In both studies, the effects were statistically significant. The relationship between blood pressure, hemoglobin level, and oxygen perfusion is complex subject to a number of feedback mechanisms that obscure cause and effect relationships.

It seems surprising that there should be a significant oxygen effect for the spinal cord radiation response. However, if one accepts that the endothelial cells are the initial radiation target, then once again we are led to conclude that it is likely that the venous endothelium is more likely to be involved than the arterial endothelium. The venous endothelial cells have been shown to have a lower oxygen tension than arterial cells. This lower oxygen tension is not truly hypoxic, but perhaps is sufficient to result in an enhancement of cell death if irradiated under conditions of high oxygen pressure. Or it may be that increasing the oxygen perfusion enhances the radiation effect in both the venous and arterial compartments. Whatever the cause, its study is more relevant to the pathogenesis of radiation myelopathy rather than the modern clinical issues since HBO treatments are no longer deployed.

Comorbid Conditions Hypertension is widely believed to be associated with increased late effects from radiation. For radiation myelopathy, the most often quoted evidence for this seems to be Asscher and Anson (1962), but Aristizabal disputes that there is evidence for such an effect in the clinical experience (Aristizabal et al. 1979). Wong et al. observed a history of hypertension in 4 of 26 cases of radiation myelopathy for whom they had data (Wong et al. 1994). In 15 cases of radiation myelopathy from M. D. Anderson Cancer Center, eight had a history of hypertension. In Asscher and Anson, hypertension was induced in rats by clipping of the right renal vein. The histopathology revealed vascular necrosis secondary to infarction of the spinal artery in nearly all cases. This is not the typical pathology of RM, and there was no arm in which hypertension was induced in unirradiated animals. There does not appear to be solid evidence implicating hypertension's association with RM. In fact, Reinhold et al. found their myelopathy cases to have significantly lower blood pressure than the non-cases (Reinhold et al. 1976). Dische et al. found a lower hemoglobin in myelopathy cases compared to similarly treated patients (Dische et al. 1986), but this was not found in the Reinhold data (Reinhold et al. 1976).

In reviews of RM, Schultheiss et al. have implicated other comorbid conditions or certain previous diseases in the development of RM (Schultheiss and Stephens 1992). These were merely observations of what seemed to be higher incidence of these factors in the RM population than one would expect or cases of RM at unusually low doses in the presence of these conditions. A history of malaria was observed in 5 out of 17 cases of RM at M. D. Anderson and this observation was confirmed by Prof. Huibert Reinhold in a series of cases from the Netherlands (Reinhold et al. 1976). Vascular diseases have been associated with RM at lower than typical doses (Laure-Kamionowska et al. 1994). Spinal malformations have been associated with low-dose RM (Weems et al. 1987; Schultheiss and Stephens 1992). There do not seem to be any reports of low-dose RM associated with disc protrusions, but it is not uncommon for patients to associate the onset of symptoms with an injury of some sort, typically a fall.

It seems reasonable to assume that the cervical and thoracic cord have the same α/β value, but differ in some other aspect that cause a separation of the responses. From a statistically conservative position, it is also reasonable to attempt to pool the data in a single analysis with the difference in responses being captured by a single parameter. Thus we would not have different values for β_0, β_1, and β_2 of Eq. (4.2a) nor different values of α/β, D_{50}, and k for Eqs. (4.5) and (4.6), but rather have all parameters in the model the same, except one.

4.5   Radiation Myelopathy in Stereotactic Radiosurgery

The continuing improvement of on-board imaging, beam tracking, and image guidance of treatment delivery has substantially increased the practicality of hypofractionated radiation treatments of 1–5 fractions per complete course. High doses per fraction to normal tissues are certainly no less toxic than before these technical advances were made in medical physics, but now when the normal tissue is separated from the target volume by even a few millimeters, the volume of dose-limiting normal tissues can be dramatically reduced and sometimes eliminated altogether. There are many applications of stereotactic body radiation therapy (SBRT) where the spinal cord is the primary sensitive normal tissue adjacent to the target volume. These applications include treatment of both benign and malignant tumors, as well as treatment with definitive and palliative intent. However, because very high dose regions and sharp dose gradients are often just millimeters away from the spinal cord, set-up and control of these treatments require the utmost vigilance.

Compared to conventional radiation treatments, there are relatively few reports of RM following SBRT or retreatment with SBRT. Regarding pathogenesis, there is nothing to distinguish RM after SBRT from RM after conventional radiation, except the lesions are probably smaller (Medin et al. 2011). The observed latent period following SBRT may be somewhat shorter, probably not because of the use of large doses per fraction, but rather as a result of the bias introduced by the generally shorter life expectancy for this patient cohort that is often being treated for metastatic disease.

Because RM is so rare and few cases have been reported following SBRT, it is currently not possible to determine the estimates of the parameters in a dose-response model of RM following SBRT. Even reliable estimates of response rates are not achievable with the available data. Moreover, statistical analysis of SBRT data is confounded by many factors: (1) Generally speaking, toxic response analysis of data with continuous independent variables, such as radiation dose, is more challenging than data that are naturally grouped. (2) In addition, since the spinal cord dose is very inhomogeneous, identifying the dose/volume combinations most predictive of RM is essentially impossible since all (dose, volume) pairs (at high dose) would show a high degree of (negative) correlation. (3) It is often the case that even if the true cord is segmented, the dosimetry that is reported in the literature applies to the OAR planning volume. Thus we may only have surrogates of the spinal cord dose to use in an analysis. Even this might be satisfactory if a consistent surrogate could be established by a consensus on dose reporting in SBRT. (4) Finally, as in conventional RT, the response of the spinal cord to SBRT appears highly idiopathic, occurring in cases where dose does not seem to be the sole cause.

A wide variety of doses has been used by NRG-RTOG protocols. At this time, there *may* be a uniform consensus of opinion regarding safe dose constraints for the spinal cord forming around the AAPM Task Group 101 guidelines (Benedict et al. 2010). Table 4.5 shows the spinal cord dose constraints by protocol for SBRT treatments currently listed for NRG/RTOG studies. For protocols specifying a maximum point dose, the volume was taken as 0.012 cm^3, the size of a $2 \times 2 \times 3$ mm^3 voxel. The single dose equivalents of fractionated doses were determined using $\alpha/\beta = 0.56$ Gy from Model 4 in Sect. 4.3. In all cases except RTOG 0631, the spinal cord segmentation was to be "based on the bony limits of the spinal canal," which is also reflected in the consensus paper authored by representatives of several study groups (Kong et al. 2011). It should be noted that the cross section of the spinal cord takes up only about 25–40% of the area of the spinal canal in the cervical and thoracic levels (Kayalioglu 2009). The volume of true cord in an 8-cm length is taken as 5.6 cm^3, as estimated using morphometry reports discussed below.

Table 4.5 Spinal cord dose constraints by protocol number

Dose (Gy)	Fractions	Volume (cm^3)	Dose per fraction (Gy)	vol/vol 8 cm	RTOG/NRG Protocol number	Equivalent single dose
10	1	0.35	10	0.0625	0915	10.0
7	1	1.2	7	0.214	0915	7.0
14	1	0.012	14	0.00536	0915	14.0
26	4	0.012	6.5	0.00214	0915	13.3
20.8	4	0.35	5.2	0.0625	0915	10.7
13.6	4	1.2	3.4	0.214	0915	7.1
22.5	3	0.03	7.5	0.00536	BR001	13.2
13	3	1.2	4.33	0.214	BR001	7.7
28	5	0.03	5.6	0.00536	BR001	12.9
22	5	0.35	4.4	0.0625	BR001	10.2
15.6	5	1.2	3.12	0.214	BR001	7.3
10	1	0.35	10	0.0625	0631	10.0
10	1	1.2	10	0.214	0631	10.0
14	1	0.03	14	0.00536	0631	14.0
18	3	0.008	6	0.00214	0618	10.6
22.5	5	0.25	4.5	0.0446	0813	10.4
13.5	5	0.5	2.7	0.0893	0813	6.4
30	5	0.01	6	0.00214	0813	13.8
8	1	1.2	8	0.214	BR002	8
8	1	0.35	10	0.0625	BR002	10
10	1	0.56 (10%)	8	0.1	BR002	8
14	1	0.03	14	0.00536	BR002	14

In Table 4.5, some of the dose constraints are actually lower than doses that are or have been used for conventional treatment volumes of the spinal cord. Generally, one would assume that this goes against common sense. However, very low dose cases of RM have been reported following SBRT (e.g., in Fig. 4.6). Even with this evidence, only the most unlikely of explanations could explain the occurrence of these cases, except for the most obvious explanation of an undetected dose delivery error.

So far, recommendations for the segmentation of the spinal cord in SBRT treatment planning have been limited to a single standard for any given trial. However, it is common to apply two definitions for the spinal cord in everyday clinical practice. That is, the spinal canal or thecal sac (dural tube) may be contoured for the purpose of minimizing the spinal cord dose, but the true cord is also contoured in order to preserve the important information comprising the actual dose distribution to the organ. Since it is not possible to understand fully the dose-volume response of the spinal cord to SBRT dose distribution without knowing the true doses and volumes to which the organ is exposed, a dual OAR definition that includes the true cord should be standard for all clinical trials.

In Fig. 4.6, the dose limits recommended in Table 4.5 are plotted as their effective single dose (ESD, Table 4.5, Column 7) versus the logarithm of the associated volume limit. These dose limits form roughly three clusters around volumes of

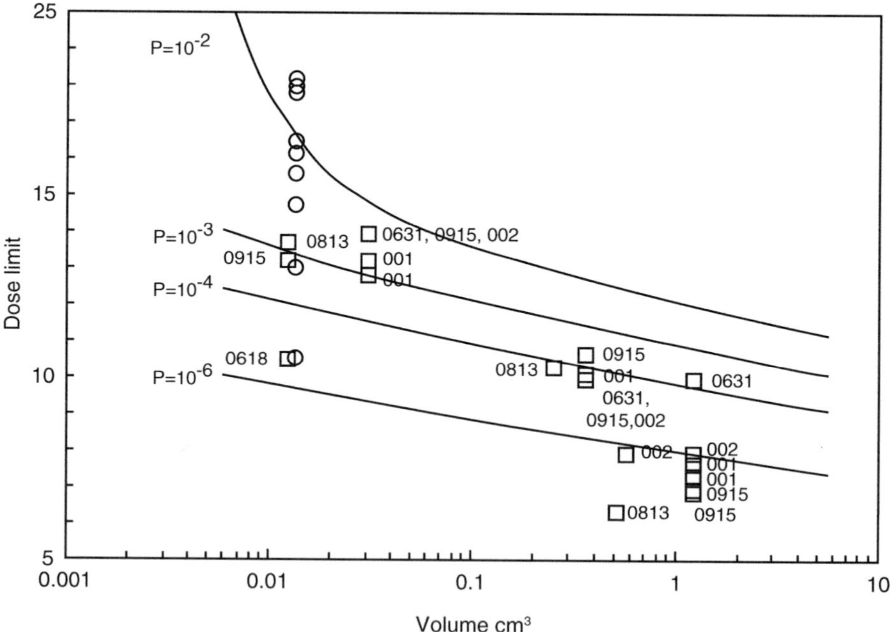

Fig. 4.6 Dose limits from clinical trials listed in Table 4.5 (squares). Also shown are D_{max} doses to the dural tube from Sahgal et al. 2013 (circles). Finally, the lines represent isoprobabilities for dose and volume according to Sect. 4.3 and assumption described in the text.

approximately 0.03 cm³, 0.3 cm³, and 1 cm³. The trend of lower-dose limits for higher volumes is obvious, reflecting the volume effect respected by writers of these clinical trials. The volume effect for RM, including its behavior at small volumes, is discussed in detail in Chap. 6.

The dose response for the spinal cord in conventional treatments has been estimated from the clinical literature. See Sect. 4.3. Combining the dose-volume response model with morphometry data, we can extrapolate the human data from the conventional volumes to SBRT volumes, *while acknowledging the loss of accuracy inherent in extrapolation.* From C3 to C8 the spinal cord volume is roughly 5.6 cm³ with an average length of 8.4 cm (Ko et al. 2004). These measurements are supported by the literature survey from Frostell et al. (2016). Furthermore, Freund et al. found cross-sectional areas from 45 mm² to 92 mm² in the cervical cord with an average of 65 mm² using MRI imaging (Freund et al. 2010). Over an 8-cm volume this yields an average spinal cord volume of 5.5 cm³. Thus we are reasonably confident in using 5.6 cm³ as the approximate reference volume for the parenchyma of the true spinal cord. Then using the critical element model as in the equation below, Fig. 4.6 shows doses for various *uniformly* irradiated volumes that elicit a given complication probability. The equation for estimating dose constraints, D (v), as a function of volume is

$$D(v) = D_{50}\left[\left(1-P\right)^{-1/v} - 1\right]^{1/k}$$

and   (4.9)

$$D_1 = \frac{\sqrt{4D(v)\left(\frac{D(v)}{N} + \alpha/\beta\right) + \left(\alpha/\beta\right)^2} - \alpha/\beta}{2}$$

where P is the incidence of RM, $D_{50} = 77.8$ Gy, $k = 11.2$, $\alpha/\beta = 0.56$ as determined by the analysis from Chap. 4, Model 4 (doses expressed in 2-Gy fractions). D (v) is the fractionated dose that results in an incidence of P, and D_1 is its corresponding dose given in a single fraction. Given that some of the dose constraints are extremely conservative and others are known to have resulted in RM (14 Gy), the extrapolation and the clinical experience appear not to conflict with each other. Finally doses for RM cases featured in Sahgal et al. (2013) are shown as circles, having been converted to equivalent single doses using $\alpha/\beta = 0.56$ Gy from Model 4. These fall above the extrapolated 1% isoeffect line for the most part. The two points that fall below this line were singled out by Sahgal for special comment.

Grimm et al. have produced a table with 60, sometimes duplicative, spinal cord dose limits with corresponding volumes for between 1 and 5 fractions (Grimm et al. 2011). An additional table gives the spinal cord definition, contouring techniques, tumor types, and delivery systems used by the data source in the dose-limits table. These tables provide compelling evidence that standardization is critically needed so that homogeneous cohorts of patients can be analyzed in a statistically uniform way. Even without standardization on dose limits, in time SBRT technology may become so sophisticated that RM becomes easily avoidable occurring only as a result of errors of treatment execution. The uncomfortable truth may be that SBRT technology has already reached or is very close to the limit of our ability to plan and deliver complex dose distributions.

Reports Referencing Dose Response for RM In conventional radiation treatments, the spinal cord dose is usually held well below that which would result in a 1% myelopathy rate. There are at least two reasons for this. An early and excessively conservative limit of 45 Gy in conventional fractionation was established as the most common spinal cord dose limit. The myelopathy rate for this dose is now understood to be much less than 0.1%. In addition, there are many thousands of patients each year whose spinal cord is irradiated as part of a definitive treatment plan. A radiation myelopathy rate of 1% would result in hundreds of cases each year, with disastrous results for patients and the field of radiation oncology. Because SBRT treatments are orders of magnitude less frequent and are often deployed with non-curative intent, a higher myelopathy rate *could* be more acceptable in the radiation community, depending of course upon the life expectancy of patients, the alternative therapies available, and the level of spinal cord in the field. In fact despite having an exhaustive survey available, it seems that RM is relatively more common in SBRT than in current definitive radiation treatment in general. Of course, this statement is speculative since we do not know what the reporting rate is in either case.

A number of publications have presented dose-response functions for RM following SBRT. Others present alternative analysis designed to indicate a dose response and to offer tolerance information without producing a dose-response analysis. Sahgal et al. (2010) presented five cases from four publications and three institutions. These cases (treated for both benign and malignant neoplasia) were compared to 19 controls from a fourth institution using analysis of variance repeated measures. The authors showed that the D_{max} to the thecal sac was larger for the cases than the controls. This extended to lower-dose/higher-volume dose measures as well. Unfortunately, because the patient sample cannot be considered a sample from a homogeneous population, the analysis serves only to show that the complication-free patients from one institution were treated with lower doses than 5 RM cases from three other institutions.

What is certain is that the tolerance was exceeded for the five cases listed. However, the authors point out one case that is especially problematic. The D_{max} for the thecal sac was reported to be 10.6 Gy in a single dose (Gerszten et al. 2008). The authors point out that Macbeth reported no myelopathies in 114 patients treated with 10 Gy in a single midplane dose in a randomized clinical trial (Macbeth et al. 1996). The Macbeth paper further states that the cord dose was generally about 5% higher than the midplane dose. Given that the maximum dose to the thecal sac probably occurs 1.5–3 mm away from the surface of the true cord, the implications are somewhat troublesome. The patient could have had an extraordinarily low radiation tolerance. Given that this case appears in multiple publications, one can only assume the dosimetry must have been checked many times and found to be correct. The real-time imaging was checked and nothing was noted that would point to a higher dose. Nonetheless, it seems likely that the dose was actually higher than recorded, possibly due to spinal cord movement within the spinal canal. As stated above, this case appears in several analyses and its existence affects tolerance estimates. No report has yet treated this case as an outlier.

In a 2012 paper, Sahgal repeated the method of taking cases from several institutions and controls from another (Sahgal et al. 2012). In this study, they investigate the implications of re-irradiation with SBRT, having five cases from four institutions and 14 controls from another institution. This process is reminiscent of Phillips and Buschke in which cases were simply added from the literature to cases and controls from their institution to obtain more dose-response data (Phillips and Buschke 1969), except in Sahgal et al., none of the controls came from the same institutions as the cases. In the study by Phillips and Buschke, the process resulted in a dose-response function that was shifted to lower doses, thereby overestimating the probability of RM and resulting in years of poor tolerance dose estimations. It will be shown in the paragraphs below that a similar effect occurs in this modern effort.

In the following year, Sahgal et al. published dose-response analysis using the same data as in 2010, supplemented with more cases and controls. Nine events, including the 5 from the 2010 paper, "were identified based on a multi-institutional and international collaboration" (Sahgal et al. 2013). In addition, 66 controls who satisfied inclusion criteria were assembled from "3 experienced academic institutions," namely M.D. Anderson Cancer Center, UCSF, and University of Toronto. The technique has some similarities with a case-control study, but again, the cases and the controls do not share common sources. Furthermore, in a case-control study, inferences about the constant in the systematic component of a GLM (viz Chap. 3) are not possible when the sampling fraction of the cases or controls is not known. This means that probability estimates are not possible. Since the data were analyzed as if they were uniformly generated from a single, complete cohort, the estimates of the probability of RM are contestable. In fact, the paper states that the dose limits at low probability do not comport "with the general understanding of radiation tolerance at the lower end of the probability spectrum," which is of course where the clinical applicability should be concentrated. It is very likely that the number of controls was not appropriate for the cases, but when faced with cases only, it is unclear how one would determine the proper number of controls and their distribution over the dose-volume space if all controls originating from the same treatment algorithm were unavailable. This problem was never overcome in the dose-response analysis of RM from external beam.

The data from this study were fitted to a linear logistic model, but the fit was assessed by the area-under-the-curve (AUC) statistic of the receiver-operator characteristic (ROC) curve, although no mention of ROC curves were made. Standard fit assessment methods were not deployed. Actually in this case it is reasonably accurate to say that there are no standard measures of goodness of fit since the data are not grouped. Nonetheless, there are some measures that are included in statistical software, the most common of which is the Hosmer and Lemeshow test (Hosmer et al. 1988; Hosmer et al. 2013). This test relies on the grouping of cases based on response rates. Of course, the most straightforward way to handle ungrouped data, which generally result from one of the variates being a continuous variable like D_{max} to the spinal cord, is to group the cases by that variable in groupings that are sufficiently large to satisfy the requisite asymptotics. Also in this study, only the doses to the RM cases were given but no description of the distribution of the doses in the control data was given beyond mean and median values. A dose-response curve was graphed, but without data points.

The AUC was used in this study to determine the fit of the model to the data, but it is not really an assessment of model fit. The AUC provides a metric for assessment of how well the model can discriminate between cases and controls. If there are m_1 cases and m_2 controls, there are $m_1 \times m_2$ unique pairs of cases and controls. The AUC gives the probability that for a pair chosen at random, the estimated probability of myelopathy is higher for the case than for the control. However, this does not determine how well the model is calibrated to the data, i.e., how well the response probabilities match the data. The global goodness of fit can be good with a poor AUC and vice versa. In this case, the model fit is indeterminant since the methodology applicable to cohort studies cannot be deployed with the data collected as described.

As stated above, the authors acknowledge that the model estimates of probability do not reflect the general understanding of RM dose response for SBRT in the low dose range. Illustrating this, we plot the logistic model from their paper with Model 4 in this chapter in Fig. 4.7. It is clear that the Sahgal et al. model has higher probability estimates for the max *point dose* to the thecal sac than the Model 4 (Sect. 4.3) has for an estimated *8 cm of fully irradiated* cervical cord. Interestingly, when the graphs are plotted for single doses rather than 2 Gy per fraction, they are much

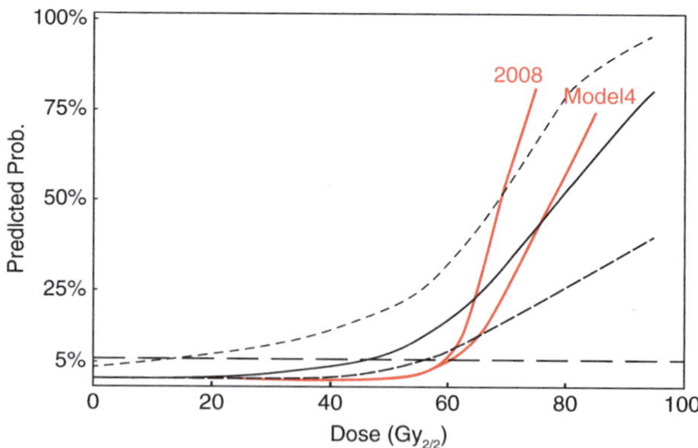

Fig. 4.7 Dose response from Sahgal (2013). Also shown are the results of the metaregression from Schultheiss (2008) and from the analysis of Sect. 4.3. Reprinted from International Journal of Radiation Oncology Biology Physics, 85 (2), A Saghal et al., Probabilities of radiation myelopathy specific to stereotactic body radiation therapy to guide safe practice. 341–347, Copyright (2013), with permission from Elsevier.

closer in the low dose range, crossing at about the 4% probability value rather than at 45%. It is unlikely that high myelopathy rates for SBRT are higher than rates for conventional radiation. The reason that the Sahgal dose-response function is located at doses that are too low is that by using an α/β ratio that was too high, the large single doses were converted into $D_{2\ Gy}$ equivalents that were too low, making it appear that the RM events occurred at lower doses than they did. This in turn shifted the estimated dose-response curve to lower doses. This will be true of all human dose-response data that use $\alpha/\beta = 2$ Gy (or higher).

The model from Sahgal et al. is overly conservative for probability estimation, and their reported doses that produce 1, 2, 3, 4, and 5% myelopathy rates are impractically low. Furthermore it is very unlikely that any practical design could be capable of distinguishing a 1% from a 5% response. In conclusion, the statistical justification of response rates presented in this study is insufficient.

Pooling data from different institutions in the manner described above exemplifies both selection bias and a type of confirmation bias. Confirmation bias in this context results from culling cases of RM from the literature and then implicitly assuming that they are representative of randomly selected cases that would have occurred in a properly designed regression analysis rather than being hand-picked. Selection bias is a result of selecting data for analysis in a non-random way. Essentially all analyses of complications from radiation suffer from selection bias since treatments are designed specifically to avoid complications. These effects can be minimized by selecting subjects for whom the event is equally likely for similar covariate patterns.

The problem with using separate sources for cases and controls should now be obvious: it is not possible to know that the number of observed cases is concordant with the number of controls. The cases from this study come from at least four and possibly six institutions. For proper calibration, all the corresponding patients without myelopathy from all these institutions should be included as controls. Presumably this would have resulted in a total greater than 66 and the estimated responses would be lower. Whether an appropriate, well-fitting model would have then resulted is impossible to say, since the case numbers are low, but at least there would have been that chance.

Obtaining the SBRT data for a viable logistic regression is an extremely difficult task. The requisite data comprise DVHs for the true spinal cord (or a consistent surrogate) and the outcomes with enough events for a meaningful fit to the data. In fact, we are now in a situation similar to the period prior to mid-1980s when estimates of response rates for radiation myelopathy were being attempted without the benefit of data from published studies and trials of the past 40 years. But the current situation for SBRT dose response of the spinal cord is made more difficult by the reduced ability to adjust the spinal cord dose versus the target dose in SBRT compared to convention XRT and the far fewer cases of radiation myelopathy resulting from the better overall understanding of cord tolerance and the improved dose delivery technology. Nonetheless, the mandates of statistical analysis remain the same.

The report by Gibbs et al. from 2007 (Gibbs et al. 2007; Grimm et al. 2016) is more typical of cohort studies that can be usefully analyzed by generalized regression model. The cohort comprised 74 patients with 102 lesions. Three patients developed RM, two of which had received prior RT. Although the prior RT doses were given, no verbal information was provided regarding the radiosurgery doses in these RM patients, but they do appear in a graph. The authors state, "The greatest permissible spinal cord dose was set at 10 Gy in a single fraction; this constraint was relaxed when the volume of spinal cord exceeding 10 Gy was less than 1 voxel." However, they also state that "The range of maximal BED3 for the spinal cord/cauda equina dose was 4.5–182 Gy." Note that the BED_3 for 10 Gy × 1 is 43.3 Gy. So we can infer that the large BED3 values come from fractionated schedules.

With only three cases of RM, it is no surprise that Gibbs et al. found no dosimetric factors that were statistically associated with the injury. The "total spinal cord volume" was "plotted against the BED_3 dose corresponding to 80% of the reference prescription dose," which simply means 80% of the prescribed dose. However, it is not clear why the total spinal cord volume (whatever that means exactly) would be plotted against anything, and it seems more likely that the volume represented the volume receiving at least the dose on the abscissa. Because of the manner in which the spinal cord doses were reported, it is difficult to combine the data with other studies or to compare the experience with that of other reports. Since all three of the RM cases were thoracic myelopathies, they repeated Wara's et al. (1975) erroneous suggestion that the thoracic could have a lower tolerance.

Grimm et al. re-analyzed the above data from Gibbs. Taking the data from the graph, they conservatively estimated being able to visualize 72 of the 90 data points out of the 102 cases that Gibbs mentioned (Grimm et al. 2016). So from the start, 30% of the cohort is not included. They then generated new doses from the dose-volume graph of Gibbs by postulating an equation from which they could determine the dose to 1 cc, 0.1 cc, and D_{max} from Gibbs' dose "to 80% of the reference prescription dose." Using these extrapolated doses they fitted the Gibbs data to a 2-parameter dose-response function, fitting the *de novo* treatments separately from the retreatments and performing separate procedures for the different dose levels. For the retreatment group, only D_{max} was used. For the *de novo* group there was one event and for the retreatment group there were two events. Since there was a single event for the *de novo* group, this data set is guaranteed to be quasi-separated and therefore maximum likelihood estimates do not exist. For the retreatment group, no information was available on the individual doses given in the initial treatment, so the effective dose and volume is unknown. Not surprisingly, a goodness-of-fit assessment was not given. The results of this analysis cannot be considered informative because, among other reasons, its information content is indeterminant.

Katsoulakis et al. present a clever and complex dosimetric analysis of cases from Memorial-Sloan Kettering (Katsoulakis et al. 2017). Although they had 228 patients with 259 sites treated, there were only two myelopathy cases. This is not enough for dose-response analysis, but they could put useful limits on the response rates with this number of cases. They deployed a technique they had previously developed and used it to create what they term a Dose Volume Atlas of Myelopathy Incidence. Imagine all 259 individual DVH's graphed on a single coordinate system. Choosing an arbitrary grid point, one can count the number of DVH's that pass above, n_a, and below, n_b, that point. (Note that the number passing above is also the number passing to the right, so it does not matter whether this operation applies to the dose axis or the volume axis.) Let the number of myelopathy cases in n_a be r_a and the number in n_b be r_b. Then using binomial statistics one can put limits on r_a/n_a and r_b/n_b. This process is then repeated for the entire dose-volume coordinate system using increments of Δd and Δv. In a similar fashion, it is also possible based on r/n to calculate the probability that the true myelopathy rate is below some value p_0.

For the data set presented in Katsoulakis et al., the calculations are more useful as a demonstration of principal rather than as clinical values of dose-volume constraints. As more data are cast in this format, the constraints become more clinically applicable. As it stands, using an example from the paper, it is not particularly helpful to know that for doses between 0 and 10 Gy given to volumes between 0 and 0.2 cc the upper 99% confidence limit of the true rate of myelopathy is 7.8%. Keep in mind, we are not talking about a 10 Gy dose, but a range of dose between 0 and 10 Gy.

In fact, the method originally described by Jackson et al. in 2006 is very useful in combining data on morbidity from different institutions (Jackson et al. 2006). Certain caveats apply, mostly concerning dose specification and the definition of the end points. Once these are made uniform, the method can be used to fill the gulf that exists between survey data or clinical experience and cohort data against which dose-response models may be tested. At present, this technique seems to be the best way of extracting dose-response information from a database for which too few events occur to successfully deploy a regression technique.

In 2019, a review from the group Hypofractionation Treatment Effects in the Clinic (HYTEC) aimed "to summarize the current understanding of the dose, volume, and outcome data for the human spinal cord specific to image guided, hypofractionated (1–5 fractions and a dose per fraction of >6 Gy) SBRT" (Sahgal et al. 2019). This paper points out many of the difficulties in estimating the incidence of RM in SBRT. In its search of the literature, only seven publications satisfied the inclusion criteria, and these included only three cases of RM. In three of the seven papers, Sahgal was the first author. Thus one can infer that these previous studies materially influence the conclusions of this paper.

In fact, the literature search does not actually form the database they ultimately use in a logistic regression effort. Rather, they use these seven papers as guidance for dose limits. There are four papers used for data generating dose-response models. The first is Sahgal et al. (2013) that has been discussed above. The next two are the papers by Katsoulakis et al. and Gibbs et al. discussed separately above. However, they pooled the *de novo* data from these two efforts for analysis. This resulted in three RM events in 278 treatments. The fact that only three cases of RM are in the primary cohort means that a statistically meaningful dose-response analysis is not achievable (Vittinghoff and McCulloch 2006). Nonetheless some useful information is contained, and the study is supplemented by other papers that did not fulfill all the inclusion criteria. The final paper they called upon for dose-response data was Kirkpatrick et al., written as part of the QUANTEC effort (Quantitative Analyses of Normal Tissue Effects in the Clinic) (Kirkpatrick et al. 2010). However, this paper presented analysis of conventional radiation treatments and not SBRT. Furthermore, it has been superseded by the analysis of Sect. 4.3 of this volume, where a more complete dataset is analyzed by the same investigator considering more alternative models.

It should be obvious by now that the chance of obtaining a well-calibrated dose-response function from logistical analysis of data with three events in 278 trials is remote. Factors that further mitigate against success in this database are: (1) Two of the events are very close to each other at about 13.5 Gy equivalent single dose (ESD) and one is at a dose so high (19 Gy ESD) that it contains little useful information on either the slope or location of the dose response. (2) The number of asymptomatic cases is difficult to discern. The data from Gibbs et al. were originally extracted from a graph where only about 70% of the data points could be visualized (72 out of 102). Katsoulakis et al. had about 14% more treatments than patients (259/228). Although it is not uncommon to include multiple treatments for patients in SBRT analyses, it is not clear that it is appropriate since multiple myelopathies could not be considered as independently occurring events in a single patient. (3) In a search for statistical significance in a 2×2 contingency table formed by varying the cut points on the dose axis, this was achievable only if 4 or fewer of the cases with the highest ESD's were matched against the 274 or more lower-dose cases. While no one should doubt the existence of a dose response for RM from SBRT, it is not clear what the authors mean when they state that "the dose-response parameter was significant" at $p = 0.015$. Presumably they refer to the coefficient of dose in the linear logistic function. However, evaluation of the model should not proceed until one can show that it fits the data. Having significant parameters is meaningless in a poorly calibrated model. Indeed, even their absolute value is uninformative in such a circumstance.

It has been suggested that model adequacy tests alone do "not suffice to demonstrate the spectrum of the predictive power of a model" (Moiseenko et al. 2021). This is obvious, but a goodness-of-fit test specifies whether the model's estimates are congruent with the data. If the model is not capable of producing estimates that agree with the data, then there is no need to explore its "spectrum of predictive power."

As a review paper, this study highlights some of the known difficult issues investigators confront when trying to analyze data from the literature or to establish meaningful protocols for defining and collecting dosimetry metrics for planning and future analysis. As with any organ at risk (OAR), it is necessary to have an accurate determination of the dose distribution over the organ. In the case of spinal SBRT, the CTV extends to and possibly into the spinal canal. Uncertainties in the spinal cord location arise from imaging resolution, windowing, technique, and patient motion. In addition there are set-up uncertainty and motion during treatment. These are routinely handled in conventional treatments, but in SBRT there is intentionally a sharp dose gradient beyond the CTV that reduces confidence in congruence between dose calculations and dose delivery. The high dose per fraction makes it especially important to treat the spinal cord conservatively.

The most common planning volumes for the spinal cord are the entire spinal canal, the dura mater, or the "true" spinal cord possibly plus a margin. In fact, the dural tube or thecal sac is often defined in as the spinal cord plus 1.5–2 mm. The spinal canal is not recommended in this review paper nor in the Ryu et al. (2007), upon which the AAPM's TG 101 report dose limits are based. Taking the dural tube has the same limitations as the spinal canal. The true cord would be ideal, especially since it is the actual organ at risk, but it is difficult to distinguish from the dura mater in imaging studies. Absent imaging uncertainty, using the true cord (+ a margin) is the least reductive method and retains the most spinal cord dosimetric information.

This review also discusses using the LQ model for converting from one fractionation scheme to another. Cited here are the papers by Brenner (2008) and Kirkpatrick et al. (2008) debating the use of the LQ model in SBRT. Much of the debate focuses on the response of the tumor cells, but normal tissue responses are also discussed. Brenner's argument is mostly based on the mechanistic nature of the LQ model. However, that is true only for cells that all respond in the same way, which implies that they must be synchronous cells since response to radiation varies significantly throughout the cell cycle. Kirkpatrick et al. argue that the LQ model is not robust enough to apply to a wide range of doses and dose rates. They both advocate for more data and better models.

Without getting distracted by the question of the applicability of the LQ model, the HYTEC paper deploys it in converting from one fractionation schedule to another using an α/β of 2 Gy. The justification of this value is "to maintain consistency" with the Sahgal 2013 paper that also used 2 Gy. That paper did explore (in supplementary material) the ramifications of using other values, but it did not address the fact that the only statistical analysis of *human* spinal cord response data has consistently found $\alpha/\beta < 1$ Gy.

In the end, the HYTEC publication adopts the recommendations from Sahgal et al. (2013), with only a slight modification. The dose limit for single-fraction treatment is given as the range 12.4–14 Gy rather than the 12.4 Gy dose from Sahgal et al. (2013). They state that the recommendations are conservative because the 66 controls in the Sahgal et al. paper were too few to represent the cohort associated with the nine cases from the literature that were selected in that paper. It is also suggested that the rate associated with these recommendations should be stated as 1–5%. Since the numbers come from Sahgal 2013 and the data in that paper did not represent an appropriately constituted cohort, the response estimates cannot be characterized as being statistically derived. Belsley et al. state "data-related problems are frequently brushed aside, all data being included without question on the basis of an appeal to a law of large numbers. But this is, of course, absurd if some of the data are in error, of *if they come from a different regime.*" (Belsley et al. 1980) (Emphasis added.)

Certainly this level of myelopathy would not be tolerated in conventional radiation treatment, but perhaps a higher rate is appropriate for SBRT if survival is expected to be short. But if that is the case, it would seem that conventional palliation could be a more appropriate treatment. A 1% myelopathy rate is usually too high to accept in a definitive treatment, but in SBRT the cord is likely to be at risk of injury from tumor progression.

Ryu et al. (2007) published an extensive analysis of the DVH's of the spinal cord at various prescribed dose levels in spine radiosurgery (Ryu et al. 2007). The authors state that "this study represents the first demonstration of the partial volume tolerance of human spinal cord to a single dose of radiation," but they defined neither "tolerance" nor "partial volume tolerance." However, the partial volume tolerance that they determined was "at least 10 Gy to the 10% spinal cord volume defined as 6 mm above and below the target." This is approximately the average $D_{v\,=\,10\%}$ for the group in which the prescribed dose was 18 Gy, the maximum prescribed dose used in the study. It appears that "tolerance" in this study has it conventional definition in radiation oncology of a dose that may be delivered safely. A single case of myelopathy was reported, originating in the group with a 16-Gy prescribed dose. In contradiction to the partial volume tolerance of 10 Gy, this patient received 9.6 Gy to 10% of the cord volume (and a maximum spinal cord dose of 14.6 Gy). How the investigators determined that 10 Gy to 10% is tolerated with their only myelopathy case violated this constraint is not clarified, but they cite the MRC study of palliation of bone metastases where 10 Gy was given at midline without any RM (Macbeth et al. 1996). It seems clear that the myelopathy case was not a result of 10 Gy to 10% of the volume but rather a result of the higher dose (up to 14.6 Gy) to a smaller volume.

In Table 4.6, some of the data from this paper are reproduced along with the confidence interval for the probability of myelopathy per lesion. (There were 230 lesions in 177 patients.) One patient developed myelopathy 13 months after a prescribed dose of 16 Gy to the 90% isodose and DVH points of $D_{v\,=\,1\%} = 13.0$ and $D_{max}=14.6$ Gy. The authors describe in some detail, the DVH points for the 86 patients who survived at least 1 year and for patients whose prescribed dose was 18 Gy. However, the prescribed dose alone is not correlated with the incidence of RM in small samples, as Table 4.6 shows. In the patient who developed myelopathy, we are given DVH points, but not the number of cases without myelopathy with these DVH points. Thus we are given insufficient information from which an incidence can be estimated. In 12-month survivors, the average doses to spinal cord volumes of 10%, 5%, 1%, and the maximum dose point were 8.6, 9.7, 10.7, and 12.2, respectively. The myelopathy cases had doses to these volume that increase from 11% higher at the 10% volume to 20% higher at the maximum dose.

Table 4.6 Myelopathy cases in treatment dose levels from Ryu et al. (2007)

Radiosurgery prescribed dose (Gy)	Number of lesions	Myelopathy cases	Exact 95% confidence interval (cases/lesion)
<10	37	0	0–9.49%
12–13	44	0	0–8.04%
14	48	0	0–7.40%
16	62	1	10^{-9} - 8.66%
18	39	0	0–9.03%
12–14	92	0	0–3.93%

Diao et al. experienced no RM out of 146 treatments in 132 patients with a D_{max}>12 Gy and a target for the spinal cord constraint of $D_{0.01cc}$ < 12.0 Gy. Median follow-up was 42 months with 60 patients alive at 24 months at the time of writing (Diao et al. 2020).

Based on the papers described above representing the most recent attempts to clarify dose limits, it should be clear that data on the spinal cord tolerance in radiosurgery treatments are insufficient to support rigorous dose-response analysis. It also seems that the field is willing to accept a higher level of RM in stereotactic treatments than in conventional RT. Most papers contain statements in their Discussion indicating that they have shown that "radiosurgery is generally safe." It is possible to agree that a 1–5% rate of myelopathy indicates that treatments are "generally safe," but this seems incompatible with attitudes toward radiation myelopathy generally expressed in the literature where radiation myelopathy is considered as both rare and devastating (Sahgal et al. 2013). The re-emergence in radiosurgery of the acceptance of such a high complication rate for RM is likely a result of an overestimation of the protection afforded by irradiating very small volumes and the lack of effective treatment alternatives.

Highly conformal SBRT and SRS are relatively recent advances in the field of radiation oncology, and a consensus regarding dose limits to organs at risk is still evolving. As with conventional RT, it is likely that robust standards of care that effectively avoid RM will be established before there are sufficient data for dose-response analysis. The standard of 45 Gy to the spinal cord in conventional RT is known to be safe for thousands of patients treated every year, but there are still a few anecdotal cases at or below this dose. For SBRT, it is probable that one or more dose-volume limits will become nearly universally established in time. Such limits will be achievable in nearly all cases, but also associated with few, if any, RM cases in the literature.

4.6   Other Altered Fractionation Schedules

In an effort to shorten the overall treatment time in head and neck radiation treatments without increasing the dose per fraction Dische and Saunders instituted a continuous, hyperfractionated, accelerated radiotherapy (CHART) protocol at Mount Vernon Hospital, Northwood, UK in 1985 (Dische and Saunders 1989). By 1989, 99 patients with advanced head and neck cancers had been treated using three fractions per day, 12 days in a row at 1.5 Gy per fraction (as of May, 1989). By late 1989, they had experienced four cases of late RM. These cases averaged spinal cord doses of about 46 Gy in 36 fractions, which one would assume would be safe. The complication rate from Eq. (4.1), assuming a denominator of 99, was 7.0%. According to Model 4 in this Chapter, this would be the equivalent of 62 Gy in 2-Gy fractions. If we equate the dose given in 1 day at three sessions to a single dose, then that dose would be 3.36 Gy per day. The total physical dose averaged 3.83 Gy per day. Thus the 6 h interval between fractions did not ultimately result in the expected repair of sublethal damage.

In 1991, Wong et al. published a report of accelerated fractionation for anaplastic thyroid carcinoma (Wong et al. 1991). The radiation therapy component of this protocol consisted of a 5 Gy initial dose to prevent upper airway obstruction followed by a regimen of four fractions per day of 1 Gy separated by 3 h. The hyperfractionation total doses were generally 30–40 Gy total. Two of the 32 patients undergoing this treatment developed RM at 7 and 12 months post treatment. The actuarial RM rate was about 20%. Using Model 4 above, this corresponds to about 68 Gy in 2-Gy fractions

The two patients who developed (and died from) RM received total tumor doses of 35 Gy in 35 fractions plus a 5 Gy initial dose and 45 Gy in 45 fractions and a 5 Gy initial boost. The latter patient's spinal cord was blocked from the posterior for the final ten fractions and the first patient's initial 5 Gy was given from the anterior only. However, simplifying matters we note that the first patient was treated on 10 days and the second on 13 days, with 85–90% of the dose delivered as four fractions per day. To further simplify, we approximate the effective doses as being the same every treatment day. Thus for the first patient, a 20% response rate would correspond to ten fractions of 3.9 Gy and the second patients treatment would correspond to 13 fractions of 3.4 Gy. Thus it seems very likely that repair of sublethal damage stopped completely at some point after the second fraction, and the subsequent fractions had a larger effect than their physical dose would indicate even with repair of sublethal damage ignored, i.e., they became supra-additive. Another case of RM after qid treatment was reported by this group, but it appears not to belong to the same treatment protocol (Wong et al. 1994). This last patient's latent period was 9 months and all patients in the thyroid protocol were followed until death or were NED at times beyond 2.5 years and any RM case would have been known to the investigators.

References

Abbatucci JS, DeLozier T, Quint R, Roussel A, Brune D. Radiation myelopathy of the cervical spinal cord. Time, dose, and volume factors. Int J Radiat Oncol Biol Phys. 1978;4:239–48.

Abramson N, Cavanaugh PJ. Short course radiation therapy in carcinoma of the lung: a second look. Radiology. 1973;108(3):685–7. https://doi.org/10.1148/108.3.685.

Ahuja CS, Wilson JR, Nori S, Kotter MRN, Druschel C, Curt A, Fehlings MG. Traumatic spinal cord injury. Nat Rev Dis Primers. 2017;3:1–21. (2056-676X (Electronic))

Aristizabal SA, Boone MLM, Laguna JF. Endocrine factors influencing radiation injury to central nervous tissue. Int J Radiat Oncol Biol Phys. 1979;5(3):349–53. https://doi.org/10.1016/0360-3016(79)91215-X.

Asscher AW, Anson SG. Arterial hypertension and irradiation damage to the nervous system. Lancet. 1962;280(7270):1343–6. https://doi.org/10.1016/S0140-6736(62)91020-6.

Atkins HL, Tretter P. Time-dose considerations in radiation myelopathy. Acta Radiol Ther Phys Biol. 1966a;5:79–94.

Atkins HL, Tretter P. Time-dose considerations in radiation myelopathy. Acta Radiol. 1966b;5(1):79–94. https://doi.org/10.3109/02841856609139546.

Belsley DA, Kuh E, Welsch RE. Regression diagnostics: identifying influential data and sources of collinearity, Wiley series in probability and mathematical statistics. New York: Wiley; 1980.

Bender E (2016) Personal communication.

Benedict SH, Yenice KM, Followill D, Galvin JM, Hinson W, Kavanagh B, Keall P, Lovelock M, Meeks S, Papiez L. Stereotactic body radiation therapy: the report of AAPM task group 101. Med Phys. 2010;37(8):4078–101.

Berkson J. A statistically precise and relatively simple method of estimating the bioassay with quantal response, based on the logistic function. J Am Stat Assoc. 1953;48(263):565–99. https://doi.org/10.2307/2281010.

van den Brenk HA, Richter W, Hurley RH. Radiosensitivity of the human oxygenated cervical spinal cord based on analysis of 357 cases receiving 4 MeV x rays in hyperbaric oxygen. Br J Radiol. 1968;41(483):205–14. https://doi.org/10.1259/0007-1285-41-483-205.

Brenner DJ. The linear-quadratic model is an appropriate methodology for determining Isoeffective doses at large doses per fraction. Semin Radiat Oncol. 2008;18(4):234–9. https://doi.org/10.1016/j.semradonc.2008.04.004.

Caplan RJ, Pajak TF, Cox JD. Toxicity criteria of the Radiation Therapy Oncology Group (RTOG) and the European Organization for Research and Treatment of Cancer (EORTC). Int J Rad Oncol Biol Phys. 1995;32(5):1547. https://doi.org/10.1016/0360-3016(95)99002-W.

Choi NCH, Grillo HC, Gardiello M, Scannell HG, Wilkins EW. Basis for new strategies in post-operative radiotherapy of bronchogenic carcinoma. Int J Radiat Oncol Biol Phys. 1980;6:31–5.

Combes PF, Daly N, Schlienger M, Humeau F. Les myelopathies radiques tardives progressives. Etudes de 27 observations. J de radiologie d'electrologie et de medecine nucleaire. 1975;56(11):815–25.

Coy P, Baker S, Dolman CL. Progressive myelopathy due to radiation. Can Med Assoc J. 1969;100(24):1129–33.

Coy P, Dolman CL. Radiation myelopathy in relation to oxygen level. Br J Radiol. 1971;44(525):705–7. https://doi.org/10.1259/0007-1285-44-525-705.

Diao K, Song J, Thall PF, McGinnis GJ, Boyce-Fappiano D, Amini B, Brown PD, Yeboa DN, Bishop AJ, Li J, Briere TM, Tatsui CE, Rhines LD, Chang EL, Ghia AJ. Low risk of radiation myelopathy with relaxed spinal cord dose constraints in de novo, single fraction spine stereotactic radiosurgery. Radiother Oncol. 2020;152:49–55. https://doi.org/10.1016/j.radonc.2020.07.050.

Dische S, Martin WMC, Anderson P. Radiation myelopathy in patients treated for carcinoma of bronchus using a six fraction regime of radiotherapy. Br J Radiol. 1981;54(637):29–35. https://doi.org/10.1259/0007-1285-54-637-29.

Dische S, Saunders MI. Continuous, hyperfractionated, accelerated radiotherapy (CHART): an interim report upon late morbidity. Radiother Oncol. 1989;16(1):65–72. https://doi.org/10.1016/0167-8140(89)90071-6.

Dische S, Saunders MI, Warburton MF. Hemoglobin, radiation, morbidity and survival. Int J Radiat Oncol Biol Phys. 1986;12(8):1335–7. https://doi.org/10.1016/0360-3016(86)90166-5.

Dische S, Warburton MF, Saunders MI. Radiation myelitis and survival in the radiotherapy of lung cancer. Int J Radiat Oncol Biol Phys. 1988;15(1):75–81. https://doi.org/10.1016/0360-3016(88)90349-5.

Duval T, Saliani A, Nami H, Nanci A, Stikov N, Leblond H, Cohen-Adad J. Axons morphometry in the human spinal cord. NeuroImage. 2019;185:119–28. https://doi.org/10.1016/j.neuroimage.2018.10.033.

Eden D, Gros C, Badji A, Dupont SM, De Leener B, Maranzano J, Zhuoquiong R, Liu Y, Granberg T, Ouellette R, Stawiarz L, Hillert J, Talbott J, Bannier E, Kerbrat A, Edan G, Labauge P, Callot V, Pelletier J, Audoin B, Rasoanandrianina H, Brisset J-C, Valsasina P, Rocca MA, Filippi M, Bakshi R, Tauhid S, Prados F, Yiannakas M, Kearney H, Ciccarelli O, Smith SA, Andrada Treaba C, Mainero C, Lefeuvre J, Reich DS, Nair G, Shepherd TM, Charlson E, Tachibana Y, Hori M, Kamiya K, Chougar L, Narayanan S, Cohen-Adad J. Spatial distribution of multiple sclerosis lesions in the cervical spinal cord. Brain. 2019;142(3):633–46. https://doi.org/10.1093/brain/awy352.

Eichhorn HJ, Lessel A, Rotte KH. Einfuss verschiedener Bestrahlungsrhythmen auf Tumor-und Normalgewebe in vivo. Strahlentheraphie. 1972;146:614–29.

Fitzgerald RH Jr, Marks RD Jr, Wallace KM. Chronic radiation myelitis. Radiology. 1982;144(3):609–12. https://doi.org/10.1148/radiology.144.3.6808557.

Fog T. Topographic distribution of plaques in the spinal cord in multiple sclerosis. Arch Neurol Psychiatr. 1950;63(3):382–414. https://doi.org/10.1001/archneurpsyc.1950.02310210028003.

Freund PAB, Dalton C, Wheeler-Kingshott CAM, Glensman J, Bradbury D, Thompson AJ, Weiskopf N. Method for simultaneous voxel-based morphometry of the brain and cervical spinal cord area measurements using 3D-MDEFT. J Magn Reson Imaging. 2010;32(5):1242–7. https://doi.org/10.1002/jmri.22340.

Frostell A, Hakim R, Thelin EP, Mattsson P, Svensson M. A review of the segmental diameter of the healthy human spinal cord. Front Neurol. 2016;7 https://doi.org/10.3389/fneur.2016.00238.

Gerszten PC, Burton SA, Ozhasoglu C, McCue KJ, Quinn AE. Radiosurgery for benign intradural spinal tumors. Neurosurgery. 2008;62(4):887–96.

Gibbs IC, Kamnerdsupaphon P, Ryu MR, Dodd R, Kiernan M, Chang SD, Adler JR Jr. Image-guided robotic radiosurgery for spinal metastases. Radiother Oncol. 2007;82(2):185–90. https://doi.org/10.1016/j.radonc.2006.11.023.

Grimm J, LaCouture T, Croce R, Yeo I, Zhu Y, Xue J. Dose tolerance limits and dose volume histogram evaluation for stereotactic body radiotherapy. J Appl Clin Med Phys. 2011;12(2): 267–92.

Grimm J, Sahgal A, Soltys SG, Luxton G, Patel A, Herbert S, Xue J, Ma L, Yorke E, Adler JR, Gibbs IC. Estimated risk level of unified stereotactic body radiation therapy dose tolerance limits for spinal cord. Semin Radiat Oncol. 2016;26(2):165–71. https://doi.org/10.1016/j.semradonc.2015.11.010.

Guthrie RT, Ptacek JJ, Hjass AC. Comparative analysis of two regimens of split course radiation in carcinoma of the lung. Am J Roentgenol. 1973;117:605–8.

Hatlevoll R, Høst H, Kaalhus O. Myelopathy following radiotherapy of bronchial carcinoma with large single fractions: a retrospective study. Int J Radiat Oncol Biol Phys. 1983;9(1):41–4. https://doi.org/10.1016/0360-3016(83)90206-7.

Hazra TA, Chandrasekaran MS, Colman M, Prempree T, Inalsignh A. Survival in carcinoma of the lung after a split course of radiotherapy. Br J Radiol. 1974;47:464–6.

Hong J, Chang A, Zavvarian MM, Wang J, Liu Y, Fehlings MG. Level-specific differences in systemic expression of pro-and anti-inflammatory cytokines and chemokines after spinal cord injury. Int J Mol Sci. 2018;19(8):doi:10.3390/ijms19082167.

Hosmer DW, Lemeshow S, Klar J. Goodness-of-fit testing for the logistic regression model when the estimated probabilities are small. Biom J. 1988;30(8):911–24.

Hosmer DW, Lemeshow S, Sturdivant RX. Applied logistic regression. Wiley series in probability and statistics. 3rd ed. Hoboken: Wiley; 2013.

Hua LH, Donlon SL, Sobhanian MJ, Portner SM, Okuda DT. Thoracic spinal cord lesions are influenced by the degree of cervical spine involvement in multiple sclerosis. Spinal Cord. 2015;53(7):520–5. https://doi.org/10.1038/sc.2014.238.

Jackson A, Yorke ED, Rosenzweig KE. The atlas of complication incidence: a proposal for a new standard for reporting the results of radiotherapy protocols. Semin Radiat Oncol. 2006;16(4):260–8. https://doi.org/10.1016/j.semradonc.2006.04.009.

Jeremic B, Djuric L, Mijatovic L. Incidence of radiation myelitis of the cervical spinal cord at doses of 5500 cGy or greater. Cancer. 1991;68(10):2138–41. https://doi.org/10.1002/1097-014 2(19911115)68:10<2138::AID-CNCR2820681009>3.0.CO;2-7.

Katsoulakis E, Jackson A, Cox B, Lovelock M, Yamada Y. A detailed Dosimetric analysis of spinal cord tolerance in high-dose spine radiosurgery. Int J Radiat Oncol Biol Phys. 2017;99(3):598–607. https://doi.org/10.1016/j.ijrobp.2017.05.053.

Kayalioglu G. The vertebral column and spinal meninges. In: The spinal cord. Elsevier; 2009. p. 17–36.

Kim YH, Fayos JV. Radiation tolerance of the cervical spinal cord. Radiology. 1981;139:473–8.

Kirkpatrick JP, Meyer JJ, Marks LB. The linear-quadratic model is inappropriate to model high dose per fraction effects in radiosurgery. Semin Radiat Oncol. 2008;18(4):240–3. https://doi.org/10.1016/j.semradonc.2008.04.005.

Kirkpatrick JP, van der Kogel AJ, Schultheiss TE. Radiation dose-volume effects in the spinal cord. Int J Radiat Oncol Biol Phys. 2010;76(3 Suppl):S42–9. https://doi.org/10.1016/j.ijrobp.2009.04.095.

Ko HY, Park JH, Shin YB, Baek SY. Gross quantitative measurements of spinal cord segments in human. Spinal Cord. 2004;42(1):35–40. https://doi.org/10.1038/sj.sc.3101538.

Kong FM, Ritter T, Quint DJ, Senan S, Gaspar LE, Komaki RU, Hurkmans CW, Timmerman R, Bezjak A, Bradley JD, Movsas B, Marsh L, Okunieff P, Choy H, Curran WJ. Consideration of dose limits for organs at risk of thoracic radiotherapy: atlas for lung, proximal bronchial tree, esophagus, spinal cord, ribs, and brachial plexus. Int J Radiat Oncol Biol Phys. 2011;81(5):1442–57. https://doi.org/10.1016/j.ijrobp.2010.07.1977.

Laure-Kamionowska M, Dambska M, Steckiewicz S, Golebiowska D. Developmental vascular anomalies of the spinal cord as a background for post radiation necrotizing myelopathy. Clin Neuropathol. 1994;13(3):117–9.

Lyman JT, Wolbarst AB. Optimization of radiation therapy, IV: a dose-volume histogram reduction algorithm. Int J Radiat Oncol Biol Phys. 1989;17(2):433–6. https://doi.org/10.1016/0360-3016(89)90462-8.

Macbeth FR, Wheldon TE, Girling DJ, Stephens RJ, Machin D, Bleehen NM, Lamont A, Radstone DJ, Reed NS, Bolger JJ, Clark PI, Connolly CK, Hasleton PS, Hopwood P, Moghissi K, Saunders MI, Thatcher N, White RJ. Radiation myelopathy: estimates of risk in 1048 patients in three randomized trials of palliative radiotherapy for non-small cell lung cancer. Clin Oncol. 1996;8(3):176–81. https://doi.org/10.1016/S0936-6555(96)80042-2.

Madden FJF, English JSC, Moore AK, Newton KA. Split course radiation in inoperable carcinoma of the bronchus. Eur J Cancer. 1979;15:1175–7.

Marcus RB Jr, Million RR. The incidence of myelitis after irradiation of the cervical spinal cord. Int J Radiat Oncol Biol Phys. 1990;19(1):3–8. https://doi.org/10.1016/0360-3016(90)90126-5.

Mathison DJ, Kadom N, Krug SE. Spinal cord injury in the pediatric patient. Clin Pediatr Emerg Med. 2008;9(2):106–23. https://doi.org/10.1016/j.cpem.2008.03.002.

McCunniff AJ, Liang MJ. Radiation tolerance of the cervical spinal cord. Int J Radiat Oncol Biol Phys. 1989;16(3):675–8. https://doi.org/10.1016/0360-3016(89)90484-7.

Medin PM, Foster RD, van der Kogel AJ, Sayre JW, McBride WH, Solberg TD. Spinal cord tolerance to single-fraction partial-volume irradiation: a swine model. Int J Radiat Oncol Biol Phys. 2011;79(1):226–32. https://doi.org/10.1016/j.ijrobp.2010.07.1979.

Miller RC, Aristizabal SA, Leith JT, Manning MR. Radiation myelitis following split-course therapy for unresectable lung cancer. (Abstr.). Int J Radiat Oncol Biol Phys. 1977;2(2):179.

Moiseenko V, Hattangadi-Gluth JA, Huynh-Le M-P, Marks LB, Grimm J, Milano MT, Jackson A, Yorke E, Pettersson N, Naqa IE. In reply to schultheiss. Int J Rad Oncol Biol Phys. 2021; https://doi.org/10.1016/j.ijrobp.2021.03.019.

dela Paz NG, D'Amore PA. Arterial versus venous endothelial cells. Cell Tissue Res. 2009;335(1):5–16.

Phillips TL, Buschke F. Radiation tolerance of the thoracic spinal cord. Am J Roentgenol Radium Therapy, Nucl Med. 1969;105(3):659–64.

Reinhold HS, Kaalen JGAH, Unger-Gils K. Radiation myelopathy of the thoracic spinal cord. Int J Radiat Oncol Biol Phys. 1976;1(7–8):651–7. https://doi.org/10.1016/0360-3016(76)90147-4.

Ryu S, Jin JY, Jin R, Rock J, Ajlouni M, Movsas B, Rosenblum M, Kim JH. Partial volume tolerance of the spinal cord and complications of single-dose radiosurgery. Cancer. 2007;109(3):628–36. https://doi.org/10.1002/cncr.22442.

Sahgal A, Chang JH, Ma L, Marks LB, Milano MT, Medin P, Niemierko A, Soltys SG, Tomé WA, Wong CS, Yorke E, Grimm J, Jackson A. Spinal cord dose tolerance to stereotactic body radiation therapy. Int J Radiat Oncol Biol Phys. 2019; https://doi.org/10.1016/j.ijrobp.2019.09.038.

Sahgal A, Ma L, Gibbs I, Gerszten PC, Ryu S, Soltys S, Weinberg V, Wong S, Chang E, Fowler J, Larson DA. Spinal cord tolerance for stereotactic body radiotherapy. Int J Radiat Oncol Biol Phys. 2010;77(2):548–53. https://doi.org/10.1016/j.ijrobp.2009.05.023.

Sahgal A, Ma L, Weinberg V, Gibbs IC, Chao S, Chang UK, Werner-Wasik M, Angelov L, Chang EL, Sohn MJ, Soltys SG, Létourneau D, Ryu S, Gerszten PC, Fowler J, Wong CS, Larson

DA. Reirradiation human spinal cord tolerance for stereotactic body radiotherapy. Int J Radiat Oncol Biol Phys. 2012;82(1):107–16. https://doi.org/10.1016/j.ijrobp.2010.08.021.

Sahgal A, Weinberg V, Ma L, Chang E, Chao S, Muacevic A, Gorgulho A, Soltys S, Gerszten PC, Ryu S, Angelov L, Gibbs I, Wong CS, Larson DA. Probabilities of radiation myelopathy specific to stereotactic body radiation therapy to guide safe practice. Int J Radiat Oncol Biol Phys. 2013;85(2):341–7. https://doi.org/10.1016/j.ijrobp.2012.05.007.

Schultheiss TE. The radiation dose-response of the human spinal cord. Int J Radiat Oncol Biol Phys. 2008;71(5):1455–9. https://doi.org/10.1016/j.ijrobp.2007.11.075.

Schultheiss TE, Higgins EM, El-Mahdi AM. The latent period in clinical radiation myelopathy. Int J Radiat Oncol Biol Phys. 1984;10(7):1109–15. https://doi.org/10.1016/0360-3016(84)90184-6.

Schultheiss TE, Kun LE, Ang KK, Stephens LC. Radiation response of the central nervous system. Int J Radiat Oncol Biol Phys. 1995;31(5):1093–112. https://doi.org/10.1016/0360-3016(94)00655-5.

Schultheiss TE, Stephens LC. Invited review: permanent radiation myelopathy. Br J Radiol. 1992;65(777):737–53.

Schultheiss TE, Thames HD, Peters LJ, Dixon DO. Effect of latency on calculated complication rates. Int J Radiat Oncol Biol Phys. 1986;12(10):1861–5. https://doi.org/10.1016/0360-3016(86)90331-7.

Scruggs H, El Mahdi A, Marks RD Jr, Constable WC. The results of split course radiation therapy in cancer of the lung. Amerjroentgenol. 1974;121(4):754–60. https://doi.org/10.2214/ajr.121.4.754.

Simpson JR, Perez CA, Phillips TL, Concannon JP, Carella RJ. Large fraction radiotherapy plus misonidazole for treatment of advanced lung cancer: report of a phase I/II trial. Int J Radiat Oncol Biol Phys. 1982;8(2):303–8. https://doi.org/10.1016/0360-3016(82)90532-6.

Taylor JM. The design of in vivo multifraction experiments to estimate the alpha-beta ratio. Radiat Res. 1990;121(1):91–7.

Ulndreaj A, Badner A, Fehlings M. Promising neuroprotective strategies for traumatic spinal cord injury with a focus on the differential effects among anatomical levels of injury. 2017;F1000Research:6:1907. https://doi.org/10.12688/f1000research.11633.1.

Vittinghoff E, McCulloch CE. Relaxing the rule of ten events per variable in logistic and Cox regression. Am J Epidemiol. 2006;165(6):710–8. https://doi.org/10.1093/aje/kwk052.

Vollmer T. The natural history of relapses in multiple sclerosis. J Neurol Sci. 2007;256(SUPPL. 1): S5–S13. https://doi.org/10.1016/j.jns.2007.01.065.

Wara WM, Phillips TL, Sheline GE, Schwade JG. Radiation tolerance of the spinal cord. Cancer. 1975;35(6):1558–62. https://doi.org/10.1002/1097-0142(197506)35:6<1558::AID-CNCR2 820350613>3.0.CO;2-7.

Weems DH, Mendenhall WM, Parsons JT, Cassisi NJ, Million RR. Squamous cell carcinoma of the supraglottic larynx treated with surgery and/or radiation therapy. Int J Radiat Oncol Biol Phys. 1987;13(10):1483–7.

Winkler EA, Sengillo JD, Bell RD, Wang J, Zlokovic BV. Blood-spinal cord barrier pericyte reductions contribute to increased capillary permeability. J Cereb Blood Flow Metab. 2012;32(10):1841–52. https://doi.org/10.1038/jcbfm.2012.113.

Winkler EA, Sullivan JS, Henkel JS, Appel SH, Zlokovic BV. Blood-spinal cord barrier breakdown and pericyte reductions in amyotrophic lateral sclerosis. Acta Neuropathol. 2013;125:111–20. (1432-0533 (Electronic))

Wong CS, Van Dyk J, Milosevic M, Laperriere NJ. Radiation myelopathy following single courses of radiotherapy and retreatment. Int J Radiat Oncol Biol Phys. 1994;30(3):575–81. https://doi.org/10.1016/0360-3016(92)90943-C.

Wong CS, Van Dyk J, Simpson WJ. Myelopathy following hyperfractionated accelerated radiotherapy for anaplastic thyroid carcinoma. Radiother Oncol. 1991;20(1):3–9. https://doi.org/10.1016/0167-8140(91)90105-P.

Experimental Studies on Rodents 5

5.1   Mouse Studies

A few studies have been performed on RM in the mouse model. Goffinet et al. irradiated 6 and 12 mm of TL spine in experiments using 20–200 Gy in 1 or 10 fractions (Goffinet et al. 1976). Dose-response curves were very shallow, generally rising from 0 to 100% over a dose range of 20 Gy or more. Latent periods were inversely related to dose, and the dose-response curves for 6 and 12 mm appeared to be parallel.

Lo et al. treated 2.2 cm of TL spine in 333 mice with single fractions of 12 to 75 Gy and analyzed the dose-response data with a log-linear logistic regression (Lo et al. 1992). About 20% of the animals died of intercurrent disease. Different dose-response functions were estimated for differing levels of morbidity, and the dose-response functions were relatively shallow. Sixteen percent of the animals receiving 28 Gy or less showed some level of permanent recovery. The latent periods and dose were inversely related. The D_{50} values at 360 days post irradiation varied from 19 to 23 Gy for the morbidity levels 1 to 4 (mild paresis to complete paralysis with incontinence). Demyelination was found in nearly all animals, with "the extent of demyelination in the cord…normally less than the nerve root." Severe vascular damage was seen at higher doses and earlier times following treatment.

Habermalz et al. treated nearly 1000 mice with 1–25 fractions of a 1-cm field to the T10-L1 spine (Habermalz et al. 1987). About 25% of these died from intercurrent disease, mostly tumors. There was an inverse relationship between dose and latency. The most remarkable finding was that there was no fractionation effect between 1 and 5 fractions, i.e., N-exponent $= 0$ in the TDF in the model and $\beta = 0$ in the LQ model.

5.2   Rat Studies

Not surprisingly, more studies on RM have been performed in the rat than any other species by far. Experimental studies have concentrated on elucidating the pathogenesis of RM; clarifying the relationship between dose, number of fractions, treatment

© Springer Nature Switzerland AG 2022   99
T. Schultheiss, *Radiation Myelopathy*,
https://doi.org/10.1007/978-3-030-94658-6_5

time, and dose per fraction in the dose response of the spinal cord; studying the effects of the sequencing of dose (top-up dose experiments); retreatments; and drug-radiation interactions. Much of classical radiobiology was focused on discovering the form of isoeffects equations for radiation response. These biomathematical models are equations containing two or more fractionation parameters (total dose, number of fractions, dose per fraction, overall treatment time, interfraction interval) whose solutions produce a constant level of effect. That level is essentially always D_{50}. The isoeffect equations accounting for nearly all the experimental studies on fractionation in RM are the NSD-like equation where a constant level of effect is given by const $= DN^aT^b$ or the LQ model where the constant level of effect is given by $E = \alpha D + \beta Dd$. (D = total dose in N fractions over T days with $d = D/N$)

There is a fundamental flaw in the majority of the dose-response experiments in radiobiology—using too few animals per dose point. With a few exceptions, the "best" designed rat experiments used eight to ten animals per dose group, but the use of only five or six animals was far too common. This also holds true in some large animal experiments. With this small number of subjects per group, the resolution is far too coarse, and a number of highly cited studies failed to reach statistical significance because of a lack of precision. In many cases, their reproducibility is poor.

For this volume, the re-analysis of many published dose-response experiments is undertaken. Generally, the dose-response experiments are appropriately character-ized as containing sparse data, that is, an inadequate number of data points (cells, covariate patterns, etc.) where the number of events and the number of non-events are not large enough to satisfy the requirements of an asymptotic χ^2 distribution (meaning model assessments and p-values are not valid). This is because too few animals were assigned to each data point or the expected value of the response is too close to 0 or 1. There is no widely accepted method for determining model ade-quacy in the case of sparse data, although there are many diagnostics that may be deployed to describe the effects of sparse data (Kuss 2002).

A lack of concern for the model fit, and therefore for the number of animals allot-ted to each dose group or covariate pattern, was probably because the investigators intended to explore the behavior of the model by determining the behavior of the D_{50}. Because the dose response was generally steep, they understood that the D_{50} could be determined or at least bracketed with relatively high precision. The behav-ior of D_{50} is not a good tool with which to investigate a model, but by the time this was finally accepted, a certain style of experimental design had been established. Even though understanding of the design elements imposed by the deployment of generalized linear models, and logistic regression in particular, eventually became widespread, dose-response experiments are still undertaken with numbers so small as to preclude successful analysis (Saager et al. 2020b).

Early studies of radiation response of the spinal cord were naturally performed in the rat model. A truly astonishing array of experiments were imagined, from pathol-ogy studies following large doses of radiation to microbeam irradiation and the effects of very low doses per fraction. Roughly, the sequence of experiments started with an emphasis on the histopathology and pathogenesis, presumably to confirm that pathogenesis in the rat and in humans was similar. Because the life span of the rat is only about 2 years, studies of latency especially as related to dose ensued. Fractionated treatments were pursued early. A special fractionation experiment involving only two fractions separated by various time intervals was performed to elucidate the repair of sublethal and long-term damage. All of these studies were initiated in just a few years starting around 1973. They were quickly followed with studies of the RBE of particle beams, studies on multiple fractions per day, studies involving putative radiation protectors and chemotherapeutic agents. The statistical analysis of early studies was sometimes unsophisticated, relying heavily on the concept of isoeffect doses that were often determined by graphical analysis. The raw data were usually in the form of dose-response experiments, but then collapsed into isoeffective doses, generally involving only D_{50}. As the shortcomings of this methodology became better understood, biomathematical models of dose response were deployed that used the raw data directly and could be assessed or validated by normative statistical methods. As usual, this process was initiated in the clinical domain, if retrospective data were available, prior to its introduction in the laboratory.

Latency as an Endpoint   Although the clinical significance is uncertain, Geraci and Mariano showed that the dose dependency of the latent period in rodents at very high doses had largely been ignored (Geraci and Mariano 1994). At single doses above 30 Gy, the latent period decreases exponentially with dose, up to at least 150 Gy. They attribute this decrease to an increase in the "functional" cells dying in interphase, i.e., from apoptosis. Brownson et al. had observed interphase cell death of oligodendrocytes in rat brain (Brownson et al. 1963a, b) and Li et al. observed it later in the spinal cord (Li et al. 1996) in both oligos and endothelial cells (Li et al. 2003, 1996).

For the first 20 years or so of investigations of RM using the rodent model, much was made of the latent period. In a large fraction of the publications, a figure would appear showing the average latent period as a function of dose. However, as early as 1978 Hubbard and Hopewell conclude that the latent period, although clearly dose dependent, cannot be used as an endpoint that is reflective of the level of damage to the spinal cord in studies that seek to determine the OER or the DMF for neutrons, and by implication, the RBE for other radiation protectors or enhancers (Hubbard and Hopewell 1978). Geraci and Mariano proposed exactly the opposite notion in 1994 (Geraci and Mariano 1994). In that study, the authors show three regions of dose dependence of the latent period. Interestingly, two of the three regions consist of dose ranges larger than what the authors call ED_{100}. As a theoretical concept, ED_{100} does not exist in most dose-response functions. As a practical matter, it would certainly depend on the number of subjects at any given dose. Assuming its practical existence, it nonetheless has little practical place in comparing responses in dose modifying experiments since doses in excess of those causing 100% paralysis can all be considered isoeffective.

We can use the data of Ang et al. to confirm the 1975 conclusion of Hubbard and Hopewell that latencies are not an adequate endpoint for bioassay (Ang et al. 1992). In the Ang study, the authors produced dose-response curves for single dose and

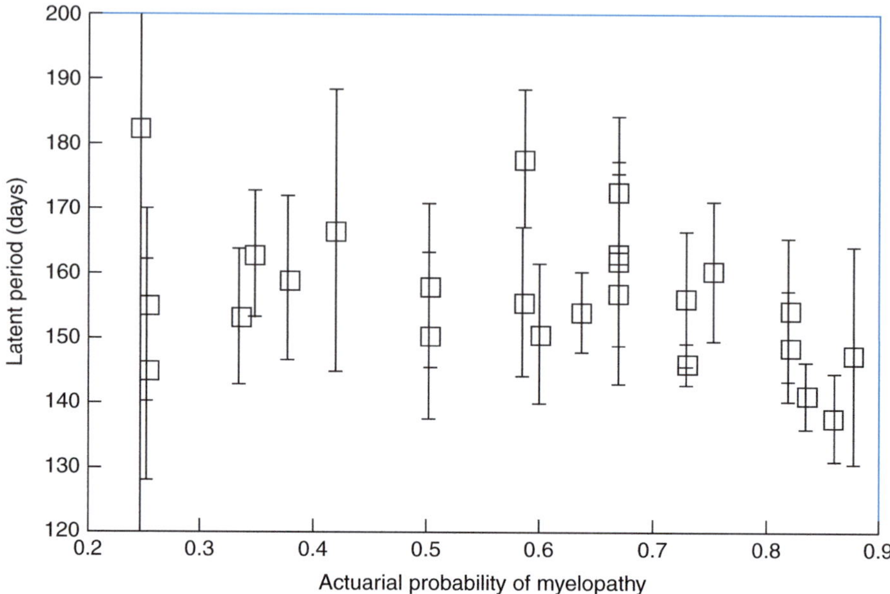

Fig. 5.1 Data from Ang et al. (1992) showing probability of RM versus latency in rats treated with 1 and 2 fractions per day. No relationship exists between latency and myelopathy rate such that latency could be used as a surrogate endpoint for RM.

fractionated treatments given at one fraction per day (plus or minus a top-up dose) and treatments given at 2 fractions per day. The interest here regarding the latent period question is whether the latent period is a reliable measure of outcome. Figure 5.1 shows the latent period as a function of the actuarial myelopathy rate. Error bars are one standard deviation. The data are restricted to responses less than 100% and to dose groups with at least three rats with paralysis. The latencies were given in days. Clearly there is no possibility of establishing a calibration curve that would reliably predict the probability of myelopathy given the average latency of the group. The correlation coefficient is −0.39 and is not significant at the 95% confidence level.

5.3   Dose Modifying Factors (DMF)

The dose modifying studies by Myers et al. and Gutin et al. have been mentioned in Chap. 2 as they relate to elucidating the pathology and pathogenesis of RM (Gutin et al. 1990; Myers et al. 1986). Other studies have addressed effects elicited by putative radioprotectors and radiosensitizers.

In 1983, van der Kogel and Sissingh investigated the effects of the hypoxic cell sensitizer, misonidazole, on the expression of RM following cervical spinal cord irradiation (van der Kogel and Sissingh 1983). Only five or six animals were used per dose group. D_{50}'s with confidence limits are given for single and fractionated doses with and without misonidazole. In addition, they investigated different anesthesias: ethrane in oxygen, Nembutal, and Nembutal plus oxygen. No hypoxia induction was observed and misonidazole had no impact on the dose response.

Later, they used the same radiation controls in a 1985 paper on the dose modifying effects of cytosine arabinoside (Ara-C) or methotrexate (MTX) in 1 or 2 fraction schedules (van der Kogel and Sissingh 1985). Treatments given in 2 fractions were separated by 4 h, 2 days, and 140 days. Intrathecal (IT) administration was used for the most part with some schedules using intravenous administration. No dose modifying effect was observed following IT MTX.

For IT Ara-C given in 1 or 2 fractions separated by 4 h, the DMF was about 1.2–1.3. For a single fraction with IV MTX or 2 fractions (IT) separated by 2 days, there was no modifying effect. For 2 fractions separated by 140 days, the effect was only seen for the WMN lesion but not the later vascular injury. Menten used a dose of Ara-C approximately 18 times higher and observed a DMF of about 1.2 for IV administration (Menten et al. 1989).

Morris et al. administered IT MTX in rats treated to field sizes of 4, 8, and 16 mm in single fractions (Morris et al. 1992). A dose modifying effect of about 1.2 was found only for the 8 mm field size with no effect for the others. In this paper as in the van der Kogel and Sissingh study, the irradiated controls were treated in a separate study, years prior.

In 1986, Spence et al. published DMF's for WR-2721 using the model of radiation-induced paralysis in Fischer 344 rats (Spence et al. 1986). They found DMF's of 1.3 and 1.6 for fore- and hind-limb paralysis, respectively. However, their method for determining the DMF was *ad hoc*, calculating it weekly rather than determining the cumulative rate of paralysis. In fact, the doses used were all higher than the D_{50} with 20 Gy being the lowest dose and with dose increments of 6 Gy. Clearly no dose-response analysis could be done with doses for which all animals would become paralyzed. This result was published again in 1988 (Spence et al. 1988). Their administration of WR-2721 did appear to delay the onset of paralysis, but to repeat, the stated DMF was not calculated as it should have been.

Also in 1986, Ang et al. studied the impact of IV-injected AZQ (aziridinylbenzoquinone), a cytostatic alkylating agent that crosses the blood-brain barrier. Using single and 2-fraction experiments, they observed no dose modifying effect with the use of AZQ, but the data were not analyzable using maximum likelihood methods since all dose-response curves were separated or quasi-separated.

In a study involving six animals per dose group, Geraci et al. found no protective effect for dexamethasone treatment following spinal cord irradiation in SPF rats. Dose grouping was relatively course, with doses being separated by 2 Gy from 18 to 30 Gy. Only at 22 Gy were the responses for the two arms not both 0% and 100%. At 22 Gy, the dexamethasone group had 2/6 myelopathies and the radiation only had 4/6—not significantly different.

Citing Fludarabine's activity against several hematological malignancies, Gregoire et al. investigated its dose enhancing effect for RM in the Fischer 344 rat (Grégoire et al. 1995). They irradiated animals in 2, 4, and 8 fractions spread evenly, they said, over 48 h. However, the interfraction intervals they reported were 24, 12, and 6 h. This was to facilitate giving the fludarabine at $t = -3$ h and $t = 21$ h for all three groups. The fact that the radiation was then actually given over periods of 24, 36, and 42 h is probably not impactful. They reported DMF's of 1.25 and slightly greater for the three different schedules—similar in its effect to Ara-C. It is interesting to note the large number of animals removed from the study due to radiation-induced sarcomas. Although the fraction censored is more than doubled in the fludarabine group (41% compared to 18%), the majority in the fludarabine group were removed after 270 days, the longest latency observed. On the other hand eight animals in the fludarabine group died on day 3 as opposed to zero in the radiation-only group. The authors reasonably suggest that "iatrogenic toxicity must be assessed closely" in clinical use of fludarabine.

Nieder et al. attempted three experiments to determine the protective effect of bFGF and IGF-1, as monotherapy or in combination (Nieder et al. 2002). No dose modifying factor could be elicited from the data but the data were sparse. Because of quasi-separation, D_{50}'s could not be estimated either (although the growth factor treatments were reported as having no protective effect in any of the three experiments). There was a hint of a longer latent period in the combined therapy group based on two of seven rats who became paralyzed. The other five died of intercurrent disease. In all three experiments, the intercurrent death rate was high—30–70%. This study can only be regarded as yielding preliminary data that show the possibility of a protective effect and that a different treatment technique should have been used to protect the esophagus since esophageal damage was responsible for the majority of intercurrent deaths.

In 2005, Nieder et al. looked at amifostine (WR-2721) in combination with IGF-1 in the retreatment setting (Nieder et al. 2005b). A second treatment of 17, 19, or 21 Gy followed 21 weeks after the initial of 16 Gy. For all three doses, the IGF-1-plus-amifostine rats exhibited longer latencies with all RM cases demonstrating vascular-based damage or vascular damage plus white matter necrosis, but none with white matter necrosis only. There was a suggestion of a DMF of 1.07, but it did not quite reach statistical significance. This was an unusual study in that it utilized a retreatment technique rather than a fractionated treatment. However, there was no full course dose response to which to compare so no repair fraction could be calculated. Although the lack of Type 1 myelopathies may confirm the suggestion made earlier that protection from WMN does not necessarily equate to protection from ultimate myelopathy, this interpretation is complicated by the fact that this is a retreatment study for which the pathogenesis could possibly vary from a single course treatment.

The last paper in this series was published in 2005 and tested different doses of IGF-1 in a 2-fraction regimen of 16 and 18 Gy given 1 day apart (Nieder et al. 2005a). The IGF-1 was given at 16 and 2 h prior to the first dose, 2 h before and 22 h after the second dose. IGF-1 doses were 1, 10, 50, and 200 µg. The 1 µg dose and the radiation-only group had indistinguishable responses, and the three higher doses of IGF-1 were also indistinguishable. Combining these groups give responses of 14/20 myelopathy cases for the 0 and 1 µg groups versus 8/34 for the 10–200 µg groups. Although these responses are highly significantly different, no dose modifying effect can be demonstrated by an experiment consisting of one radiation dose level. The IGF-1 latencies were significantly longer at the $p = 0.04$ level, but only if the variance of the latencies is considered the same in both groups. As has been seen above, it is not unusual to see significant delays in the expression of injury without significant shifts in the dose-response function.

Andratschke et al. performed an interesting set of experiments using PDGF as a biologic agent to reduce the incidence of spinal cord injury in rats (Andratschke et al. 2004). The rationale was that through various mechanisms on endothelial cells and oligodendrocytes (and possibly other targets), the expression of radiation damage in the spinal cord could be reduced. Indeed, they found a statistically significant dose modifying factor of 1.05 and a significant extension of the latent period. An interesting aspect of their analysis is that in the fitting of the two dose-response curves they allow the slope of the radiation-plus-PDGF curve to vary independently of the radiation-only curve, thus using a 4-parameter model to fit the data. See Fig. 5.2. It should be noted that the authors acknowledge that an examination should be performed at lower dose per fraction than was used in this study (16 Gy and higher). What was not stated was that because the curves were allowed to have different slopes, they will in fact cross at lower probabilities of injury. Our re-analysis shows this probability to be 0.9%, that is where the radiation + PDGF has a higher probability of injury than radiation alone. Using a single slope for both curves (that is, using a 3-parameter model with parallel dose-response curves) obviates this problem and does not impact the significance of the result. In fact, it is preferable according to the AIC metric. Thus we can see that the exact shape of the distribution of tolerances in a population could have an important impact in the estimate of the DMF at clinical doses. Of course, in most experimental settings, it is impossible to distinguish one distribution from another, that is, to examine the response at very low levels, because too many subjects would be required to be practically and ethically feasible.

Data from Nieder et al. using IGF-1 (above) are also plotted on Fig. 5.2. The control data agree well, but somewhat surprisingly, the data point for IGF-1 after a total dose of 34 Gy falls almost on top of the data point for PDGF. Having these two experiments support each other lends confidence to the overall result as well as the general collaborative efforts of Professors Nieder and Andratschke. At this point there are at least three lines of investigation that seem important to pursue. First, does the putative protection of PDGF and IGF-1 extend to longer times following irradiation and therefore protect against the RM vascular lesion as well? Otherwise it might transpire that the net protection would be zero. Second, does the DMF change (increase) with fractionation?

Finally, as in Nieder et al. (2005b), the retreatment setting should be explored. It seems likely that this field has reached a point where RM can be avoided as long as the radiation treatment is designed and delivered properly. However, patients face unpredictable risks in the retreatment setting. Therefore it is probable that the most essential deployment of radioprotectors will be in retreatment of the spinal cord.

Treatment with the ace inhibitor ramipril has been shown to reduce VEGF-A expression and demyelination without impacting the incidence of RM (Clausi et al. 2018). Saager et al. attempted to assess the dose modification effect in photons and carbon ion irradiation, but with only four animals per dose group and doses grouped

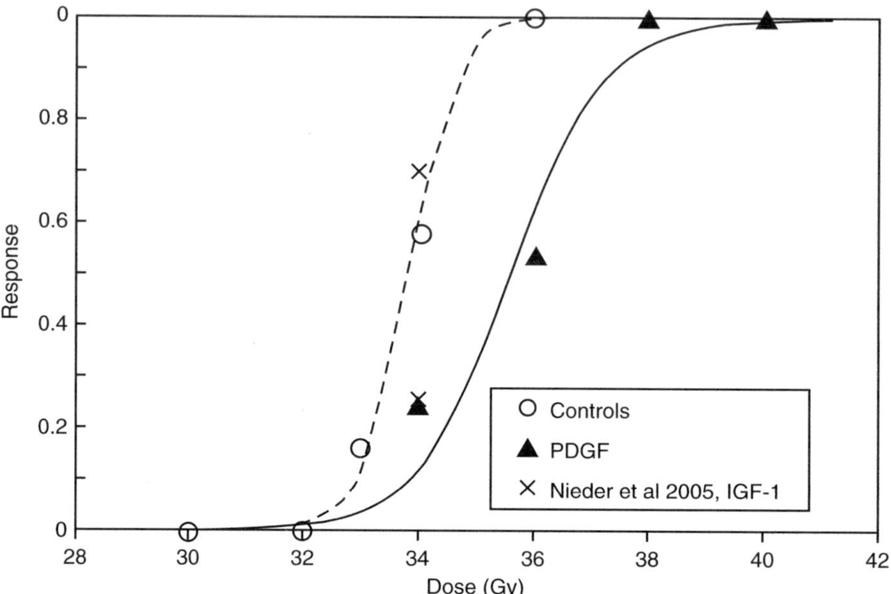

Fig. 5.2 Protective effect of PDGF and IGF-1 from Andratschke et al. (2004) and Nieder et al. (2005a).

in intervals not less than 2 Gy, they were not successful (Saager et al. 2020b). They performed dose-response studies for both radiations. In re-analysis, logistic regression was used to model the dose response for the pooled data and for the photon and carbon data pooled separately. When all the data were pooled, the model was rejected. When pooled by irradiation, the carbon data could not be analyzed since both with and without ramipril the data were completely separated. The photon data did not fit the model. Looking at only the dose points with a difference in the response of ramipril versus no ramipril, the incidences were not statistically significant by Fisher's exact test. The design of this experiment virtually excluded the possibility of an analyzable result.

Although some efforts toward mitigating radiation spinal cord damage have been promising, there has not yet been a clearly successful strategy. Nearly all efforts have been directed at the amelioration of the early lesion of white matter necrosis. Even a successful strategy against this injury component leaves open the path of continued progressive vascular damage leading to the later development of paralysis from massive vascular injury.

5.4   Fractionation Studies in the Rat Up to 1990

Starting in the mid-1970s and continuing for more than 20 years, the rat model was studied extensively to quantify dose, time, and fractionation response of the spinal cord. Of course, human observations were published contemporaneously with the rat studies, and undoubtedly informed what sort of experiments would be useful. Experiments included dose response for various numbers of fractions, 2-fraction dose response where the fractions were separated by increasing lengths of time, and dose response with a large single dose at the end of treatment, referred to as top-up dose experiments. Later the top-up dose became a fractionated dose rather than a single dose and was given alternatively at the beginning or at the end of treatment. Top-up dose experiments were designed to facilitate investigating small doses per fraction in rodents; they were not thought to be of intrinsic or clinical interest. RBE experiments were performed to ascertain the magnitude and the effect of fractionation of particle treatment compared to photon treatment. Nearly all of these experiments shared a common trait—they were significantly underpowered. This fact manifests itself in the large proportion of published dose-response experiments from this period that exhibit complete separation or quasi-separation. This will be addressed as this section unfolds.

5.4.1 Transition from the Power-Law Model to the LQ Model

It is probably fair to say that most of the fractionation studies were performed in order to determine a mathematical formula (model) that could predict how the rate of RM would change in response to changes in dose per fraction, number of fractions, the interval between fractions, the overall treatment time, or in general, the schedule of dose delivery. By determining such a model, it would be possible to alter fractionation patterns, based on a treatment known to be safe and not exceed the spinal cord tolerance. This was the search for isoeffect formulas.

In 1974, van der Kogel and Barendsen first published slope values for rat spinal cord tolerance data plotted on a log-log scale ($\log D$ versus $\log N$) (van der Kogel and Barendsen 1974). (This was referred to as a Strandquist-type curve at the time even though Strandquist always used overall time rather than fraction number in his analysis; many also referred to this as the "power law" model.) "Tolerance" in this paper was defined by the largest dose for a given N for which no paralysis was observed (clearly dependent on the number of animals allocated to that dose level). The slope (exponent of N) for the plot was given as 0.44. Thus the isoeffect dose was proportional to $N^{0.44}$. Similar publications by Madsuda et al. (thoracic cord) and White and Hornsey (lumbar cord) followed in 1977 and 1978 (Masuda et al. 1977; White and Hornsey 1978); van der Kogel also published a follow-up paper appearing in the issue of Radiology immediately following Masuda et al. (van der Kogel 1977). van der Kogel found a slope of 0.44, Masuda et al. found 0.44, and White and Hornsey found "about" 0.4 (current re-analysis gives 0.395).

In neither the 1974 nor the 1977 papers by van der Kogel were the data presented in a way that allowed re-analysis, so the slope of 0.44 could not be independently verified. However, the data of White and Hornsey *were* reported such that a re-analysis here was feasible and their conclusions could be tested and validated. Their experiments consisted of two sets of dose-response experiments—one with variable fraction number and the other with 2 fractions with variable interfraction intervals.

The first set had dose-response curves for 1, 4, 8, 15, and 30 fractions to the lumbar cord over 6 weeks. The two-fraction studies had dose-response curves for 1-, 4-, 8-, 16-, and 32-day interfraction intervals. Because the 2-fraction data that were separated by 4, 8, and 16 days all had the same dose response, these experiments were combined in the re-analysis with the first series to serve as the data for 2-fraction dose response. So the re-analysis had 1, 2, 4, 8, 15, and 30 fractions. It should be noted that complete or quasi-separation existed in some of these sets of dose-response data (see Chap. 3). All the data were fitted simultaneously using three different 3-parameter models rather than getting a D_{50} value from each set by a separate analysis. The three models used were:

Model 1

$$P = \frac{1}{1 + \left(\dfrac{D_{50}}{D}\right)^k} \tag{5.1}$$

where D_{50} is the single-fraction median tolerance and D is the effective single-fraction dose from the LQ model.

Model 2

$$P = 1 / \left(1 + \exp\left(\beta_0 - \beta_1 D - \beta_2 Dd\right)\right) \tag{5.2}$$

where $\alpha/\beta = \beta_1/\beta_2$.

Model 3

$$P = \exp\left\{-\exp\left[\beta_0 - \beta_1\left(\ln(D) - \beta_2 \ln(N)\right)\right]\right\} \tag{5.3}$$

where β_2 is the exponent of N in a TDF-type (power-law) model. Note that the signs of some of the terms have been adjusted for the convenience of having positive coefficients.

The first model provided the best fit, but of course the second was very close, being a similar form of the LQ model. They are both logistic models, but in the first model, dose is on a logarithmic scale whereas the second is dose on a linear scale. The third (power-law) model did provide an acceptable fit, not as good as either of the LQ models. Also, to obtain an acceptable fit for the power-law mode, a log-log link function was used because the logistic link did not fit. Figure 5.3 shows the first and third models (LQ and power law) along with the D_{50}'s fitted by eye by the original authors. Perhaps the most interesting aspect of this re-analysis is that despite the authors' purely visual determination of D_{50}'s and a number of separated dose-response curves, the statistical analysis here gave results that are quite close to the authors' conclusions. Furthermore, visual examination would seem to support the notion that Model 3, the power-law model, should fit as well as Models 1 and 2, *but it does not*. This figure also clearly illustrates that the seeming conformity of the D_{50} values to some functional dependence does not mean that a model fits the

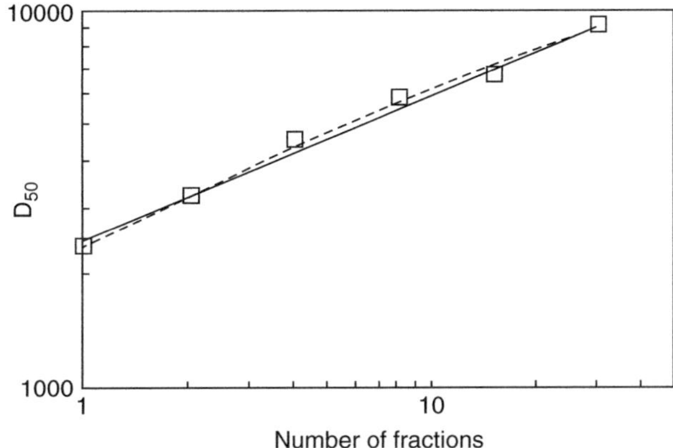

Fig. 5.3 Data from White and Hornsey (1978) with D_{50} values fitted by eye. Dashed line is the fit to the raw data using Model 1 and solid line is from Model 3.

Table 5.1 D_{50} values from White and Hornsey (1978) and Models 2 and 3 above

N	D_{50} Authors	D_{50} LQ (MLE)	D_{50} power-law (MLE)
1	2400	2369	2444
2	3270[a]	3229	3189
4	4600	4334	4159
8	5900	5697	5425
15	6800	7128	6904
30	9200	8823	9006

[a]This value was the average of the D_{50}'s for $N = 2$ treatments given 24 h to 16 days apart

dose-response data. *D_{50} is not a good metric for model assessment*, and plotting data on a log-log scale may elicit overall trends but suppresses nuance.

The D_{50} values were calculated based on the MLE of the three variates and the fraction number. These, along with the authors' graphical estimates, are given in Table 5.1. The data are graphed on a log-log scale in Fig. 5.3. The dashed line represents the LQ model whose parameters were determined in the re-analysis ($\alpha/\beta = 4.6$ Gy). The curvature is obvious, but it would be much less so without the line put there to guide the eye. The solid line is the power-law model. Of course, the fact that these lines are continuous is somewhat misleading since treatment fractions are restricted to integer values only.

In radiation biology, as in all fields, models of processes are invented to organize observations (data) as efficiently as possible using the fewest possible explanatory parameters. Fitting five dose-response functions with separate D_{50}'s and slope parameters requires ten constants. In this re-analysis, fitting all five dose-response curves simultaneously with one location, one slope, and one fractionation parameter required only three parameters total and was done according to normative statistical methods. Finally, it is important to note that the fit of the models to the data can be assessed by normative methods.

The data from this paper were re-analyzed in detail to illustrate several points vital to the interpretation of published data. First, the data were provided in sufficient detail to perform a re-analysis. It is important to provide original data whenever practicable because the reader should be able to satisfy themselves that the data were properly analyzed and that the analysis was faithfully reported. Second, if the paper suggests that the data follow a mathematical pattern, i.e., a model, then it is absolutely necessary to demonstrate that the model fits the data. "Strictly speaking, an assessment of the adequacy of the fitted model should precede an attempt at interpreting it" (Hosmer and Lemeshow 1989). Of course it would be pointless to discuss in depth the interpretation of the values of α, β, and α/β ratio of the LQ model derived from data that were in fact not adequately fitted by the model. After all, the LQ model has been extensively applied to clinical situations requiring conversion from one fractionation scheme to another. Finally, and this point may seem contradictory, it should be noted that just because a poor job was done assessing the performance of the model or even if the model does not fit the data, that does not mean that the model is not useful or not applicable. In fact in such a case, readers should consider the possibility that the experimental method or the acquisition of the data were exceptional in some way. For example, many clinical RM reports come from thoracic treatments delivered with one field per day, and erroneous conclusions were reached regarding the dose response of the thoracic cord.

The re-analysis of this paper and other papers in this volume is not a criticism of the original work. Although maximum likelihood methods had been used in analysis of radiation dose responses (Best 1959; Frome and Beauchamp 1968; Herring 1980; Gerber 1982; Metz et al. 1982; Potish et al. 1983; Frome and DuFrain 1986), modern analytical methods applied to binomial data became widespread as computers in radiation oncology departments were deployed against iterative algorithms. Leith et al. seem to be the first to use maximum likelihood methods in logistic regression in the analysis of radiation myelopathy (Leith et al. 1982).

Interestingly, Hornsey and White later extended the fractionation schedule to 60 fractions (Hornsey and White 1980). However, this was done by irradiating the rats twice daily, approximately 10 h between fractions. Their previous work was 1 fraction per day, and apparently there was no treatment schedule that calibrated the two experiments against each other. In this study, the 60-fraction D_{50} was stated to be 112 Gy and indeed the log-log plot did not appear to show a decrease in slope at the higher fraction number. However, the single-fraction dose was now significantly below the isoeffect line whose slope they estimated to be 0.37.

A re-analysis was attempted for the data of Masuda et al. but could not be accomplished because of many internal inconsistencies in the presentation of the data. Doses and numbers of paralyzed animals were internally inconsistent in the two graphical presentations. Based on the information given in the text (ten animals per dose group) and the number of animals used to calculate the average latency (never more than 6 in a dose group no matter how high the dose), it appears that at least 40% of the animals died of intercurrent disease. Finally, because of outliers at lower doses, the fit to the data was rejected by the χ^2 statistic. However, this paper subtly contributes to the advancement of statistically methodology in radiation dose-response analysis. Dose-response curves were given for 1, 2, and 4 fractions. D_{50}'s were determined by "a logit regression line" fitted to each fractionation schedule separately. This was presumably an unweighted linear regression of $-\ln(1/P\text{-}1)$

versus ln (dose). Although this is not the equivalent of logistic regression as it is now understood and has a number of problems (0% and 100% responses cannot be used, the fit is rejected in this case, and the weights vary with P), it is an improvement over "fit by eye" *if* it is applied appropriately (Schultheiss et al. 1990).

As a reminder of what was being published in the human literature about this time, Phillips and Buschke had published a value for the N-slope in the power-law model of 0.5 for the human thoracic cord in 1969, with the value being determined by a sort of graphical analysis (Phillips and Buschke 1969). A scattergram of log (D) versus log (N) was created where the D and N were obtained from treatments of individual lung cancer patients. The cases with and without RM were separately identified in the graph and a straight line drawn through the two most extreme cases with myelopathy. This was akin to a graphical version of discriminant analysis. A somewhat more scientifically determined value of 0.44 was published by Wara et al. (Wara et al. 1975). Although the slopes for the human data were not determined using isoeffect data as in the rat study, they yielded the same value. Because the rat experiments were published after the human data for spinal cord and also for lung, it is probable that it was the human data that inspired the rodent investigations.

5.4.2 Studies from van der Kogel

The collective early publications of van der Kogel were summarized by him in invited reviews in 1986 and 1991 (van der Kogel 1986, 1991). His early studies were seminal works, often being the first in the field. Summarized in this section are the fractionation parameters in the studies prior to 1985. For the WR rat, the α/β ratio for the C-cord was given as 2.4 Gy with an N-exponent of 0.43; for the L-cord these were 2.5–4.7 Gy and 0.42, respectively. The tolerance dose expressed as ED_{100} were 21 and 24 Gy for the WR and BN rats, respectively and did not differ significantly for the different levels (van der Kogel 1986). These results were originally published in a 1977 paper and the 1979 PhD dissertation (van der Kogel 1977, 1979). The dose-response analysis for both works was limited to a graphical analysis. In fact, more than 80% of the x-ray dose-response curves in the thesis were completely or quasi-separated. This was a result of too few animals per dose point. As long as D_{50} values were estimated graphically and not with biostatistical procedures, separated dose-response data did not present a methodological problem, but statistically normative, quantitative estimates were not possible. In a paper in 1985, van der Kogel addressed this problem, stating "To carry out this analysis legitimately, at least two points in between 0% and 100% response are needed" (van der Kogel and Sissingh 1985). Although this fact was widely appreciated, it did not seem to make a strong impression on researchers who continued to fit "by eye" well into the 1990s and beyond.

van der Kogel's 1979 thesis is a highly cited and authoritative work and seems to be the first instance of applying the LQ model to spinal cord data. Leith appears to be the first peer reviewed study to give α/β values for the spinal cord (Leith et al. 1981). Although some of the data that were re-analyzed to determine α/β values for the cord came from van der Kogel's thesis, van der Kogel's deployment of the LQ model was not cited. Admittedly, van der Kogel's discussion of the LQ model in his dissertation was brief—comprising three paragraphs only and receiving no mention in the abstract. Only 3 months after the appearance of Leith's 1981 paper, Thames et al. published the famous "spaghetti plot," showing for the first time that the α/β ratio for late effects was smaller than for acute effects and tumors (Thames et al. 1982).

Interestingly, it seems that the 1981 Leith paper was the first to suggest 2 Gy as the value for α/β. Although this value was initially suggested only for the cervical cord, it eventually became the default value for both the thoracic and cervical cord but sometimes a different value was suggested for the lumbar cord because of the fact that a different pathogenesis was found for the lumbar cord. Initially at least, these values were reached based on a flawed analytic method—the Fe-plot.

5.4.3　The Fe-Plot

The Fe-plot was first introduced by Douglas and Fowler as a technique for determining the α/β ratio based on isoeffect data (Douglas and Fowler 1976). The basic LQ equation cell survival equation

$$S = e^{-\alpha dN - \beta d^2 N} \tag{5.4}$$

can be rearranged as

$$E = \alpha D + \beta D d$$

or

$$\frac{E}{D} = \alpha + \beta d \tag{5.5}$$

where D is the total dose, d is the dose per fraction, and E is a constant level of effect. Thus when pairs of (D,d) that are assumed to be isoeffective are plotted as $(1/D, d)$, then the ratio of the slope to the intercept gives an estimate of α/β. Since this exercise depends on the use of linear regression, which in turn assumes an independent stimulus variable (x) and a dependent response variable (y), a case *cannot* be made for the legitimacy of the estimate obtained by this method since $1/D$, the inverse of total dose, is certainly not independent of d, the dose per fraction. Another issue with the use of the Fe-plot is that it deploys derivatives of the original data, i.e., isoeffect doses rather than the data themselves. It is likely that this construct arose because of the level of understanding and penetration of statistical methods into radiobiological research at the time and because of the relatively primitive software and hardware available to analyze dose-response data. Certainly the use of the Fe-plot diminished dramatically in the 1990s due to better tools for dose-response analysis and the understanding of the considerable shortcomings of the Fe methodology (Herbert 1989; Herbert et al. 1993; Taylor and Kim 1989).

In addition to the questionable α/β estimates from the Fe-plot, it was also asserted that the straightness of the graph of $1/D$ versus d was evidence of the validity of the

LQ model itself (Douglas and Fowler 1975, 1976). The straightness of this graph does not indicate that the LQ model fits the data. Arguments supporting this are strong (Herbert et al. 1993). However, the theoretical arguments are probably less persuasive than a demonstration. To this end, a Monte Carlo simulation was performed to mimic a dose-response experiment that was designed and analyzed under the assumption that the LQ model was the underlying biology behind the dose response, but in fact the data were generated according to the power-law model with the N-exponent $= 0.4$ and the T-exponent $= 0.02$. The D_{50} for 2 fractions was defined as 30 Gy. 1000 histories of the experiment were replicated. Five fractionation schemes were used, $N = 1, 2, 4, 8$, and 20. Five dose levels were used at each value of N. The experiment was designed to allow for some random variation of the parameters of the experimental design itself. Thus, the "investigator" randomly guessed that the D_{50} for $N = 1$ was between 22 and 23 Gray, in increments of 0.1 Gy. The investigator guessed that the value of α/β was between 2 and 4 Gy in increments of 0.5 Gy. (The guesses were produced by using a rectangular distribution between the limits stated.) Using the "guessed" value of α/β, the investigator estimated the D_{50}'s for the other values of N and added two dose points above and below these D_{50} values. The true value of P was calculated at these doses and the number of animals who exhibited RM was determined based on Bernoulli statistics. Maximum likelihood estimates were obtained for α/β, and the $N = 1$ values for D_{50} and k (Eq. 5.1). Based on the responses of the individual values of N, maximum likelihood estimates were obtained for D_{50}, and Fe-plot generated, and fitted using linear regression. From this an α/β was calculated using the ratio of the slope to intercept of the Fe-plot.

Of the 1000 histories of separate experiment simulations, 31.5% of them rejected the LQ model by the χ^2 statistic. (A value of $P(\chi^2) > 0.15$ was taken to imply a satisfactory fit to the data.) However, *all* experiments produced Fe-plots with statistically significant straight lines, even the subset of 315 that were poorly described by the LQ model. In fact, the average R^2 value of the Fe-plot for the *rejected* experiments was 0.98—not just straight, but very straight. Of the experiments that were not rejected by χ^2, we have α/β values from both the logistic regression and the Fe-plot. The correlation coefficient of these values was only 0.57. For Fe-plot analysis to be believable, the results should at least be highly correlated with results from normative statistical methods.

To help visualize this process, Figure 5.4 shows one example replication. The data points show the responses to the doses "designed" by the experimenter, and each fractionation schedule has two associated dose-response curves. The solid curves are the dose-response functions fitted to the individual values of N, from which was generated an Fe-plot. The dashed curves are the 3-parameter logistic regression fitted to the entire dataset. Obviously, the solid lines were generated using 10 (two per curve) rather than three parameters. This dataset was not well fitted by the LQ model with $p(\chi^2) \sim 0.06$. The Fe-plot is shown in Fig. 5.5.

To conclude, unless the experimental RM data are very poorly behaved, the Fe-plot cannot help but be a straight line. This straightness in no way indicated a good fit of the LQ model to the data, and the estimates for the α/β ratio could not be relied upon. While it is true, as stated above, that the Fe-plot fell out of use by the

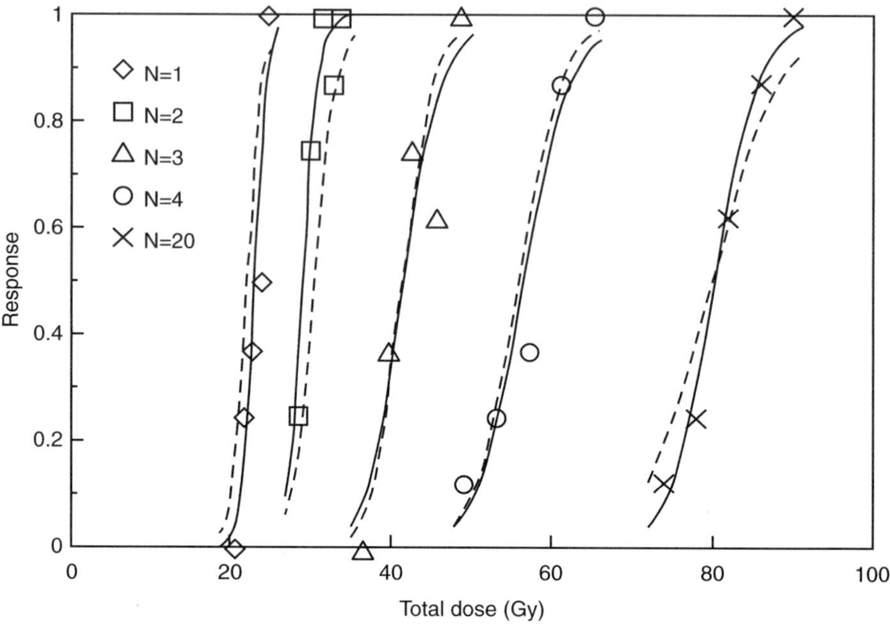

Fig. 5.4 Data generated in a Monte Carlo simulation and fitted by pooling the data and assuming the LQ model of dose response.

mid-1990s, one should be aware of its impressive weaknesses when reading the older literature where it was deployed.

Thus, we can conclude that the fractionation parameters (α/β = 2 Gy; N-exponent = 0.4, etc.) for experimental RM upon which countless tolerance dose estimates were made were not derived from statistical analysis. This does not mean that these parameters were incorrect. Indeed, no statistical analysis can disclose the "correct" values of model parameters. It means that the degree to which they should have been relied upon and the limits of their validity were not established. In the years that followed until the late 1990s, many dose-response experiments were performed in rats to elucidate the biomathematics of various processes, such as fractionation, retreatment, long-term recovery, etc. Many of these dose-response experiments were characterized by an excess of extreme responses resulting from a lack of understanding of where the dose-response curve being probed should lie on the dose axis. Perhaps if experiments with more statistical power had been performed earlier, better quantitative estimates of underlying response parameters would have yielded experiments more tightly trained on the sigmoid portion of the dose-response curve.

5.4.4 Altered Fractionation and Top-Up Experiments

As it became clear that damage to late responding organs such as the spinal cord could be decreased by lowering the dose per fraction, the use of multiple fractions

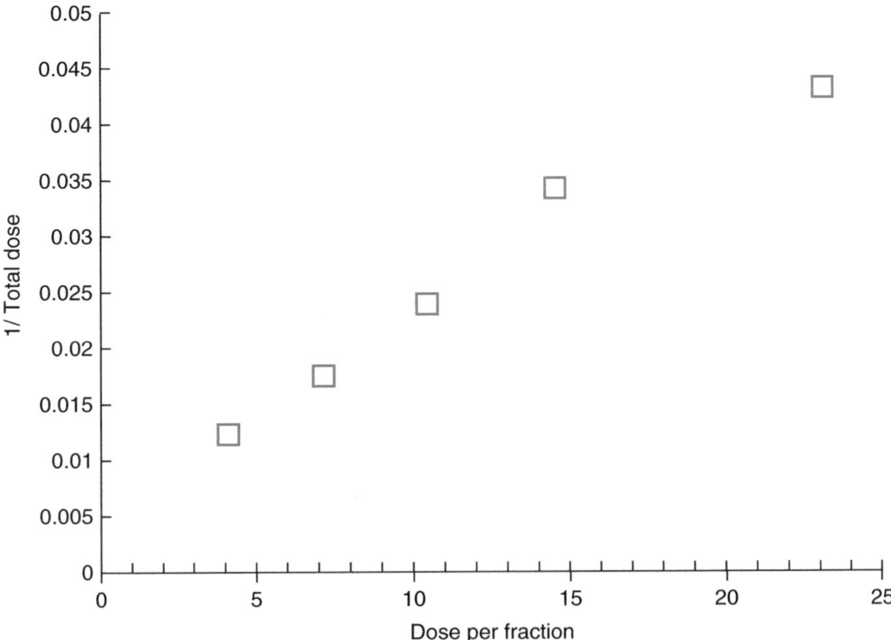

Fig. 5.5 The Fe-plot for Fig. 5.4. The clearly straight lines do not imply a good fit to the data.

per day (MFD) again became an attractive alternative to daily fractionation. This seemed especially true for certain head and neck cancers whose proliferative response to treatment induced the deployment of a more rapid course afforded by multiple fractions per day. The first cases of myelopathy following MFD had been treated with three fractions per day (van den Bogaert et al. 1986; Dische and Saunders 1989). Two fractions per day to the spinal cord treated according to a clinical trial protocol rarely if ever resulted in RM (Parsons et al. 1997).

Although rats were the animal of choice for experimental studies in RM, it was difficult, tedious, and expensive to perform fractionated experiments where the fractionation schedules were similar to clinical schedules, i.e., with doses per fraction of 1.5 to 3 Gy or so. Making this even more difficult was the fact that the D_{50} for rats in this region was considerably higher than the human, thus requiring even more fractions. Very early attempts approached this problem by deploying multiple fractions per day (van der Kogel and Sissingh 1983), but the problem of incomplete interfraction repair soon became apparent. In 1983, Ang et al. published the top-up dose technique where a significant number of small fractions was replaced by a single large fraction, thereby reducing the overall number of treatments (and set-ups, anesthesias, etc.) (Ang et al. 1983). This technique had its origins in the mouse foot skin reaction studies (Denekamp 1973; Denekamp et al. 1984). In Ang's 1983 study, the top-up dose was one fraction of the 2-fraction D_{50} dose, thereby halving the number of fractionated treatments required. Wong later modified this technique by using 3 fractions of the 4-fraction D_{50}, allowing him to go to even smaller doses per fraction and thereby mimicking more total fractions (Wong et al. 1993). Ang et al. referred to this concept as "partial tolerance," a term for which they gave no attribution but that was introduced in 1969 by Frank Ellis for use in an NSD calculation (Ellis 1969). It amounts to dividing a fractionated treatment into two or more fractionation schedules that achieve the same level of response as a known schedule. In the top-up experiments, the division was made such that the top-up dose was a fraction of a dose schedule (initially one half) with a known outcome (50% response), and the remainder of the dose was delivered as the fractionated test dose being studied.

The objective of the top-up paper was to verify that the concept of partial tolerance was valid. That meant, for example, if the top-up dose was 1 fraction of the 2-fraction D_{50}, then the study dose that would elicit a 50% response would be half of the number of fractions of a fractionated D_{50}, say 10 of the 20-fraction D_{50}. To demonstrate this effect, four dose schedules were chosen for dose-response curves with and without a top-up dose. They were 1, 2, 4, and 10 fractions with no top-up, and 2, 5, 10, and 20 fractions followed by a top-up dose as described above. In theory, the 2- and 5-fraction schedule with a top-up would have one half the D_{50} of the 4- and 10-fractions schedule without the top-up. Unfortunately, issues in the execution prevented the statistical validation of the concept, allowing only a graphical analysis of D_{50} values. These limitations were not recognized at the time.

Although the authors state the 16 animals were irradiated simultaneously, this does not necessarily mean that all 16 received the same dose. Six of the eight dose-response curves were completely or quasi-separated, including all of the schedules without top-up doses. Six of the eight dose-response curves contained only 3 dose points, and two thirds of the 27 dose points had extreme responses of 0% or 100%. Although it was stated that the "curves were constructed by computer probit analysis," it was clear that the D_{50}'s were in fact determined by eye since maximum likelihood estimates for separated dose-response curves are undefined and computer algorithms would fail. Also, 95% confidence limits were shown, but their determination was not specified. Of course, no D_{50} confidence limits for separated data could be determined by standard methods. Neither the D_{50} values nor their confidence limits were given in tabular form.

Once the D_{50}'s were determined, they were plotted on a log-log plot to estimate the N-exponent of an NSD-like (power law) model. The top-up dose points were plotted as the (N, D) pair using $(2N, 2D_{50})$ where N is the fraction number of the top-up experiment ignoring the 15 Gy top-up dose and D_{50} is the visually estimated D_{50}, because "the use of the top-up dose reduces the ED_{50} for the fractionated treatment with the same fraction size with exactly 50%, as had been predicted." All points lie on a straight line with a slope of 0.45, and the points for $N = 4$ and $N = 10$ are, as reported, overlapping with the $N = 2$ and $N = 5$ point from the top-up component.

The mathematics of the NSD-type model was well understood at this time (Orton and Ellis 1973). Orton and Ellis simplified the use of the power-law model by constructing a dose metameter that is additive (unlike the NSD) since it is directly proportional to N. They called it the TDF, standing for time-dose factor. Following this method, it is possible to create a logistic dose-response model using a modification of Orton's TDF. By combining all the data in the top-dose experiment, separation problems are averted and the power-law model's fit to these data can be assessed. As stated in the chapter on statistics, no further data analysis involving implications of a model should be undertaken if the model fails to fit the data by appropriate goodness-of-fit metrics. Re-analyzing the data from this paper shows that the model fails to fit the data, assuming that there were in fact 16 animals per dose group. The model's values of the N-exponent and the D_{50}'s cannot be used for further analysis that would serve to validate the partial tolerance concept because it cannot be shown that the model fits the data. The model fitting exercise was repeated using the LQ model, which also failed to fit the data as well. Thus the authors were using parameters, the values of D_{50}, obtained by an *ad hoc* method when the standard model assessment rejected the model and consequently the parameters it contained.

It should be further noted that validation used in the paper itself relies upon the D_{50}'s from the top-up dose regimens with 2- and 5-fractions overlapping the D_{50}'s from the 4- and 10-fraction data without top-up doses. However, all four of these D_{50} values were indeterminant since the dose-response data were separated or quasi-separated. Thus, this paper also illustrates what has been said previously, namely that distilling dose-response points down to D_{50} values and then applying a graphical analysis to them is a blunt instrument at best and very misleading much of the time.

It is in fact true that according to both the power-law model and the LQ model, the top-up paper should have shown exactly what it claimed: the validity of the partial tolerance concept. In both models, the dose metameter is proportional to N times a function of dose per fraction that is independent of N. Ultimately, this means that the incremental damage elicited by an incremental dose is completely independent of the fractionation schedule used in eliciting the initial damage. Thus the concept of partial tolerance is closely related to the fundamental hope of radiobiology: that equal fractions result in equal effects independent of the sequence in which the fractions are delivered. However, the conclusions from this paper were not validated by the analysis of the data and cannot be taken as support for the partial tolerance concept.

5.4.5 Additional Ang Studies

Ultimately Ang and colleagues from M.D. Anderson performed a series of experiments primarily involving the study of incomplete repair (Ang et al. 1992). However, included in the study were dose-response data for 1, 2, 5, and 10 fractions without top-up along with 10 and 20 fractions and 2 and 4 Gy per fraction with top-up. Although these data were not analyzed with validating the partial tolerance in mind (because they assumed the matter was settled), the data were re-analyzed here based on the LQ model. By deleting one or two of the 38 points as outliers, the model was found to fit the data, and the concept of partial tolerance was determined not to be in conflict with the data. This is a very important assumption in dose-response analysis carried out in the subsequent decades.

Out of concern that a significant increase in the spinal cord tolerance dose was being predicted for humans in clinical studies using doses per fraction less than 2 Gy, Ang et al. designed a rat study to test the assumption (Ang et al. 1985). They irradiated rats at constant doses per fraction (plus a top-up dose) of 1.3, 1.5, 1.8, and 2 Gy. Unfortunately, in order to shorten overall treatment times, different percentages of animal at each dose per fraction level were treated twice a day with 4 h intervals. How this percentage was distributed over the dose levels within the various dose-per-fraction groups was not disclosed. Therefore an erroneous (or at least unsupportable) conclusion was reached that no further sparing was achieved by using doses per fraction smaller than 2 Gy. Thames et al. showed that incomplete repair with mono-exponential kinetics would explain the observation of lack of sparing (Thames et al. 1988). The estimated value for α/β was 2.7 Gy, with a half-time for repair of 1.6 h. However, it was not just the data from the 1985 paper that were re-analyzed; data from 3 additional papers were included. Based on the Figure 1 of that paper, depicting the dose-response data, the fit to the data from the 1985 paper was quite poor, although no model assessment was offered. The only real conclusion that can be reached with respect to the 1985 study is that it served as a preliminary effort in exploring our understanding of repair kinetics and the dose response for the spinal cord.

Arguing that "repair of sublethal damage is the main factor determining the tolerance of the CNS to fractionated radiotherapy" in the clinical setting, Ang et al. designed a series of experiments to explore the kinetics of repair of sublethal damage to the spinal cord using the rat model. In a 1984 publication, he studied the effect of 2- and 4-fraction regimens on the incidence of myelopathy when these fractions were separated by intervals of 20 min to 24 h (Ang et al. 1984). He demonstrated that the repair appeared to follow first order kinetics and furthermore, that the rate of cellular repair is faster after smaller fractions. These studies were done without recourse to the use of a top-up dose. A potential issue with the study was that the 2-fraction data were obtained using 20 MeV electrons while the 4-fraction data used 18 MV photons. An RBE for electrons of 1.09 was obtained from single-fraction dose-response experiments with both modalities and of course "this cannot influence the obtained results concerning repair kinetics" since the modality was constant within each fraction number. However, the repair kinetics was not measured directly. The kinetics was inferred by introducing the fraction of dose that was recovered between radiation treatments and attributing the dose recovery to the "kinetics of cellular repair of sublethal damage." Cellular damage was not measured nor was the type of cell specified.

They found the half-time of repair was "about 110 min [1.83 h] for 2 fractions (11–15 Gy/fraction) and … about 85 min [1.42 h] for 4 fractions (7–11 Gy/fraction)." Neither the LQ model nor the concept of incomplete repair was directly referenced in the analysis. These fractionation schemes, with all fractions being separated by a constant interfraction interval rather than two or three fractions given daily for several days, were exactly the schedules that Thames' IR model was designed to model.

Subsequently an additional study was performed using 10 equally spaced fractions followed by a top-up dose of 15 Gy (Ang et al. 1987). The data from the first study were combined with these later data and analyzed together using maximum likelihood estimation with a double exponential dose-response function that incorporated the LQ model and the mathematics of incomplete repair (Thames et al. 1984). The pooled data were re-analyzed for this book with closely matching results. The overall model fit the data well according to the χ^2 statistic, even with trimming the superfluous degrees of freedom at the extreme response values.

The results of the original analysis were $\alpha/\beta = 3.4$ Gy and $t_{1/2} = 1.55$ h, and there was no need to deploy a second exponential component of repair. However, when the fractionation schedules were analyzed separately (along with the single dose data), α/β ratios of 2.0 and 2.7 were obtained. It seems reasonable that the pooled α/β ratio should have fallen within the range of the separately analyzed data. Of course, using only a single fractionation schedule plus single dose data is not a robust method for obtaining α/β ratios. Nonetheless, one cannot say that this study gives very precise estimates of α/β, even the context of a single study with multiple values for the estimate. In general, this is not uncommon in the radiobiological studies of RM. In fact, it appears that precise estimates of α/β are not of great concern, at least not in animal models. Although these values are reported whenever possible, discussions of the values or the values relative to other animals or experimental situations are not common.

5.5   Dose Response After 1990

By 1990, it was not uncommon, but not universal, for normative statistical procedures to be applied to dose-response data. These procedures included maximum likelihood estimation of parameters and, much less commonly, assessment of the model fit, usually through a χ^2 statistic. These procedures were not always applied appropriately, and one of the more common errors was the use of an excessive number of degrees of freedom associated with the value of χ^2. As discussed in Chap. 3, superfluous high-dose points with 100% response or low-dose points with 0% response should not be used in the count of degrees of freedom since the parameters estimates do not really depend on these data points. In the re-analysis done for this volume, the number of degrees of freedom will reflect only those the points with non-extreme responses plus one value of 0% or 100% on either side of these. This will be termed the "trimmed degrees of freedom."

Part of the original impetus for the fractionation studies of Ang, discussed above, was to explore fractionation effects at low doses per fraction compared to high dose per fraction courses. Certainly it was important to perform experiments at clinically relevant doses per fraction in the neighborhood of 2 Gy, but clinical hyperfractionation protocols necessitated pushing the limit even further. For this reason Wong et al. designed experiments to explore the dose response for RM in rats down to approximately 1 Gy per fraction (Wong et al. 1992, 1993, 1992, 1993). Fischer 344 rats (a strain independent of the Wistar strain) were given daily doses using protocols of 1, 2, 4, 8, and 40 fractions, 5 days per week except the 40 fraction group were treated through the weekends (Wong et al. 1992). Another set of rats was given 1, 2,

10, 20, 30, or 40 fractions following a top-up dose of 3 × 9 Gy. This was the first report of a spinal cord top-up dose that was fractionated. The top-up dose represented 3 fractions of the 4-fraction "tolerance" dose. Note that none of the doses were given twice daily and all were given in less than about 6 weeks.

The "full course fractionation" (no top-dose component of the experiment) was reported to be well fitted by the LQ model with $\alpha/\beta = 2.41$ Gy, despite most of the schedules (3 of 5) being completely or quasi-separated. The full course fractionation portion of the experiment consisted of 40 dose points, only 7 of which had responses greater than 0 but less than 100%. As a consequence, there were 22 points that were trimmed to correct the number of degrees of freedom. When this is done, the model does *not* fit the data *unless* 2 two additional points (unfortunately from the 7 non-extreme responses) are treated as outliers. As stated in Chap. 3, having many of these extreme responses adds to the number of degrees of freedom without adding appreciably to the sum of the χ^2 values. When the *df* are thus adjusted, one finds that the χ^2 statistic rejects the model.

The top-dose component could not be fitted by the LQ model, even when outliers were removed. We can only conclude that the experiment was flawed in some way that cannot be guessed or that the fractionated top-up up dose given prior to the metered doses that were the primary thrust of the experiment caused a failure to fit the model. Potential problems with the experimental data could include: a variation of sensitivity to radiation in different groups of animals, a dose delivery issue or a dose calibration issue, or an unrecognized change in the experimental conditions. Of course, it may also be that the LQ model is not appropriate.

The fractionated top-up dose was characterized as 3 fractions of the "ED_{50} for the 4-fraction experiment." The problem is that the data of the 4-fraction experiment were quasi-separated with only the lowest dose, 34.8 Gy, having a nontrivial response (1/9). All other responses were 100%, and the next lowest dose was 36.4 Gy. Thus the MLE of the D_{50} was indeterminant, but likely between 34.8 and 36.4 Gy. More critically, the top-up dose accounted for the great majority of damage with the metered doses being only a minority of the stimulus. Whether or not this was responsible for the failure of the model to fit the data, it was nonetheless a potentially problematic design.

However, the objective of this study was not to validate the LQ model, but to assess the radiation response of the rat spinal cord at daily low doses per fraction. Within the context of the experimental conditions, the thesis announced in the title was true in this case—the LQ model *did* underestimate the sparing effect of small daily doses, but the relevance of this statement is diminished since the LQ model did not fit the data. Furthermore, this conclusion was put in doubt from a later Wong paper (Wong et al. 1995).

In 1993, as part of this impressive series of papers that extended over 20 years, Wong et al. tested the placement of the top-up dose (Wong et al. 1993). They did so by initiating an experiment to complement the 1992 paper discussed above. In the 1993 paper, the 3 × 9 Gy top-up dose was placed at the end of the course rather than at the beginning as in 1992. As in 1992, the treatment schedules deployed 1, 10, 20, 30, and 40 fractions, but there was no 2-fraction schedule as there had been in 1992. Generally there were nine animals per dose group. The data exhibited no separation and the degrees of freedom did not need trimming because of excess extreme responses.

The authors reconfirmed that the 1992 data with the initial top-up doses yielded $\alpha/\beta = 0.97$ Gy but the LQ model did not fit the data. The conclusion from the analysis of the new data was that the α/β value was 1.23 Gy, but the LQ model also did not fit. Interestingly, re-analysis agreed with the α/β value, but it was determined by the same re-analysis that a different link function had been used compared to the 1992 analysis. In 1992, the log-log link function was used, but in 1993, a logit link function was used. Using a log-log link function for the 1993 data, it was found that $\alpha/\beta = 1.00$ Gy in surprising agreement with 1992, but the LQ model still does not fit the data. Despite the fact that the LQ model fits neither data set, the authors go on to discuss the result in terms of the LQ model, stating that the α/β ratio decreases in both data sets with increasing fraction. In fact, this statement serves to indicate further that the LQ model fails to fit. Since the LQ model fits neither data set, *it* cannot inform us regarding the importance of the sequence of the top-up dose placement, nor can meaningful statements be made regarding the α/β ratio. Furthermore, if the LQ model fits, then α/β is a constant.

To utilize the data without recourse to the LQ model, a different approach was taken here. Each fractionation schedule was analyzed using simple linear logistic regression with varying values of the location parameter but assuming a common slope. Thus the model was

$$z = \beta_0 + \beta_1 \ln\left(\text{dose}\right) + \beta_2\, \delta_{\text{top-up after}} \tag{5.6}$$

and

$$P_{\text{after}} = \text{logit}\left(z\right) \tag{5.7}$$

where β_2 is the log odds ratio for top-up after relative to top-before, $\delta_{\text{top-up after}}$ is 1 for top-up after and 0 for top-up before, β_1 is common for all fractionation schedules, and β_0 varies with the fractionation schedule. Thus there are a total of 7 variates for the entire data set. Fitting this model to the entire data set produces an acceptable fit with $p\,(\chi^2) \sim 0.2$ and $\beta_2 = 0.747$ or the odds ratio $= 2.11$ with 95% confidence interval of 1.26 to 3.58. Thus this analysis shows that the placement of the top-dose *does* matter and elicits more damage when placed after the graded doses.

It is important to remember that we showed in Chap. 3 that in linear logistic regression for dose response, the odds ratio can be recast into a dose modifying factor with

$$\text{DMF} = \exp\left(\beta_2 \,/\, \beta_1\right). \tag{5.8}$$

In this case, the DMF $= 1.05$, a value that can be difficult to detect in a dose-response experiment.

The subject of the placement of the top-up dose was addressed again in a paper by Kim et al. from the same group (Kim et al. 1997). Again in this study the top-up dose was 3×9 Gy, but the graded doses were given BID in 1 Gy fractions, testing the importance of the top-up dose placement on hyperfractionation. The hyperfractionated doses were separated by time intervals of 0, 1, 2, 4, 8, and 24 h. The 0 h separation represents a QD fractionation with 2 Gy per fraction and the 24 h separation represents a QD treatment with 1 Gy per fraction. The top-up doses were given before or after the graded treatments.

The authors analyzed the data using D_{50}'s for each dose-response curve as well as maximum likelihood estimation for all initial top-up data, all final top-up data, and all data combined. Both mono- and bi-exponential incomplete repair models were assessed. In Table 3 of their paper, they provide the results of -2 times log likelihood, so the likelihood ratio was easily calculated using a saturated model. In no case, did the model fit the data as assessed by the likelihood ratio. Nonetheless, comparisons of model parameters were made and tested with a z-test. This test assumes a known true and single standard deviation for the populations being compared. In no instance were any of the estimated parameters (α/β or repair half-times) different for the two groups. Since (1) the confidence limits for all parameter estimates included 0 so a statistical difference would be highly unlikely and (2) the models did not fit, the parameter estimates were meaningless. The authors concluded, based on lack of statistically significant differences in the parameters for the initial versus final top-up doses, that the placement of the top-up dose was immaterial. However, their analysis did not support this conclusion.

A re-analysis was performed similar to that above for the Wong 93 paper that assessed top-up dose placement relative to daily fractionated treatments. Again a linear logistic regression was used to assess the significance of odds ratio of the before versus after placement of the top-up dose. Unfortunately, even this relatively simple model fails to fit the data and cannot be used to assess the placement of the top-up dose.

The data can be analyzed for top-dose placement irrespective of whether the LQ model with incomplete repair or any other model fits the data. These data are naturally amenable to a simple analysis involving multiple 2×2 contingency tables. Cox and Cox and Snell go into considerable detail on this type of analysis using the framework of a logistic response function (Cox 1972; Cox and Snell 2018). Using the number of responders and the number of subjects, a statistic representing the log odds ratio can be determined along with its significance. For the top-up data of Kim et al. it was found that the MLE of the log odds ratio was 0.708. This yields an odds ratio of 2.23 with a 95% confidence interval of 1.38 to 3.65 with the odds ratio applying to the dose response of the group with the final top-up dose. This value of the odds ratio is in remarkably good agreement with that found on the re-analysis of the 1993 paper.

It is worth noting that in all BID treatments the D_{50}'s for the final top-up group were higher than for the initial top-up, indicating higher probability of RM and an odds ratio greater than one. So even in the paper there is some statistical evidence that a final top-up is associated with increased probability of myelopathy. The log odds ratio is approximately the difference in the D_{50} values for appropriately paired dose-response curves. The average difference in the stated D_{50} values of all treatments was 0.62 Gy and the log of the odds ratio was 0.80.

The implication of a significant odds ratio greater than 1 for top-up doses following rather than preceding the graded doses is that relatively more radiation damage is done when there is damage present already. This contradicts the common and virtually essential radiobiological assumption of equal effects from equal fractions.

In 1995, Wong et al. returned to the question of fractionation sensitivity at small doses per fraction (Wong et al. 1995). Up to this point, attempts to characterize the fractionation sensitivity for doses per fraction of less than 2 Gy were unsuccessful because the models deployed failed to fit the data. Although this fact was generally acknowledged by the authors, they nonetheless reached conclusions regarding the value of α/β. It is somewhat oxymoronic to discuss α/β values for data that are not well described by the LQ model. The 1995 experiment was inspired by reports that cells *in vitro* exhibited unexpectedly large radiosensitivity to single doses below 1 Gy (Marples and Joiner 1993). To investigate this possibly *in vivo*, the authors again used a large top-up dose (3 × 10.25 Gy) followed by graded fractionated doses from 0.4 to 1.5 Gy per fraction, with an additional group of 3 × 10.25 Gy plus graded single fractions. In this case they were essentially looking to see if the LQ model would fail below 1 Gy with an apparently larger value of α/β, assuming that β would remain the same.

In this experiment, the data were well described by the LQ model (p (χ^2) = 0.26) and α/β = 2.07 Gy. Re-analysis gave only slightly different values of the parameters and the χ^2 p-value. The α/β value for doses per fraction at or above 1 Gy was higher than that obtained for doses per fraction of 0.4 to 0.8 Gy. However the difference was not significant.

In 1998 and 1999, Niewald published a two-armed experiment using either 30 daily (total dose 45–120 Gy) or 60 BID fractions (total dose 45–160 Gy) to treat Wistar rats with 100 kV x-rays (Niewald et al. 1999, 1998). A linear logistic regression was performed using only total dose and a categorical variable representing BID treatment. However, since all BID treatments were given in 60 fractions and all daily treatments were given in 30 fractions, this variable confounds the effects of BID and the effects of conventional 30 versus 60 fractions. No dose × dose per fraction variable, which would have facilitated an estimate of α/β, was included. Although it was possible to reproduce the estimates of the values of the variates, it was not possible to reproduce the quoted values of the Hosmer and Lemeshow statistic. Why this statistic was used at all is questionable since it was designed for use in binary data or sparse binomial data, but it is not typically used in situations where the data are grouped in groups of 9 to 15 subjects. Nonetheless, the authors reported a $\hat{C}$ value of 8.012 with 8 degrees of freedom. Because they used SAS for their analysis, an attempt was made to reproduce this statistic using the current SAS algorithm for $\hat{C}$, and a value of >30 was found with 7 degrees of freedom. (Eight degrees of freedom would have implied 10 groups of data, but the current algorithm would have the data grouped in 9 groups.) This value of $\hat{C}$ clearly indicates a poor fit, as does the Pearson χ^2 resulting from the estimated dose response. Unfortunately, this paper does not contain useful information on hyperfractionation.

Continuous Low-Dose Rate Radiation In 1985, Dale published the formalism to handle continuous radiation in the context of the LQ model, assuming mono-exponential repair (Dale 1985). The response metameter in this case is given by

$$z = \beta_0 - \beta_1 D - \beta_2 D^2 \cdot g\left(\mu t\right) \tag{5.9}$$

where z is the systematic component of the GLM (generally logistic regression) and g is analogous to $(1 + h)$ term for incomplete repair in fractionated treatment. This is given by.

$$g(\mu t) = 2\left[\mu t - 1 + \exp(-\mu t)\right] / (\mu t)^2. \tag{5.10}$$

In 1989, Scalliet et al. published results on continuous radiation effect on the spinal cord (Scalliet et al. 1989). Dose rates of 2.0, 3.9, 14.7, and 107.6 cGy per hour were used. The Wag/Rij strain was irradiated using a Co-60 unit with the dose rate varied by adjusting the SSD and adding lead attenuators. The entire cervical cord was irradiated. The very low dose rates required very long anesthesia that resulted in loss of a majority of the animals in the lowest dose group. Extensive dosimetry was performed.

The data were analyzed using a *graphical* method described by Dale (Dale 1985). This method is reminiscent of the Fe-plot and has the same shortcomings. No regression method was applied to the complete data set. A re-analysis was performed using the LQ model with incomplete repair for continuous radiation as described by Dale in the same publication. The model failed to fit the data. However, because the data were well behaved, a heuristic model was able to provide a good fit to the data. Since this model has no biological basis, it is not reported here. The conclusion of the authors was that there is a significant dose rate effect across the dose rates studied. This is certainly correct. However, the radiobiological parameters they report for the Dale model cannot be confirmed because that model did not fit the data.

Pop et al. published results on continuous radiation effect on the spinal cord using either a pair of Ir-192 wires or parallel catheters for the travel of a high-dose rate Ir source with various dwell positions (Pop et al. 1997). These were very difficult experiments to perform and verify. Dose rates of 120.6 cGy per hour (HDR) and 0.49 or 0.96 cGy per hour (LDR) were achieved. These authors also used the graphical analysis of Dale et al. which does not produce statistically useful information. However, in addition they used maximum likelihood estimation of the variate in the incomplete repair model. This produced a good fit to the data that was verified by re-analysis, although the re-analysis resulted in somewhat different parameter values, especially the α/β ratio. The value of this parameter obviously depends upon the normalization dose in the inhomogeneous dose distribution produced in a brachytherapy procedure. Normalizing to the minimum dose to the ventral surface of the cord, the authors estimate $\alpha/\beta = 1.45$ Gy and the $T_{1/2} = 1.76$ h. It appears that they used a logit link function rather than a log-log even though they interpret the constant as a log N, where N is the number of critical units (Pop et al. 2000). They state that the addition of a second repair pathway (bi-exponential repair) was not supported by the data. Re-analysis confirmed this as well.

In a more extensive examination of the dose rate effect for paresis, the same group used a similar set-up, either IR wires or HDR catheters, to a longer segment of the thoracolumbar cord. Dose rates examined were 0.53, 0.9, 1.63, 2.55, 4.4, and 9.9 using LDR wires, and 120 Gy/h using HDR (Pop et al. 1998). Not all D_{50}'s could be calculated independently because the data were separated for some of the dose rates. Therefore in a re-analysis, a common value of the constant and the value of α were used for all dose rates and β was varied for each dose rate. This provided D_{50}'s very similar to the authors' graphical estimates (where the data were separated). The importance of D_{50} values obtained independently for each dose rate will become apparent in the following paragraphs.

The substantial value of this study is the range of dose rates that were covered and the consistency of the data. Again using Dale's LQ+incomplete repair model, a good fit to the data was obtained. For mono-exponential repair, Table 5.2 shows the values obtained by the authors. Re-analysis produced a very good fit to the data and the parameter values are in general agreement with those of the authors. The variate estimates for the bi-exponential model are also given in Table 5.2. The authors state that the difference in the log likelihood values indicated that the bi-exponential model should be preferred ($p \sim 0.024$), whereas re-analysis shows that the mono-exponential model is preferable ($p \sim 0.1$). Nonetheless, the original analysis should be considered sound.

In their book entitled *Fractionation of Radiotherapy*, Thames and Hendry show a power-law plot of isoeffect doses versus overall treatment time for continuous irradiation, with both axes on a log scale (Thames and Hendry 1987). This is somewhat similar to the power-law plot of log dose versus log N but should not be confused with a plot showing the overall time exponent of an NSD-like model. In Fig. 5.6, log D_{50} as determined above (varying only the β term in a logistic regression of the individual dose rates) is plotted versus log D_{50}/dose rate (treatment duration) for the Pop et al. (1998) data. Also shown is the theoretical curve. As explained in Thames and Hendry (Sect. 2.3), the exponent goes to zero at very long and very short treatment durations and is somewhat sigmoidally shaped. The rising portion of the curve is where the dose rate effect is operative. Also shown in this figure are data from Pop et al. (1997).

Since g depends only on μ and t, in theory it should be possible to create the curve in Fig. 5.6 using μ determined from fractionated data. In fact, it cannot be said that the values of μ reported from fractionated data substantially disagree with those reported by Pop et al. however, the confidence intervals are rather broad and therefore not especially probative. Of course, the situation is made more complicated by the inclusion of both slow and fast repair in the bi-exponential model.

A cautionary issue associated with the data from Pop et al. and any studies involving the lumbar segments of the spinal cord in the rat is that of pathogenesis.

Table 5.2 Parameters for incomplete repair model from Pop et al. (1998). Since the upper limit of the half-times of repair are indeterminant and the upper limit for the fraction of slow repair exceeds 1, if the bi-exponential model is statistically preferable as the authors state, that may be only an arithmetic advantage

Fitted parameters	Mono-exponential	Bi-exponential
α/β	4.0 (2.9–5.1) Gy	2.7 (1.5–4.4) Gy
$T_{1/2}$	1.60 (1.31–2.04) h	2.42 (1.51–) h
$T_{1/2}$	–	0.13 (0.03–) h
Fraction of slow repair	–	0.59 (0.37–82)

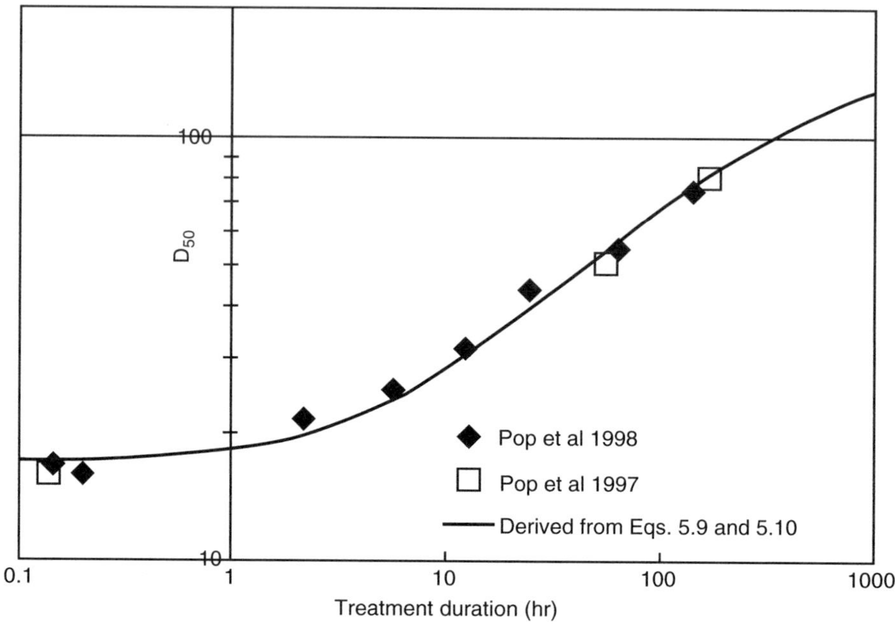

Fig. 5.6 Data from Pop et al. on paresis after continuous irradiation with the theoretical curve for D_{50} versus treatment time.

Pop specifically noted that the pathology of paresis in their studies is primarily that of nerve root demyelination and degeneration. This is simply not the same thing as white matter necrosis of the spinal cord. It does not involve the same glial cell populations or even the spinal cord, and one cannot assume that if the target cell theory is correct, the same target cells are involved in white matter necrosis and nerve root damage. So how applicable the Pop data are to the cervical and thoracic levels of the spinal cord is speculative, although the work of van der Kogel and others do not seem to indicate that a major difference exists. It is a matter of uncertainty.

5.6 Relative Biological Effectiveness (RBE)

The relative biological effectiveness (RBE) is the ratio of the dose from a reference photon beam to the dose from a particle beam that achieves the same biological effect. In the literature, the reference beam is generally either Co-60 gamma rays or 250–300 kVp x-rays. Megavoltage photon beams may also serve as the reference radiation. The main factor in determining which reference beam was used is almost certainly the equipment available in the lab. For most *in vitro* endpoints, orthovoltage x-rays were used. RBE's using orthovoltage should be multiplied by 1.11 to be equivalent to megavoltage/C0–60.

It is well known that the RBE is a complex function of dose, fractionation, LET, end point, and other factors, even for a given particle beam. Nonetheless, it is very often treated as a constant. RBE seems quite straightforward when we first learn about it. We are given two cell survival curves, one for photons and one for particle radiation. Because neutrons and charged particles are more densely ionizing than photons, they exhibit less recovery from sublethal damage than do photons, and their cell survival curve lies to the left and beneath the curve for photons. The RBE in this case is the ratio of the dose of photons to the dose of particles that elicit the same level of cell survival for both. Since these cells survival curves do not have the same shape, the RBE will depend on dose, even for this simple situation.

If the dose is fractionated, the effect on cell survival is not the same for photons and particles and therefore the RBE will depend on the dose per fraction of each beam and the number of fractions. If the biological effect is not simply cell survival, but a clinically measured end point, the situation becomes more complex. Finally, particle beams change energy and therefore LET as they penetrate tissue, and the RBE changes as well.

Thus one cannot simply assign an RBE to a particle beam. The exact circumstances of its measurement must be stated. However, more often than not, the RBE is depicted as a constant. However as a concept, "the RBE for protons," for example, is no more specific than "the tolerance dose for protons."

Since clinical data on RBE are rare and experimental data cannot encompass all the relevant factors that affect RBE, the following discussion of RBE for the end point of radiation myelopathy will tend toward listing the values and the conditions of the experiment.

The particles used in radiation therapy include both charged particles and neutrons. Thus far these include a range from pions to carbon-12 nuclei. Except of course for neutrons, these all exhibit a Bragg peak at the end of their range. The particles lose energy at an almost constant rate once they enter an absorber, until their kinetic energy approaches that of the binding energy of the electrons in the medium. Then they interact strongly with the orbital electrons losing their remaining energy over a short range. The same is also true of electrons, except owing to their low mass, they are scattered at larger angles than more massive particles and the abrupt energy loss does not occur solely in the direction of increasing depth.

5.6.1 Neutrons

Starting roughly in the 1960s, there was great clinical interest in high energy (as opposed to thermal or epi-thermal) neutrons as a treatment modality because treatment failure was often attributed to radioresistant hypoxic tumor cells and neutron radiation killed these cells at essentially the same rate as it killed oxygenated tumor cells. Early neutron treatment delivery systems were technologically inferior to the photon delivery systems of the same era. Late effects from these treatments were excessive because of low beam energy, poor collimation, a fixed beam configuration, and use of high doses per fraction necessitated by long treatment sessions. Dedicated cyclotrons built for single institutions had higher energies, better collimation, and were capable of isocentric treatment. However, interest waned (by about 1990) as it became clear that for nearly all cancers a therapeutic advantage was not afforded by the neutron beam.

In 1974, Geraci et al. reported RBE values for single doses of 8 MeV neutron irradiation compared to 250 kVp x radiation (Geraci et al. 1974). They used the ratio of the average latent periods as the metric for the RBE. This measure was not then regarded as an unreliable quantitative measure of biological response. They found that the RBE decreased with increasing neutron dose from 1.75 at 9 Gy to 0.9 at 45 Gy. The shape of the RBE versus dose relationship was bilinear. In 1978, again using the latent period as an endpoint, Geraci et al. reported an RBE of 3.5 for neutrons given in 10 fractions as compared to 250 kVp x-rays in 10 fractions and a neutron dose per fraction of 1 Gy (Geraci et al. 1978). White matter necrosis, vascular damage, and nerve root necrosis were seen in addition to gray matter injury at high doses.

Also in 1974 van der Kogel and Barendsen published dose-response curves for the lumbar spinal cord treated with 200 kV x-rays or 15 MeV neutrons for 1 and 5 fractions (van der Kogel and Barendsen 1974). The neutron dose rate was between 6 and 20 cGy/min. Both sets of curve were completely separated for both schedules, but they interpolated to get the D_{50} values. From these they determined the approximate RBE's to be 1.1 and 1.7 for single doses and 5 fractions, respectively. The data were later supplemented to include non-extreme responses and refined values of 1.2 and 2.1 were given (van der Kogel 1979). Because of the low dose rate, these values are lower than typically reported.

By 1990, neutron treatment was becoming extinct and very few new experimental reports of neutron RBE values were being published. Hornsey gives an excellent summary of findings in her chapter, "Experimental central nervous system injury from fast neutrons," in Gutin et al. *Radiation injury to the nervous system* (Hornsey 1991). Their Figure 3, reproduced here as Fig. 5.7 effectively summarizes the experimental experience regarding the neutron RBE for different neutron energies and animal species as a function of dose per fraction of photons. For neutron energies above 50 MeV the RBE decreases from just below 5 to about 2 as the dose per fraction increases from 0.65 to 20 Gy. The values are reasonably consistent across species.

In the third of three reports from M.D. Anderson on RM in rhesus monkeys, Schultheiss et al. look at the RBE as applied to clinical dose schedules (Schultheiss et al. 1992). Animals in the photon groups were all treated with 2.2 Gy per fraction from a Co-60 source and the animals in the neutron groups were treated with 9 fractions (13, 14.25, and 15 Gy) or 12 fractions to 14.4 Gy. The change in fraction number was the result of a change in the clinical regimen. In addition, the 14.4 Gy group was treated on a different cyclotron. The photon data were gathered several years later as part of large animal radiobiological studies of dose and volume responses of the spinal cord, so there was no designed experiment. Clearly the neutron arm could have benefitted from more subjects per group.

The results were that the latent periods for neutrons were shorter than those for photons. The slope of the neutron dose-response curve was steeper and the D_{50}'s for neutrons and photons were 14.6 Gy (14.3, 15.1 C.I.) and 75.8 Gy (73.2, 80.8 C.I.), respectively. The RBE as determined by the ratios of $D\gamma$ to D_n that resulted in the same probability of RM decreased from 5.2 at 50% to 4.2 at 1% below which it decreased rapidly. Of course, this is well below the probabilities for which there were any data.

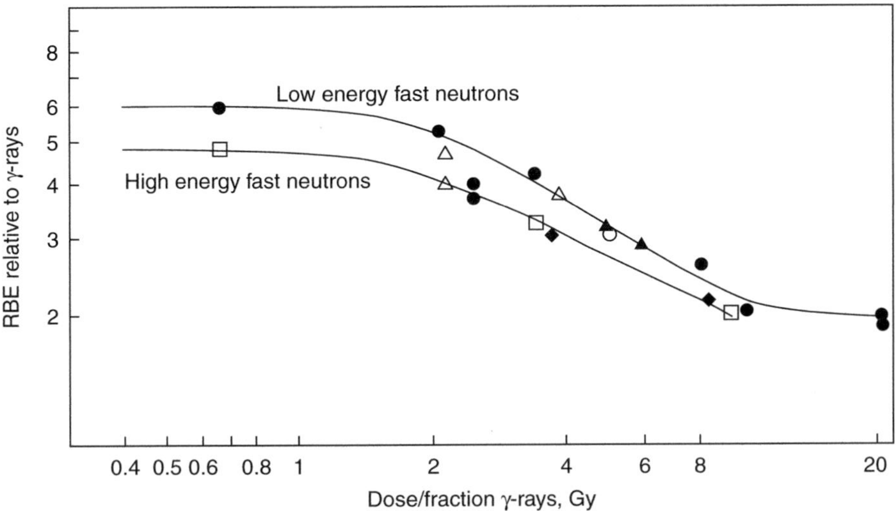

Fig. 5.7 The RBE relative to ^{60}Co γ rays as a function of γ-ray dose/fraction. ● d[16]Be, rat (Hornsey et al. unpublished); ▲ d[22]Be, mouse (Geraci et al. 1974, 1978) neutrons and for □ p[62]Be, rat (Hornsey et al. unpublished); ◆ d[0.27]3H, rat (van der Kogel 1979); ○ d[50]be, monkey (Stephens et al. 1983); △ d[50] Be neutrons, monkey (Peters et al. 1986). From "Experimental central nervous system injury from fast neutrons," in Gutin et al. *Radiation injury to the nervous system* (Hornsey 1991).

In a report on 76 cases from the Fermilab and an abbreviated review of the literature (limited to three studies) in 1985, Cohen et al. suggested a dose limit of 12.5 Gy and an RBE of 4 (equivalent to a 50 Gy dose limit for x-rays) (Cohen et al. 1985). Other cases of RM had been reported at lower doses, but were not cited. Laramore attributes these myelopathies at lower doses to facility differences, which include fractionation schedules, beam energy, photon contamination, etc. (Laramore et al. 1986). He converted doses from older equipment to the higher energy machine of the time and concluded that 10.5–11 Gy represented a safe and conservative spinal cord dose. This certainly seems reasonable, considering an RBE in excess of 4.

Thus for neutrons, as with any new clinical treatment technique, the emphasis was not so much on determining precise biological effect parameters as it was on discovering doses or dose schedules that were safe for patients.

5.6.2 Leith et al. Charged Particles

In a series of papers from 1975 to 1982, Leith et al. reported on the spinal cord response to neon, helium, and carbon ions at Lawrence Berkeley Laboratory (Leith et al. 1975a, b, 1982). The RBE's are given in Table 5.3. The carbon ions were accelerated at 400 MeV/nucleon. The helium and neon do not have any extant beam to which comparisons can be made but carbon ions are the subject of research and development currently. The T12 to L1 region was treated (approximately 1.1 cm).

Table 5.3 Summary of RBE values from Leith et al. (1975a, b, 1982)

Particle	# of fractions	RBE[a]
He	1	1.07
C-12 plateau region	1	1.45
C-12 spread peak	1	1.48
C-12 plateau region	4	1.31
C-12 spread peak	4	1.95
Neon plateau	1	1.46
Neon spread peak	1	1.86
Neon plateau	4	1.8
Neon spread peak	4	2.18

[a]Compared to 230 kV x-rays whose RBE $\approx$ 1.1

The histopathology for symptomatic rats was demyelination and white matter necrosis with little to no vascular damage mentioned. The dose-latent period relationship was fundamentally similar to that of x-rays, with the expected difference being a shift of the curve occurring along the dose axis.

In the 1970s and 1980s, it was recognized that the RBE was not a constant, but the literature reads very much as if there were a search for *the* RBE. For example, a value of 3 was often used when neutrons were given in 9–12 fractions, with an exception generally being made for the spinal cord whose RBE was estimated at 4–5. Being a low LET radiation, protons have an RBE only slightly larger than 1, the exact value of which has not really captured the enthusiasm of the field and generally has not been found to be significantly different from unity.

5.6.3 Carbon Ions

The only other charged particle that has reached widespread clinical use is the fully ionized carbon-12 ion. Because the LET of the carbon beam changes significantly as it penetrates an absorber, the RBE changes in a similar fashion. Models based on *in vitro* cell survival have been conceived to account for the variation in RBE as a function of LET and therefore as a function of depth in tissue (Carante et al. 2020; Scholz and Elsässer 2007). Possibly this concept could be a reasonable way forward for malignant cells, but modeling the RBE for late clinical changes based a cell survival model is doomed to failure. The pathogenesis of RM is clearly complex, involving interactions among endothelial cells, essentially all glial cell types, inflammatory cells, cytokines, and growth factors. RM does not occur because a specific cell type is reduced below a specific level of cell survival. These cell-survival-based models derive the values for α and β, without specifying the cell type or the critical level of survival. Experimental confirmation at the microscopic level is impossible and inferring confirmation from clinical results is inapposite.

Radiation myelopathy studies in rats were performed at carbon ion facilities in Germany and reported in a series of papers from 2003 to 2020 (Debus et al. 2003; Karger et al. 2006a, b; Saager et al. 2014, 2016, 2020a, b). Unfortunately, only five animals (Saager et al.) or six animals (Debus et al. and Karger et al.) were used per dose group, which resulted in a number of separated or quasi-separated dose-response curves. Of course this resulted in ML estimates of D_{50} being undefined. Since RBE is defined as a ratio of D_{50}'s, the observed RBE is thereby also indeterminant. The reason these experiments were undertaken was primarily to determine the RBE as a function of LET, but also to use this information to inform RBE modeling (as described above). Two figures served to illustrate the importance of these studies and the success or lack thereof of modeling efforts.

The first of these figures is from Carante et al. (2020). The authors state, "Based on the BIANCA biophysical model, a radiobiological database was constructed to predict RBE values for head-and-neck tumor radiation with C-ions or protons," and "The RBE values predicted by BIANCA were in line with experimental RBE values obtained by irradiating the rat spinal cord at different depth positions along C-ion or proton SOBPs." This prediction is illustrated in Fig. 5.8.

The second figure (Fig. 5.9) depicts the modeled and observed results of the local effect model (LEM) (Scholz and Elsässer 2007), which is currently in its fourth iteration, LEM IV. Shown here are the RBE values as a function of LET and equivalently, tissue depth in the spread out Bragg peak (Saager et al. 2020a). The original LEM model (LEM I) and LEM IV are shown. LEM I clearly does not fit the data. LEM IV also does not intersect with the data except at the highest LET value, 99 kev/μm. Nonetheless, the RBE versus depth, which also shows the depth dose for the SOBP, convincingly demonstrates the importance of understanding the LET behavior of the RBE for this particle. In the peak region,

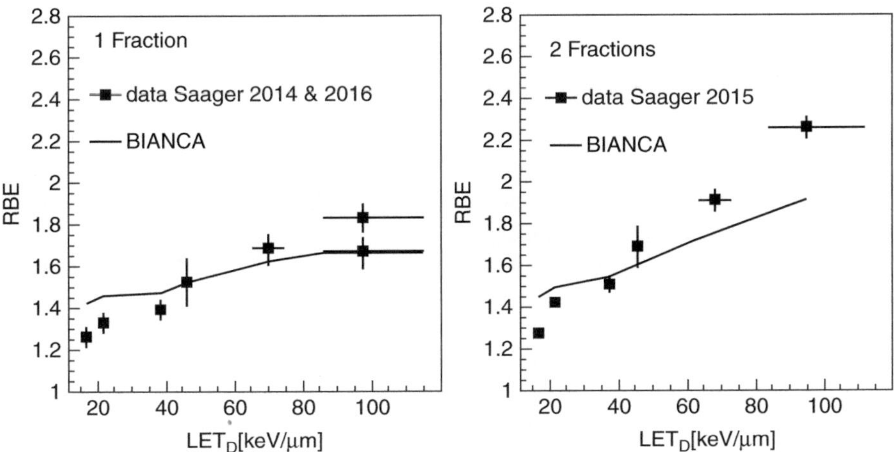

Fig. 5.8 "RBE as a function of dose-averaged LET for single-fraction (Panel a).or two-fraction (Panel b) irradiation of the rat spinal cord at different positions within the carbon-ion Spread-Out Bragg Peak (SOBP). The lines are predictions obtained by BIANCA, the points are experimental data taken from [12, 14] (panel a), or [13] (panel b). In panel a, at 99 keV/um the higher experimental value is from a repetition experiment [14] and is considered as more reliable." Although no model assessment was offered for the BIANCA model, it is easy to see that the curves that define the model and the data points somewhat intersect, but do not follow have the same functional behavior. This article is distributed under the terms of the Creative Commons Attribution 4.0 International License (http://creativecommons.org/licenses/by/4.0/), which permits unrestricted use, distribution, and reproduction in any medium.

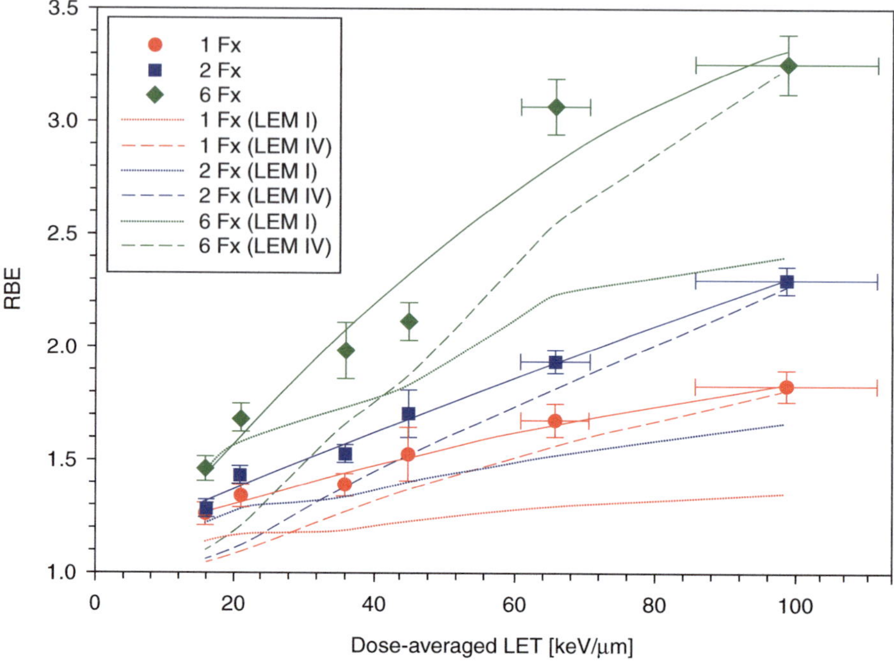

Fig. 5.9 The solid lines represent the authors fit to a second order polynomial, but the dashed and dotted curve are said to be the LEM I and IV fits. These latter models interconnect with the data only in a very loose fashion. This article is distributed under the terms of the Creative Commons Attribution 4.0 International License (http://creativecommons.org/licenses/by/4.0/), which permits unrestricted use, distribution, and reproduction in any medium.

ostensibly corresponding to the target region, the RBE in six fractions varies by a factor of two.

The conclusion one draws is that the RBE dependence is important but not yet well described by biophysical models.

The series of experiments mentioned above consisted of animals irradiated with x-rays and carbon ions using 1, 2, and 6 fractions. The particle beam was modulated so that the spinal cord was at 6 different effective depths with consequent LET's of 16, 21, 36, 45, 66, and 99 KeV/μm. The x irradiations were done as part of a different set of experiments where the treatments were given in 1, 2, 6, and 18 fractions and the carbon ion treatments were either on the plateau region or the peak region of the beam with 13 and 125 KeV/μm, respectively. Table 5.4 shows the data sources.

Of the 3 x-ray dose-response curves, one was quasi-separated ($N = 6$). Of the 6 carbon ion dose-response curves, 5 were completely or quasi-separated. So if each dose-response curve were analyzed separately, then RBE's for $N = 6$ could not be calculated because there would be an undefined reference dose; for $N = 1$, and 2, four of the 12 RBE's could not be calculated because of the undefined D_{50} for carbon due to separation of the dose-response data. Nonetheless, the authors seemed to obtain estimates for the D_{50}'s from the statistical software, acknowledging only that the standard errors of separated data were unreliable. Because of separation, but more importantly in general to reduce the number of unknown variables to be

Table 5.4 Source of data for RBE experiments with number of fractions and LET of carbon ions

N	x-ray	13 KeV/μ (plateau)	16–99 KeV/μ (plateau, 6 beams)	125 KeV/μ (peak)
1	Debus et al. '03	Debus et al. '03	Saager et al. '14, Saager et al. '16	Debus et al. '03
2	Debus et al. '03	Debus et al. '03	Saager et al. '15	Debus et al. '03
6	Karger et al. '06	Karger et al. '06	Saager et al. '20a, Saager et al. '18	Karger et al. '06
18	Karger et al. '06	Karger et al. '06		Karger et al. '06

estimated, the data must be combined and the D_{50}'s calculated using the coefficients in the model. The math for this procedure follows.

Re-analysis of the Carbon Ion Data of Saager (Table 5.4) By combining all data for an LET value into a single group, a simple analysis is achievable. Using data from $N = 1, 2,$ and 6 for photons and carbon ions, the systematic component of linear logistic models become,

$$z_x = \lambda_x - \alpha_x D_x - \beta_x D_x d_x \tag{5.11}$$

and

$$z_c = \lambda_c - \alpha_c D - \beta_c Dd. \tag{5.12}$$

where the z's are constructed such that the unknown variates should all be positive. Then

$$P_x = \left(1 + \exp\left(-z_x\right)\right)^{-1} \tag{5.13}$$

and

$$P_c = \left(1 + \exp\left(-z_c\right)\right)^{-1} \tag{5.14}$$

where P_c and P_x are the probabilities of RM for carbon ions and photons, respectively, λ is the constant in the linear logistic model, α and β have their usual meaning in the LQ model, D and d are the total dose and dose per fraction, the doses for photons are subscripted with "x," and the doses for carbon ions are unsubscripted.

Thus there are only three parameters for each LET value no matter how many fractionation schemes were investigated. Once the parameters are estimated, d and d_x are replaced by their respective terms of D/N, we set $z = 0$ (for $P = 0.5$) and the resultant quadratic equation is solved for D or D_c. The general ratio of D_x to D is the RBE and is given by

$$\mathrm{RBE} = \frac{\sqrt{\left(\dfrac{\alpha}{\beta}\right)_x^2 + 4\dfrac{\left(\lambda_x - z\right)}{\beta_x}\dfrac{1}{N}} - \left(\dfrac{\alpha}{\beta}\right)_x}{\sqrt{\left(\dfrac{\alpha}{\beta}\right)_c^2 + 4\dfrac{\left(\lambda_c - z\right)}{\beta_c}\dfrac{1}{N}} - \left(\dfrac{\alpha}{\beta}\right)_c} \tag{5.15}$$

This is the LQ-based *clinical* model for RBE that should replace the LQ-based cell survival models (LEM I, LEM IV, BIANCA) if the LQ model is going to be used to plan and assess the clinical outcomes of particle therapy. One should not mix or conflate both the clinical and cell survival LQ models in a single theory of RBE. As we will see, the LQ RBE model gives a concise theory of RBE and is consistent with the data.

In fact, Eq. (5.15) is not limited to determining RBE values at D_{50} only. Since in Eqs. (5.11) and (5.12), $z = \ln(1/P-1)$, and equal values of P are required for the RBE determination, (5.15) is the general RBE equation for any P. For the RBE at D_{50}, $z = 0$ and the RBE is given by

$$\text{RBE} = \frac{\sqrt{\left(\dfrac{\alpha}{\beta}\right)_x^2 + 4\dfrac{\lambda_x}{\beta_x}\dfrac{1}{N}} - \left(\dfrac{\alpha}{\beta}\right)_x}{\sqrt{\left(\dfrac{\alpha}{\beta}\right)_c^2 + 4d\left(\left(\dfrac{\alpha}{\beta}\right)_c + d\right)} - \left(\dfrac{\alpha}{\beta}\right)_c} \tag{5.16}$$

where the denominator has been written with its dependency on d rather than N. Fitting the 1, 2, and 6 fraction data produces the results shown in Fig. 5.10 and given in Table 5.5. Dose-response functions were fitted for each LET separately with the $N = 1$, 2, and 6 carbon ion data being combined with the corresponding x-ray data. For all LET sets, good fits were obtained as determined by the Pearson χ^2 value. A total of 21 parameters were estimated, three parameter each for the x-ray data and for the 6 LET data sets. Had all D_{50}'s been estimated separately, the number of parameters for the LET data would have been 3 for each value of N times 7 LET+x-ray sets times 2 parameters in each curve, giving 42 parameters in total if no dose-response data exhibited separation. Thus pooling data not only obviates the separation problem, it halves the number of parameters. Pooling of the data was done in Saager (2020a) but not in the previous papers. Interestingly, it was done solely for BED analysis and not for determining estimates of D_{50} and RBE.

Figure 5.10a shows the RBE plotted versus dose per fraction of photons. The data points are those given in the Saager papers of Table 5.4. In Saager et al. (2014), the RBE for LET = 99 KeV/m and $N = 1$ seemed to be an outlier, therefore that component was repeated and published in Saager et al. (2016). The solid lines in this figure are a result of the LQ RBE theory of Eq. (5.16). These lines are entirely a function of the fitted parameters of the dose-response data. Presumably the very close fit of the D_{50}'s from Saager et al. is a result of good agreement between them and the D_{50}'s calculated here from the pooled data. It is clear from this figure that grouping RBE's according to the fraction number, N, is more edifying than grouping them according to the LET.

Figure 5.10b shows the RBE versus LET. It seems apparent in this figure that the RBE is, within experimental error, a linear function of LET, as noted by Saager et al. from 2014 onward. The straight lines in this figure represent a regression of the data with the slope varying as a function of the number of fractions, but all going through the same point at low LET. Given the definition of RBE and the fact that RBE versus LET can be expressed as $\text{RBE} = \text{LET}_0 + m \cdot \text{LET}$, we can easily find the dependence of $D_{50,\text{C-ion}}$ on LET is

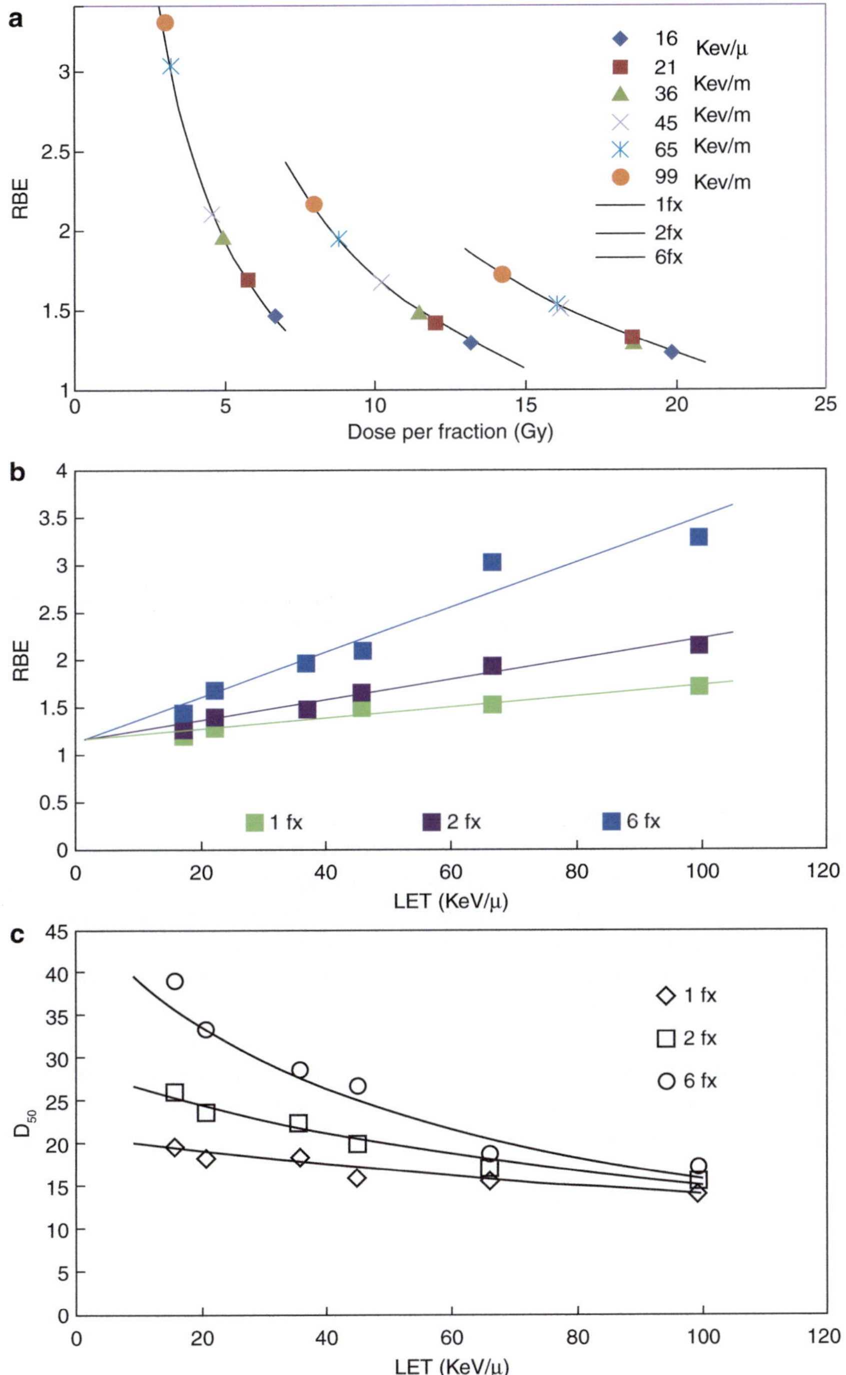

Fig. 5.10 Results of re-analysis of Saager et al. data using LQ model defined by Eq. (5.11)–(5.16). RBE's reflect Saager et al. estimates of D_{50}. (**a**) RBE versus dose per fraction. (**b**) RBE versus LET. (**c**) D_{50} versus LET with Saager et al. data.

Table 5.5 MLE parameters of the model defined by Eqs. (5.11)–(5.14)

LET (KeV/μm)	λ	β_1	β_2	$\beta_1/\beta_2 = \alpha/\beta$
x-ray	15.5	0.04	0.024	1.8
16	14.8	0.19	0.028	6.9
21	24.0	0.45	0.046	9.8
36	43.1	1.20	0.061	19.7
45	33.7	0.93	0.072	12.8
66	35.4	1.80	0.026	68.2
99	26.9	1.47	0.030	49.0

$$D_{50,c} = \frac{D_x / \mathrm{LET}_0}{1 + \dfrac{m}{\mathrm{LET}_0}\mathrm{LET}} \tag{5.17}$$

The values of $D_{50,\text{C-ion}}$ are plotted versus LET in Fig. 5.10c along with Eq. (5.17) above. The agreement seems reasonable, with the possible exception of 66 KeV/μm for $N = 6$. It should be noted that Fig. 5.10b, c are essentially the same data expressed in slightly different ways. It is reasonable to assume that much of the dispersion in the data could have been reduced by using a larger number of animals per dose point.

In general, the RBE is not usually understood as a linear function of LET. However, that understanding is based on cell survival models. In the clinical situation, where many cell types interact over long periods before an outcome is manifest, the relationship between RBE and LET is not known. If the data suggest a linear model, there is no compelling reason to assume otherwise. In fact, this is also the assumption made by Saager with regard to protons (Saager et al. 2018).

To conclude, it is important to understand the dependence of RBE on LET and therefore on depth in tissue. If this dependence is represented in treatment planning systems by a model, it is vital that the model fits the data (and that the data are reliable). Heuristic models may serve well to represent the data.

It is likely that treatment with carbon ions will ultimately be guided, not by modeled values of RBE, but rather by clinical experience as is currently the case with SRS. Whether through errors, accidents, or statistical variations in clinical tolerance, eventually safe doses for irradiating the spinal cord with carbon ions will be established either by successful definitive or curative treatments without cases of RM or by discovering doses at which RM might occur. There is good reason to hope that the former situation is as a result of the dosimetric and radiobiological advantages heavy ion treatments enjoy over photon treatment.

5.6.4 Protons

A similar study was performed by Saager et al. using a proton beam at the Heidelberg Ion Beam Therapy Center. The cervical spinal cord was irradiated "at four different positions (35, 100, 120, and 127 mm) of a 6 cm SOBP ranging from 70 to 130 mm

(Fig. 1a) water-equivalent depths (energy range 97.5–135.8 MeV). The corresponding dose-averaged LET values were 1.4, 2.7, 3.9, and 5.5 KeV/mm, respectively." Five animals were irradiated at each dose level (255 animals total) and about 20 died or were removed due to intercurrent disease. This resulted in only four animals being available for evaluation in 6 of the 17 non-extreme responses.

Re-analysis generally supports the most what the authors point to as the most important result of this study, namely a "general trend of an increasing RBE with LET." However, the features of this study diminish the certainty of specific conclusions. Of course, the number of animals per dose group is the overriding concern.

A feature of the physical data is that the proton treatment field is much less homogeneous than the carbon ions. This makes it difficult to establish exactly what proton dose to designate as the independent variable in the dose-response analysis.

It is surprising that only two of the 8 proton dose-response data sets were separated. However, the data had other unusual aspects. When all the proton data are pooled for analysis using the LQ dose-response model (Eqs. 5.11–5.14), the MLE of the coefficient of the dose term (α) is negative and a poor fit is obtained. Adding an LET term improves the fit, but it is still poor. Pooling the data by LET results in very good fits for the two lowest LET values but with negative α's. A marginally good fit is achieved for the 3.9 KeV/μm LET, but this time with $\alpha > 0$. Both $N = 1$ and $N = 2$ datasets for the highest LET value, 5.5 KeV/μm, were quasi-separated and could not be analyzed with MLE. Clearly it cannot be said that the LQ model fits these data.

Given the issues with the data in this paper, it is hard to have much confidence in the linear dependence of the RBE on LET, except of course, it was seen in carbon ion treatments. However, the value of the proton RBE of 1.1 was supported by the observations in this study. The low value of the RBE is consistent with clinical assumptions that the RBE for clinical treatment is 1.1 or lower. We also see that this may not be true at the end of the proton range (Paganetti 2014; Wouters et al. 2015). Clinical cases of RM have been reported after proton treatment, but there are insufficient numbers to assess the RBE (Chowdhry et al. 2016; Marucci et al. 2004). As with carbon ion treatment, one should be cautious when the spinal cord is near the end of the particle range; the combination of slight set-up error and higher distal Bragg peak RBE could result in greater than anticipated spinal cord damage.

5.7 In Closing

One may legitimately question the appropriateness of small animal models of RM. Other than the general questions regarding the pertinence of rodent models in medical research, there are very specific issues related to RM that could cause some doubt regarding their clinical relevance. By far the major issue is the fact that experimental models and human data exist on very difference response levels. Nearly all response percentages in experimental models are measured in the double digits, while most reports of clinical RM are anecdotal. Of the 20 reports used to assess the human dose response in Chap. 4, only 4 had crude response rates in excess of 10%. The consequence of this is that experience with human RM nearly always arises from the low probability tail of the dose-response curve where the dose response is relatively insensitive to changes in dose and volume and possibly other causative factors. Whether the changes in the location of the dose-response curve of rodents when subjected to variations in experimental conditions can be taken as being applicable to humans could be considered speculative.

The fractionation schedules used in animal experiments are frequently conducted at higher doses per fraction than are typically relevant to human treatments. Of course there are many examples where this has been overcome either by brute force or by deploying the top-up dose technique. Nonetheless, many experiments, especially those involving dose modification factors, are limited to single doses. Furthermore, one may ask a more fundamental question of whether a rodent Gray = human Gy, that is, are the biochemical changes brought about in human cells following a small dose d, similar in magnitude and consequence to the changes in rodent cells at the same dose.

Of course field sizes in small animals are necessarily much smaller than comparable (measured by number of vertebrae) field sizes in humans. It is quite likely that in addition to the relative length of spinal cord being important in dose response, the absolute length also plays a role. Certainly the size of the smallest lesions associated with symptoms does not scale according to the size of the animal.

It is difficult to know how the differences in human and rodent life expectancies impact the late radiation response. Are the time scales involved in expression and repair of damage similar? And what changes take place in the human over time scales beyond the lifespan of the rodent?

Although there is considerable room for debate, it seems that the fractionation sensitivity of experimental animals is not substantially different from humans, at least as measured by the α/β ratio. It is certainly true that RM is more dependent on changes to dose per fraction than acute effects. Clearly the α/β ratio for RM is on the low end of the spectrum of values reported for all tissue responses, both experimentally and clinically.

Finally it is reasonable to assume that the pathogenesis of RM is similar in most experimental animals and humans. An apparent exception to this is the role that nerve root degeneration seems to play in specific experimental situations (rat lumbar cord, pig). But the fundamental pathogenesis of the response of the true cord has only relatively minor differences in rodents versus humans. One area where some differences are likely to exist is neuro-inflammation. However, this field is relatively young and perceived differences in the role of neuro-inflammation in rodents and humans may diminish with time.

No matter what differences may exist between small animal models and humans with regard to RM, these models will remain a useful and indeed necessary tool for future research. The ethical use of experimental animals requires that experiments are designed to maximize the efficient allocation of animals. In dose-response experiments, this necessitates the use of an adequate number of animals to discriminate sufficiently the important aspects of the dose-response model while at the same time minimizing the pointless sacrifice of animals at the extreme responses of 0 and 100%.

References

Andratschke NH, Nieder C, Price RE, Rivera B, Tucker SL, Ang KK. Modulation of rodent spinal cord radiation tolerance by administration of platelet-derived growth factor. Int J Radiat Oncol Biol Phys. 2004;60(4):1257–63. https://doi.org/10.1016/j.ijrobp.2004.07.703.

Ang KK, van der Kogel AJ, Van Der Schueren E. The effect of small radiation doses on the rat spinal cord: the concept of partial tolerance. Int J Radiat Oncol Biol Phys. 1983;9(10):1487–91. https://doi.org/10.1016/0360-3016(83)90323-1.

Ang KK, van der Kogel AJ, Van Dam J, van der Schueren E. The kinetics of repair of sublethal damage in the rat cervical spinal cord during fractionated irradiations. Radiother Oncol. 1984;1(3):247–53. https://doi.org/10.1016/S0167-8140(84)80007-9.

Ang KK, van der Kogel AJ, van der Schueren E. Lack of evidence for increased tolerance of rat spinal cord with decreasing fraction doses below 2 Gy. Int J Radiat Oncol Biol Phys. 1985;11(1):105–10. https://doi.org/10.1016/0360-3016(85)90368-2.

Ang KK, Thames HD Jr, van der Kogel AJ, Van Der Schueren E. Is the rate of repair of radiation-induced sublethal damage in rat spinal cord dependent on the size of dose per fraction? Int J Radiat Oncol Biol Phys. 1987;13(4):557–62. https://doi.org/10.1016/0360-3016(87)90071-X.

Ang KK, Jiang GL, Guttenberger R, Thames HD, Stephens LC, Smith CD, Feng Y. Impact of spinal cord repair kinetics on the practice of altered fractionation schedules. Radiother Oncol. 1992;25(4):287–94. https://doi.org/10.1016/0167-8140(92)90249-T.

Best JB. Maximum likelihood analysis of ionizing radiation induced mortality in whole body fractionated dose experiments. Health Phys. 1959;2(2):139–56. https://doi.org/10.1097/00004032-195904000-00003.

Brownson RH, Suter DB, Diller DA. Acute brain damage induced by low dosage x-irradiation. Neurology. 1963a;13(3):181–91. https://doi.org/10.1212/wnl.13.3.181.

Brownson RH, Suter DB, Oliver JL, Diller DA. Acute brain damage induced by x-irradiation with special reference to rate and recovery factors. Neurology. 1963b;13(12):1011–20. https://doi.org/10.1212/wnl.13.12.1011.

Carante MP, Aricò G, Ferrari A, Karger CP, Kozlowska W, Mairani A, Sala P, Ballarini F. In vivo validation of the BIANCA biophysical model: benchmarking against rat spinal cord RBE data. Int J Mol Sci. 2020;21(11) https://doi.org/10.3390/ijms21113973.

Chowdhry VK, Liu L, Goldberg S, Adams JA, De Amorim Bernstein K, Liebsch NJ, Niemierko A, Chen YL, DeLaney TF. Thoracolumbar spinal cord tolerance to high dose conformal proton-photon radiation therapy. Radiother Oncol. 2016;119(1):35–9. https://doi.org/10.1016/j.radonc.2016.01.002.

Clausi MG, Stessin AM, Tsirka SE, Ryu S. Mitigation of radiation myelopathy and reduction of microglial infiltration by Ramipril. ACE inhibitor Spinal Cord. 2018;56(8):733–40. https://doi.org/10.1038/s41393-018-0158-z.

Cohen L, Ten Haken RK, Mansell J, Yalavarthi SD, Hendrickson FR, Awschalom M. Tolerance of the human spinal cord to high energy P(66)Be(49) neutrons. Int J Radiat Oncol Biol Phys. 1985;11(4):743–9. https://doi.org/10.1016/0360-3016(85)90306-2.

Cox DR. The analysis of multivariate binary data. Appl Stat. 1972:113–20.

Cox DR, Snell EJ. Analysis of binary data. Routledge; 2018.

Dale RG. The application of the linear-quadratic dose-effect equation to fractionated and protracted radiotherapy. Br J Radiol. 1985;58(690):515–28.

Debus J, Scholz M, Haberer T, Peschke P, Jäkel O, Karger CP, Wannenmacher M. Radiation tolerance of the rat spinal cord after single and split doses of photons and carbon ions. Radiat Res. 2003;160(5):536–42. https://doi.org/10.1667/3063.

Denekamp J. Changes in the rate of repopulation during multifraction irradiation of mouse skin. Br J Radiol. 1973;46(545):381–7. https://doi.org/10.1259/0007-1285-46-545-381.

Denekamp J, Joiner MC, Maughan RL. Neutron RBEs for mouse skin at low doses per fraction. Radiat Res. 1984;98(2):317–31. https://doi.org/10.2307/3576239.

Dische S, Saunders MI. Continuous, hyperfractionated, accelerated radiotherapy (CHART): an interim report upon late morbidity. Radiother Oncol. 1989;16(1):65–72. https://doi.org/10.1016/0167-8140(89)90071-6.

Douglas BG, Fowler JF. Fractionation schedules and a quadratic dose-effect relationship. Br J Radiol. 1975;48(570):502–3.

Douglas BG, Fowler JF. The effect of multiple small doses of X rays on skin reactions in the mouse and a basic interpretation. Radiat Res. 1976;66(2):401–26. https://doi.org/10.2307/3574407.

Ellis F. Dose, time and fractionation: a clinical hypothesis. Clin Radiol. 1969;20(1):1–7.

Frome EL, Beauchamp JJ. Maximum likelihood estimation of survival curve parameters. Biometrics. 1968;24(3):595–605. https://doi.org/10.2307/2528320.

Frome EL, DuFrain RJ. Maximum likelihood estimation for cytogenetic dose-response curves. Biometrics. 1986;42(1):73–84. https://doi.org/10.2307/2531244.

Geraci JP, Mariano MS. Relationship between dose and the latent period for radiation myelopathy in rats. Radiat Res. 1994;140(3):340–6. https://doi.org/10.2307/3579111.

Geraci JP, Thrower PD, Jackson KL, Christensen GM, Parker RG, Fox MS. The relative biological effectiveness of fast neutrons for spinal cord injury. Radiat Res. 1974;59(2):496–503. https://doi.org/10.2307/3573998.

Geraci JP, Jackson KL, Christensen GM, Thrower PD, Mariano M. RBE for late spinal cord injury following multiple fractions of neutrons. Radiat Res. 1978;74(2):382–6. https://doi.org/10.2307/3574897.

Gerber GB. Interactions: dose effect relationships and isoeffect curves. Radiat Environ Biophys. 1982;20(4):235–43. https://doi.org/10.1007/BF01323749.

Goffinet DR, Marsa GW, Brown JM. The effects of single and multifraction radiation courses on the mouse spinal cord. Radiology. 1976;119(3):709–13. https://doi.org/10.1148/119.3.709.

Grégoire V, Ruifrok ACC, Price RE, Brock WA, Hittelman WN, Plunkett WK, Ang KK. Effect of intra-peritoneal fludarabine on rat spinal cord tolerance to fractionated irradiation. Radiother Oncol. 1995;36(1):50–5. https://doi.org/10.1016/0167-8140(95)01563-V.

Gutin PH, McDermott MW, Ross G, Chan PH, Chen SF, Levin KJ, Babuna O, Marton LJ. Polyamine accumulation and vasogenic oedema in the genesis of late delayed radiation injury of the central nervous system (CNS). Acta Neurochir Suppl. 1990;51:372–4.

Habermalz HJ, Valley B, Habermalz E. Radiation myelopathy of the mouse spinal cord--isoeffect correlations after fractionated radiation. Strahlenther Onkol. 1987;163(9):626–32.

Herbert D. Reflections on the LQ model. Does it 'fit'? (does it matter?). In: Paliwal B, Fowler J, Herbert D, Kinsella T, Orton C, editors. Prediction of response in radiation therapy: analytical models and modelling. New York: American Institute of Physics Inc; 1989. p. 400–516.

Herbert D, Schultheiss TE, Feldman A, Krishnan E, Orton CG, Ovadia J, Paliwal B, Shrivastava P, Smith A, Stovall M, Cohen L. Quality assessment and improvement of dose response models: some effects of study weaknesses on study findings "C'est magnifique?". Madison, WI. 1993. https://doi.org/10.37206/42.

Herring DF. Methods for extracting dose response curves from radiation therapy data, I: a unified approach. Int J Radiat Oncol Biol Phys. 1980;6(2):225–32. https://doi.org/10.1016/0360-3016(80)90042-5.

Hornsey S. Experimental central nervous system injury from fast neutrons. In: Gutin P, Leibel S, Sheline G, editors. Radiation injury to the nervous system. New York: Raven Press; 1991.

Hornsey S, White A. Isoeffect curve for radiation myelopathy. Br J Radiol. 1980;53(626):168–9. https://doi.org/10.1259/0007-1285-53-626-168.

Hornsey S, Myers R, Joiner M. RBE for damage to the spinal cord with d(16)BE and p(62)Be fractionated neutron irradiations. unpublished.

Hosmer DW, Lemeshow S. Applied logistic regression. Wiley series in probability and mathematical statistics Applied probability and statistics. New York: Wiley; 1989.

Hubbard BM, Hopewell JW. The dose-latent period relationship in the irradiated cervical spinal cord of the rat. Radiology. 1978;128(3):779–81. https://doi.org/10.1148/128.3.779.

Karger C, Peschke P, Sanchez-Brandelik R, Scholz M, Debus J. Radiation tolerance of the rat spinal cord after single and fractionated doses of photons and carbon ions. Radiother Oncol. 2006a;81:S82.

Karger CP, Peschke P, Sanchez-Brandelik R, Scholz M, Debus J. Radiation tolerance of the rat spinal cord after 6 and 18 fractions of photons and carbon ions: experimental results and clinical implications. Int J Radiat Oncol Biol Phys. 2006b;66(5):1488–97. https://doi.org/10.1016/j.ijrobp.2006.08.045.

Kim JJ, Hao Y, Jang D, Wong CS. Lack of influence of sequence of top-up doses on repair kinetics in rat spinal cord. Radiother Oncol. 1997;43(2):211–7. https://doi.org/10.1016/s0167-8140(97)01928-2.

Kuss O. Global goodness-of-fit tests in logistic regression with sparse data. Stat Med. 2002;21(24):3789–801.

Laramore GE, Bauer M, Griffin TW, Thomas FJ, Hendrickson FR, Maor MH, Griffin BR, Saxton JP, Davis LW. Fast neutron and mixed beam radiotherapy for inoperable non-small cell carcinoma of the lung. Results of an RTOG randomized study. Am J Clin Oncol. 1986;9(3):233–43. https://doi.org/10.1097/00000421-198606000-00012.

Leith JT, Lewinsky BS, Woodruff KH, Schilling WA, Lyman JT. Tolerance of the spinal cord of rats to irradiation with cyclotron-accelerated helium ions. Cancer. 1975a;35(6):1692–700. https://doi.org/10.1002/1097-0142(197506)35:6<1692::AID-CNCR2820350631>3.0.CO;2-M.

Leith JT, Woodruff KH, Lewinsky BS, Lyman JT, Tobias CA. Tolerance of the spinal cord of rats to irradiation with neon ions. Int J Radiat Biol. 1975b;28(4):393–8. https://doi.org/10.1080/09553007514551181.

Leith JT, Keith Dewyngaert J, Glicksman AS. Radiation myelopathy in the rat: an interpretation of dose effect relationships. Int J Radiat Oncol Biol Phys. 1981;7(12):1673–7. https://doi.org/10.1016/0360-3016(81)90191-7.

Leith JT, McDonald M, Powers-Risius P, Bliven SF, Howard J. Response of rat spinal cord to single and fractionated doses of accelerated heavy ions. Radiat Res. 1982;89(1):176–93. https://doi.org/10.2307/3575694.

Li YQ, Jay V, Wong CS. Oligodendrocytes in the adult rat spinal cord undergo radiation-induced apoptosis. Cancer Res. 1996;56(23):5417–22.

Li YQ, Chen P, Haimovitz-Friedman A, Reilly RM, Wong CS. Endothelial apoptosis initiates acute blood-brain barrier disruption after ionizing radiation. Cancer Res. 2003;63(18):5950–6.

Lo YC, McBride WH, Rodney Withers H. The effect of single doses of radiation on mouse spinal cord. Int J Radiat Oncol Biol Phys. 1992;22(1):57–63. https://doi.org/10.1016/0360-3016(92)90982-N.

Marples B, Joiner MC. The response of Chinese hamster V79 cells to low radiation doses: evidence of enhanced sensitivity of the whole cell population. Radiat Res. 1993;133(1):41–51.

Marucci L, Niemierko A, Liebsch NJ, Aboubaker F, Liu MCC, Munzenrider JE. Spinal cord tolerance to high-dose fractionated 3D conformal proton-photon irradiation as evaluated by equivalent uniform dose and dose volume histogram analysis. Int J Radiat Oncol Biol Phys. 2004;59(2):551–5. https://doi.org/10.1016/j.ijrobp.2003.10.058.

Masuda K, Reid BO, Withers HR. Dose effect relationship for epilation and late effects on spinal cord in rats exposed to gamma rays. Radiology. 1977;122(1):239–42. https://doi.org/10.1148/122.1.239.

Menten J, Landuyt W, van der Kogel AJ, Ang KK, van der Schueren E. Effects of high dose intraperitoneal cytosine arabinoside on the radiation tolerance of the rat spinal cord. Int J Radiat Oncol Biol Phys. 1989;17(1):131–4. https://doi.org/10.1016/0360-3016(89)90380-5.

Metz CE, Tokars RP, Kronman HB, Griem ML. Maximum likelihood estimation of dose-response parameters for therapeutic operating characteristic (TOC) analysis of carcinoma of the nasopharynx. Int J Radiat Oncol Biol Phys. 1982;8(7):1185–92. https://doi.org/10.1016/0360-3016(82)90066-9.

Morris GM, Hopewell JW, Morris AD. The influence of methotrexate on radiation-induced damage to different lengths of the rat spinal cord. Br J Radiol. 1992;65(770):152–6. https://doi.org/10.1259/0007-1285-65-770-152.

Myers R, Rogers MA, Hornsey S. A reappraisal of the roles of glial and vascular elements in the development of white matter necrosis in irradiated rat spinal cord. Br J Cancer. 1986;53(Suppl. 7):221–3.

Nieder C, Price RE, Rivera B, Andratschke N, Ang KK. Experimental data for insulin-like growth factor-1 (IGF-1) and basic fibroblast growth factor (bFGF) in prevention of radiation myelopathy. Strahlenther Onkol. 2002;178(3):147–52. https://doi.org/10.1007/s00066-002-0897-8.

Nieder C, Andratschke N, Price RE, Rivera B, Ang KK. Evaluation of insulin-like growth factor-1 for prevention of radiation-induced spinal cord damage. Growth Factors. 2005a;23(1):15–8. https://doi.org/10.1080/08977190500055919.

Nieder C, Price RE, Rivera B, Andratschke N, Ang KK. Effects of insulin-like growth factor-1 (IGF-1) and amifostine in spinal cord reirradiation. Strahlenther Onkol. 2005b;181(11):691–5. https://doi.org/10.1007/s00066-005-1464-x.

Niewald M, Feldmann U, Feiden W, Niedermayer I, Kiessling M, Lehmann W, Abel U, Berberich W, Staut W, Büscher E, Walter K, Nieder C, Nestle U, Deinzer M, Schnabel K. Multivariate logistic analysis of dose-effect relationship and latency of radiomyelopathy after hyperfractionated and conventionally fractionated radiotherapy in animal experiments. Int J Radiat Oncol Biol Phys. 1998;41(3):681–8. https://doi.org/10.1016/s0360-3016(98)00079-0.

Niewald M, Feldmann U, Feiden W, Kiessling M, Berberich W, Lehmann W, Abel U, Staut W, Buscher-Rolles E, Walter K, Nieder C, Nestle U, Deinzer M, Schnabel K. Radiomyelopathy after conventionally fractionated and hyperfractionated radiotherapy - experimental data and clinical consequences. In: Wiegel T, Hinkelbein W, Brock M, Hoell T, editors. Controversies in neuro-oncology, Frontiers of radiation therapy and oncology, vol. 33. Basel: Karger; 1999. p. 293–304.

Orton CG, Ellis F. A simplification in the use of the NSD concept in practical radiotherapy. Br J Radiol. 1973;46(547):529–37.

Paganetti H. Relative biological effectiveness (RBE) values for proton beam therapy. Variations as a function of biological endpoint, dose, and linear energy transfer. Phys Med Biol. 2014;59(22):R419.

Parsons JT, Mendenhall WM, Stringer SP, Cassisi NJ, Million RR. An analysis of factors influencing the outcome of postoperative irradiation for squamous cell carcinoma of the oral cavity. Int J Radiat Oncol Biol Phys. 1997;39(1):137–48. https://doi.org/10.1016/S0360-3016(97)00152-1.

Peters LJ, Schultheiss T, Maor MH. Tolerance of human spinal cord to high energy neutrons. Int J Radiat Oncol Biol Phys. 1986;12(2):292–3. https://doi.org/10.1016/0360-3016(86)90117-3.

Phillips TL, Buschke F. Radiation tolerance of the thoracic spinal cord. Am J Roentgenol Radium Therapy, Nucl Med. 1969;105(3):659–64.

Pop LAM, van der Plas M, Skwarchuk MW, Hanssen AEJ, van der Kogel AJ. High dose rate (HDR) and low dose rate (LDR) interstitial irradiation (IRT) of the rat spinal cord. Radiother Oncol. 1997;42(1):59–67. https://doi.org/10.1016/s0167-8140(96)01862-2.

Pop LAM, van der Plas M, Ruifrok ACC, Schalkwijk LJM, Hanssen AEJ, van der Kogel AJ. Tolerance of rat spinal cord to continuous interstitial irradiation. Int J Radiat Oncol Biol Phys. 1998;40(3):681–9. https://doi.org/10.1016/S0360-3016(97)00852-3.

Pop LAM, Millar WT, van der Plas M, van der Kogel AJ. Radiation tolerance of rat spinal cord to pulsed dose rate (PDR-) brachytherapy: the impact of differences in temporal dose distribution. Radiother Oncol. 2000;55(3):301–15. https://doi.org/10.1016/s0167-8140(00)00205-x.

Potish RA, Twiggs LB, Adcock LL, Prem KA. Logistic models for prediction of enteric morbidity in the treatment of ovarian and cervical cancers. Am J Obstet Gynecol. 1983;147(1):65–72. https://doi.org/10.1016/0002-9378(83)90086-8.

Saager M, Glowa C, Peschke P, Brons S, Scholz M, Huber PE, Debus J, Karger CP. Carbon ion irradiation of the rat spinal cord: dependence of the relative biological effectiveness on linear energy transfer. Int J Radiat Oncol Biol Phys. 2014;90(1):63–70. https://doi.org/10.1016/j.ijrobp.2014.05.008.

Saager M, Glowa C, Peschke P, Brons S, Grun R, Scholz M, Huber PE, Debus J, Karger CP. The relative biological effectiveness of carbon ion irradiations of the rat spinal cord increases linearly with LET up to 99 keV/mu m. Acta Oncol. 2016;55(12):1512–5. https://doi.org/10.108 0/0284186x.2016.1250947; https://doi.org/10.1016/j.radonc.2015.07.006.

Saager M, Peschke P, Brons S, Debus J, Karger CP. Determination of the proton RBE in the rat spinal cord: is there an increase towards the end of the spread-out Bragg peak? Radiother Oncol. 2018;128(1):115–20. https://doi.org/10.1016/j.radonc.2018.03.002.

Saager M, Glowa C, Peschke P, Brons S, Grün R, Scholz M, Debus J, Karger CP. Fractionated carbon ion irradiations of the rat spinal cord: comparison of the relative biological effectiveness with predictions of the local effect model. Radiation oncology (London, England). 2020a;15(1):6. https://doi.org/10.1186/s13014-019-1439-1.

Saager M, Hahn EW, Peschke P, Brons S, Huber PE, Debus J, Karger CP. Ramipril reduces incidence and prolongates latency time of radiation-induced rat myelopathy after photon and carbon ion irradiation. J Radiat Res. 2020b;61(5):791–8. https://doi.org/10.1093/jrr/rraa042.

Scalliet P, Landuyt W, van der Schueren E. Repair kinetics as a determining factor for late tolerance of central nervous system to low dose rate irradiation. Radiother Oncol. 1989;14(4):345–53. https://doi.org/10.1016/0167-8140(89)90147-3.

Scholz M, Elsässer T. Biophysical models in ion beam radiotherapy. Adv Space Res. 2007;40(9):1381–91. https://doi.org/10.1016/j.asr.2007.02.066.

Schultheiss TE, Cohen L, Mansell J. Normal tissue reactions and complications following high-energy neutron beam therapy II: complication rates adjusted for censoring. Int J Radiat Oncol Biol Phys. 1990;18(1):165–71. https://doi.org/10.1016/0360-3016(90)90280-W.

Schultheiss TE, Stephens LC, Ang KK, Jardine JH, Peters LJ. Neutron RBE for primate spinal cord treated with clinical regimens. Radiat Res. 1992;129(2):212–7. https://doi.org/10.2307/3578159.

Spence AM, Krohn KA, Edmondson SW, Steele JE, Rasey JS. Radioprotection in rat spinal cord with WR-2721 following cerebral lateral intraventricular injection. Int J Radiat Oncol Biol Phys. 1986;12(8):1479–82. https://doi.org/10.1016/0360-3016(86)90198-7.

Spence AM, Krohn KA, Steele JE, Edmondson SE, Rasey JS. WR-2721, WR-77913 and WR-3689 radioprotection in the rat spinal cord. Pharmacol Ther. 1988;39(1–3):89–91. https://doi.org/10.1016/0163-7258(88)90044-7.

Stephens LC, Hussey DH, Raulston GL, Jardine JH, Gray KN, Almond PR. Late effects of 50 MeV d→Be neutron and cobalt-60 irradiation of rhesus monkey cervical spinal cord. Int J Radiat Oncol Biol Phys. 1983;9(6):859–64. https://doi.org/10.1016/0360-3016(83)90012-3.

Taylor JMG, Kim DK. The poor statistical properties of the fe-plot. Int J Radiat Biol. 1989;56(2):161–7. https://doi.org/10.1080/09553008914551311.

Thames HD, Hendry J. Fractionation in radiotherapy. London, Philadelphia: Taylor & Francis; 1987.

Thames HD, Rodney Withers H, Peters LJ, Fletcher GH. Changes in early and late radiation responses with altered dose fractionation: implications for dose-survival relationships. Int J Radiat Oncol Biol Phys. 1982;8(2):219–26. https://doi.org/10.1016/0360-3016(82)90517-X.

Thames HD, Wither HR, Peters LJ. Tissue repair capacity and repair kinetics deduced from multifractionated or continuous irradiation regiments with incomplete repair. Br J Cancer. 1984;49(Suppl. 6):263–9.

Thames HD, Ang KK, Stewart FA, Van Der Schueren E. Does incomplete repair explain the apparent failure of the basic LQ model to predict spinal cord and kidney responses to low doses per fraction?? Int J Radiat Biol. 1988;54(1):13–9. https://doi.org/10.1080/09553008814551461.

van den Bogaert W, van der Schueren E, Horiot J-C, Chaplain G, Devilhena M, Raposo S, Leonor J, Schraub S, Chenal C. Early results of the EORTC randomized clinical trial on multiple fractions per day (MFD) and misonidazole in advanced head and neck cancer. Int J Radiat Oncol Biol Phys. 1986;12(4):587–91.

van der Kogel AJ. Radiation tolerance of the rat spinal cord: time dose relationships. Radiology. 1977;122(2):505–9. https://doi.org/10.1148/122.2.505.

van der Kogel AJ. Late effects of radiation on the spinal cord; dose-effect relationships and pathogenesis. RBI; 1979.

van der Kogel AJ. Radiation-induced damage in the central nervous system: an interpretation of target cell responses. Br J Cancer. 1986;53(Suppl. 7):207–17.

van der Kogel AJ. Central nervous system radiation injury in small animal models. In: Gutin PH, Leibel SA, Sheline GE, editors. Radiation injury to the nervous system. New York: Raven Press; 1991.

van der Kogel AJ, Barendsen GW. Late effects of spinal cord irradiation with 300 kV X rays and 15 MeV neutrons. Br J Radiol. 1974;47(559):393–8. https://doi.org/10.1259/0007-1285-47-559-393.

van der Kogel AJ, Sissingh HA. Effect of misonidazole on the tolerance of the rat spinal cord to daily and multiple fractions per day of X rays. Br J Radiol. 1983;56(662):121–5. https://doi.org/10.1259/0007-1285-56-662-121.

van der Kogel AJ, Sissingh HA. Effects of intrathecal methotrexate and cytosine arabinoside on the radiation tolerance of the rat spinal cord. Radiother Oncol. 1985;4(3):239–51.

Wara WM, Phillips TL, Sheline GE, Schwade JG. Radiation tolerance of the spinal cord. Cancer. 1975;35(6):1558–62. https://doi.org/10.1002/1097-0142(197506)35:6<1558::AID-CNCR2 820350613>3.0.CO;2-7.

White A, Hornsey S. Radiation damage to the rat spinal cord: the effect of single and fractionated doses of X rays. Br J Radiol. 1978;51(607):515–23. https://doi.org/10.1259/0007-1285-51-607-515.

Wong CS, Minkin S, Hill RP. Linear-quadratic model underestimates sparing effect of small doses per fraction in rat spinal cord. Radiother Oncol. 1992;23(3):176–84. https://doi.org/10.1016/0167-8140(92)90328-R.

Wong CS, Minkin S, Hill RP. Effect of small doses per fraction in rat spinal cord: influence of initial vs. final top-up doses. Radiother Oncol. 1993;28(1):52–6. https://doi.org/10.1016/0167-8140(93)90185-B.

Wong CS, Hao Y, Hill RP. Response of rat spinal cord to very small doses per fraction: lack of enhanced radiosensitivity. Radiother Oncol. 1995;36(1):44–9. https://doi.org/10.1016/0167-8140(95)01558-X.

Wouters BG, Skarsgard LD, Gerweck LE, Carabe-Fernandez A, Wong M, Durand RE, Nielson D, Bussiere MR, Wagner M, Biggs P, Paganetti H, Suit HD. Radiobiological intercomparison of the 160 MeV and 230 MeV proton therapy beams at the Harvard Cyclotron Laboratory and at Massachusetts General Hospital. Radiat Res. 2015;183(2):174–87. https://doi.org/10.1667/RR13795.1.

Volume Effect Studies in Rodents 6

In 1987 Hopewell et al. were the first to explore the dose response for different lengths of cervical spinal cord. SD rats were irradiated with 250 kV x-rays to 4, 8, and 16 mm of the cervical cord (Hopewell et al. 1987). This interesting paper shows a clear and substantial relationship between incidence of paralysis and field size for latencies less than 30 weeks. In this time period, the "predominant histological lesion" for RM in all animals was white matter necrosis. Gray matter changes and nerve root demyelination and necrosis were seen at the higher doses for the smaller fields, lesions not previously reported in the rat cervical cord. Additionally, the paper also shows a clear relationship between latency for vascular damage, which occurred after 30 weeks, in both dose and field size. Interestingly, the relationship between latency for WMN and dose was almost nil below about 55 Gy single dose for all field sizes. This single dose experiment was well described by a linear logistic dose-response function in dose and volume. Later in this chapter, it will be shown that a better fit is obtained with the critical volume model of Jackson et al. (Jackson and Kutcher 1993).

In 1987, van der Kogel published in abstract form a volume effect study on rats using field sizes from 2 to 60 mm (van der Kogel 1987). This was later published in graphical form without a table of values of the D_{50}'s (van der Kogel 1991, 1993). The actual dose-response data were not published. Bijl et al. published a partial list of van der Kogel's D_{50} values in 2002 (Bijl et al. 2002). Using those values it is possible to calibrate a captured image of van der Kogel's graph of D_{50}'s and thereby determine the D_{50} values that Bijl et al. did not include. The larger field sizes of these data encompass both the cervical and thoracic cord (van der Kogel 1987). Where the fields were centered is unknown and the experiment itself has not been described in the literature.

Starting in 2002, volume effect studies were published from two groups in the Netherlands under the lead authorship of Hendrik Bijl, Peter van Luijk, and Marielle Philippens with Albert van der Kogel as senior author in all cases. A number of these papers authored by Bijl et al. studied the cervical cord. Philippens et al. later

© Springer Nature Switzerland AG 2022
T. Schultheiss, *Radiation Myelopathy*,
https://doi.org/10.1007/978-3-030-94658-6_6

published somewhat similar papers studying the thoracolumbar and the cervical cord and an additional paper that involved SLD repair in the cervical cord.

The first paper in this series was a straightforward, yet difficult study of the volume effect in the cervical cord (Bijl et al. 2002). (In most cases, the volume effect will refer to the responses elicited by varying the homogeneously irradiated length of spinal cord, but can include inhomogeneous irradiations of the cord in either the longitudinal or transverse directions. Unless otherwise specified, it shall have the former meaning.) This study deployed proton irradiation for 2-, 4-, 8-, and 20-mm lengths of the cervical cord centered at C4–5. While the 8- and 20-mm lengths were relatively uniform in dose longitudinally, the 2- and 4-mm lengths were less uniform over the irradiated length. The inhomogeneities in the dose distributions were not accounted for in the analysis; nor were they addressed, other than being presented in the graph reproduced in Fig. 6.1. The experiment was underpowered at six animals per dose group and for unstated reasons, the 4-mm group was repeated, not using the same doses, but using doses interleaved between the original doses. The RBE of the plateau region of the unmodulated proton beam was believed to be unity, but because the data were not compared directly to photon treatments the point is moot.

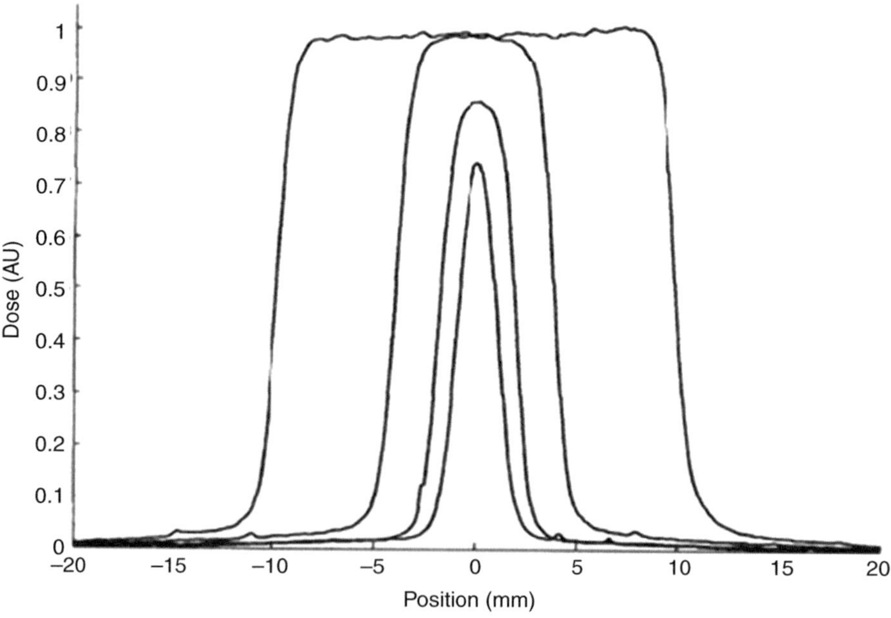

Fig. 6.1 Beam profiles for 2, 4, 8, and 20 mm proton beams used in Bijl et al. (2002). Diminishing peak values simply shows the field-size effect. In practice, all treatments were normalized to the peak value. Reprinted from International Journal of Radiation Oncology*Biology*Physics, Vol 51/1, Hendrik P Bijl, Peter van Luijk, Rob P Coppes, Jacobus M Schippers, Antonius W.T Konings, Albert J van der Kogel, Dose-volume effects in the rat cervical spinal cord after proton irradiation, 205–211, Copyright (2002), with permission from Elsevier.

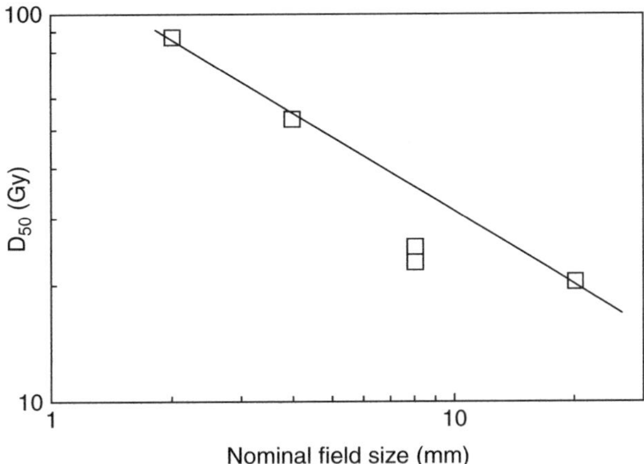

Fig. 6.2 D_{50} versus field size from Bijl et al. (2002, 2006). The solid line is added here to prejudice the eye into believing that the D_{50}'s for the 8 mm field size might be problematic. The explanation for this deceptive maneuver is fully disclosed in the text.

A dramatic length effect was shown using D_{50} as a metric. This is reproduced in Fig. 6.2, where the D_{50} versus field length is plotted on a log-log scale. The dose-response data were analyzed using probit analysis according to the paper's Methods section. As is common with low-powered, single dose experiments, the 20-mm group's data were quasi-separated, so the D_{50} estimate cannot have been produced by probit analysis that deploys maximum likelihood estimation. A MLE of D_{50} could have been produced assuming a normal distribution of tolerances (same as probit) *if* a value for the slope of the dose-response curve was assumed, but there was no mention of this technique in the paper. Also shown in this figure is the D_{50} for an 8-mm study published later (Bijl et al. 2006). The latent period also showed a significant length effect, with shorter lengths having shorter latencies; of course a much higher dose had to be used at shorter lengths in order to elicit the same response rates. The histopathology was as expected based on the previous studies of Hopewell, except that "necrosis of the gray matter and nerve root was not observed."

Hopewell et al. attributed the field-size effect to the *possible* migration of cells from unirradiated tissue into the irradiated area and thereby ameliorating the radiation damage. van der Kogel was not circumspect about the nature of these cells stating, "This indicates a critical dependence on migration of glial precursor cells from the unirradiated borders for regeneration of the early demyelinating lesions." This theory was substantially amended and elaborated in subsequent publications from van der Kogel and colleagues. See below. A recurring argument made in the Bijl series was that Hinks et al. (2001) "showed that a single dose of 40 Gy to a 7-mm segment of rat thoracic spinal cord almost completely depleted the oligodendrocyte progenitor cell[s]," yet "This depleted segment was repopulated slowly to normal levels by OPCs from flanking unirradiated areas." It is worth observing that this dose results in paralysis nearly 100% of the time and therefore the repopulation of OPC's cannot have had any long-term beneficial effect.

The van der Kogel volume data are in good agreement with Bijl et al. (discussed in detail below) and in qualitative agreement with Hopewell et al., as shown in Fig. 6.3 (Hopewell et al. 1987). They show essentially no volume (length) response

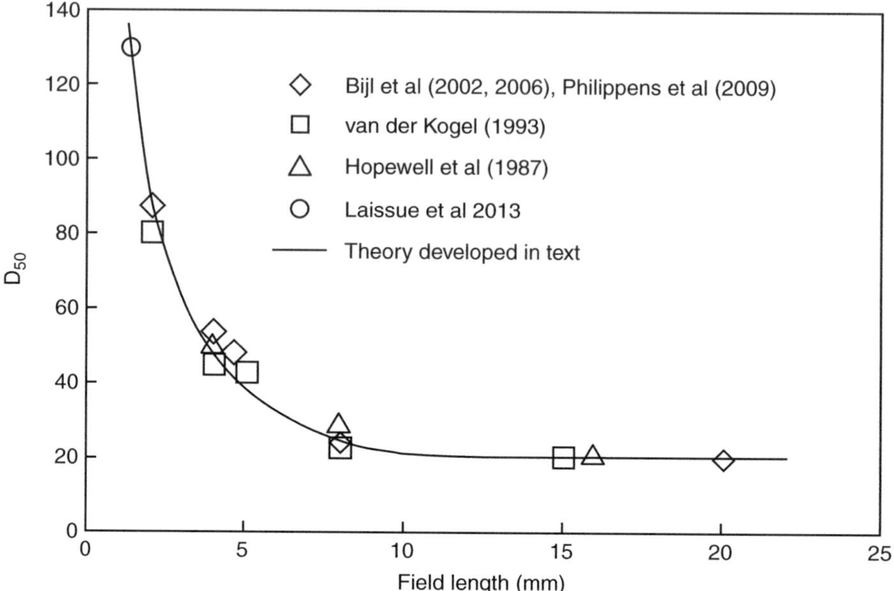

Fig. 6.3 Data points from small field irradiation in rats.

from 1 to 6 cm and a very dramatic volume response below 8 mm. The solid line is a model that will be addressed in some detail below. First, a description of several papers from the labs of van der Kogel follows with first authors of van Luijk, Bijl, and Philippens.

Modeling the volume effect was not part of the Bijl et al. (2002) study. However, in a 2005 publication, van Luijk et al. fitted these data to 14 different volume effect models, finding that none fit the data (van Luijk et al. 2005). That paper is discussed in more detail below. One of the reasons that van Luijk could not fit these data to dose-response models is that some of the data are poorly behaved. First, for reasons not stated in the paper and as alluded to above, the 4-mm data were repeated (with somewhat different results). Referred to in the paper as groups I and II, the D_{50}'s for groups I, II, and the combination were 51.8, 55.6, and 53.8 Gy, respectively. However, group II are completely separated and could not have produced a probit estimate of D_{50}. The group I data, when analyzed by itself rejects the probit model. When the data are combined, the model is again rejected because of a low-dose (43-Gy) outlier with a response of 1 out of 3, but excluding this point it produces a satisfactory fit when all the data are pooled. (This fit is demonstrated in Fig. 6.6 and will be discussed below, along with the modeling.)

An additional observation is that fitted separately, the 8- and 20-mm dose-response functions would cross each other, although this is almost certainly a result of too few animals per dose group and too large a spacing between groups. The 20-mm data are quasi-separated. Thus it is no surprise that van Luijk could not find a satisfactory fit to all the data simultaneously.

A log-log plot of the Bijl D_{50} data seems to indicate that the 8-mm D_{50}, and therefore the underlying dose-response data, are spurious (Fig. 6.2). The D_{50} seems too low. However, as shown in Fig. 6.3, the D_{50}'s from 60 mm down to 2 mm all fall along a single curve. From 1987 onward, van der Kogel maintained the position that in the rat, there was no volume effect (for single doses) above field lengths of 1 cm, but there was a very significant effect for smaller field lengths. In effect, there was an abrupt transition just below 1 cm. Figure 6.3 and the analysis below demonstrate that van der Kogel's understanding of the shape of volume effect in the rat is well founded, but cell migration is an unlikely explanation.

Bijl et al. published two important follow-up papers—one about 6 months later and another in 2006 (Bijl et al. 2003, 2006). In these studies, inhomogeneous dose distributions were generated to probe the spinal cord response for a better understanding of the volume effect. In fact, these papers are as important with respect to their implications for the pathogenesis as they are for their discoveries regarding the volume effect.

The 2003 paper introduced the "bath and shower" technique. With this technique, a larger úniform field is applied to the spinal cord and then, too quickly for repair of SLD, a smaller field is applied on top of the first field. If the fields had very sharp dose fall off, this technique would produce two uniformly irradiated regions in the spinal cord. (In practice, the dosimetry is somewhat more complicated.) For this study, the investigators used a 20-mm field for the "bath" field and a 4-mm field for the "shower." In addition, they deployed two 4-mm fields separated from each other by either 8 or 12 mm. They also used the results of the 8-mm from the 2002 study discussed above for comparison to the split-field irradiation with the two 4-mm fields. Finally, the authors used an asymmetric bath and shower where the bath dose field was shortened to 12 mm and its edge was aligned with the shower dose, thereby creating a one-sided bath effect. These are illustrated in Fig. 6.4.

Although the authors never explicitly define the bath-and-shower effect, in the abstract of the 2006 paper they describe it succinctly as "the spinal cord tolerance of relatively small volumes (shower) [being] strongly affected by low-dose irradiation (bath) of adjacent tissue." To refine this slightly one might say "very small volumes" rather than "relatively small volumes" and elaborate that the tolerance was significantly reduced by this mechanism. In what follows below, the absence of a bath effect is taken to mean that there is no out-of-field effect that requires toxins or protectants to migrate into or outside the radiation field.

The results of the split-field study are given in tabular form in the 2003 publication. Applying the analysis of multiple 2×2 contingency tables (developed by Cox' 66), we find that the 8- and the 12-mm splits are not statistically different from each other (Cox 1966). If one then combines the data for a single logistic regression, a value of 43.3 Gy is obtained for the MLE of D_{50}. The 8-mm field from the 2002 study had a D_{50} of about 25 Gy. The authors stated that "the same total tissue volume [was] irradiated" with the 8-mm field as the two 4-mm fields. Although this is not strictly true, as the DVH for the two 4-mm fields is not identical to an 8-mm field, reconstruction shows that the average dose over the nominal field length is only

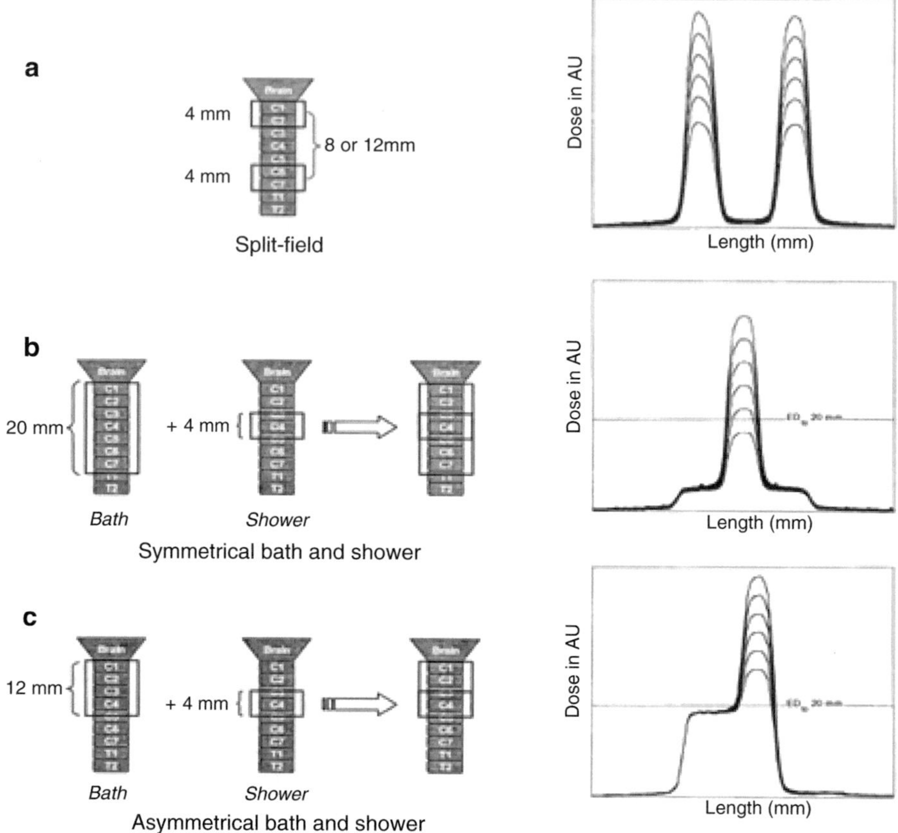

Fig. 6.4 "Schematic outline of the present experiments. The rat cervical spinal cords with the used combination of irradiation fields are depicted on the left. The corresponding dose distributions are shown on the right: split field (**a**), symmetrical bath and shower (**b**), and asymmetrical bath and shower (**c**)." Reprinted from International Journal of Radiation Oncology*Biology*Physics, 57 (1), Bijl HP et al., Unexpected changes of rat cervical spinal cord tolerance caused by inhomogeneous dose distributions, 274–281, Copyright (2003), with permission from Elsevier.

about 4% higher for the 8-mm field than the combined 4-mm fields. This is further support for a novel sort of volume effect being operative at these small field sizes.

In 2006, the authors published another paper that supplemented data from their previous two studies. In this paper, they again used the B&S technique with $D_{bath} = 4$ Gy and the shower field sizes of 2 and 8 mm. The 8-mm experiment from the 2002 paper ($D_{bath} = 0$) was repeated; a 2-mm field B&S treatment was given with $D_{bath} = 18$ Gy; and a 2-mm field was used with an asymmetrical bath of 18 Gy. These experiments logically expanded the data obtained in earlier studies.

At this point, it is useful to introduce some results from Philippens et al. (2009). In this paper, the 4-mm B&S technique was deployed using megavoltage photons. As would be expected, the 4-mm field had a significant penumbra. Also, the descriptor of "4 mm" was used for what was actually reported as a 4.6-mm field. The intent of the paper was to investigate the effect of the time interval between the bath dose and the shower dose, with the bath given first. Although the authors state that for their photon irradiations the B&S effect for the 4.6 cm shower and 4 Gy bath was less than that for protons, when their photon points are included in the data of Fig. 6.3, they follow the pattern of the proton data. All the data in this figure are plotted using the field sizes and doses as reported in the papers, i.e., not effective fields size, not equivalent uniform dose, and not recalculated D_{50}'s.

The data from the three Bijl et al. studies are presented in Table 6.1. Only the D_{50}'s as given by the authors are reproduced here, not the entire dose responses, which were reported in the individual papers. The point should be made here that Bijl et al. expressed results in terms of ED_{50}, but as measured by the shower dose alone without including the bath dose. This leads to some confusion and statements that are essentially erroneous. For example, the D_{50} of the 4 mm shower regimen is 53.7 Gy. The authors state that "applying a bath dose of only 4 Gy" results in "an iso-effective dose (ED_{50}) value of only 39 Gy, compared with the 53.7 Gy for the 4 mm single field." The error is that 39 Gy is not isoeffective with the 53.7 Gy, but rather the 39 Gy shower plus 4 Gy bath is the D_{50}. Clearly with a significant effect occurred. However, the authors then state that a "further increase of the bath doses to 12 … and 18 Gy … results in a further, but relatively smaller, decrease to ED50 values of 33.4 and 31.3 Gy, respectively." In fact, the D_{50} doses increase to 45.4 and 49.3 Gy, respectively, when the total dose is used, and the effect is the *opposite* of what they suggest. Keep in mind that it is not the bath dose alone that is responsible for the tissue damage, but the total dose. For this work, all values of D_{50} will be understood as the total dose of the bath and shower.

We shall take it as axiomatic that a bath dose, either symmetric or asymmetric, should reduce the shower D_{50} by an amount at least equal to the bath dose, otherwise the bath would have a protective effect in the shower region. In fact, it is useful to clarify the consequences of the bath dose with the following equations.

$$D_{50}\left(D_{bath}=0\right)=D_{50}\left(D_{bath}\right)\Rightarrow \text{no out-of-field effect from } D_{bath}; \tag{6.1a}$$

$$D_{50}\left(D_{bath}=0\right)>D_{50}\left(D_{bath}\right)\Rightarrow \text{enhanced}\left(\text{toxic}\right)\text{out-of-field effect from } D_{bath}; \tag{6.1b}$$

$$D_{50}\left(D_{bath}=0\right)<D_{50}\left(D_{bath}\right)\Rightarrow \text{protective out-of-field effect from } D_{bath}. \tag{6.1c}$$

The assumption of the Bijl et al. papers was that Inequality (6.1b) holds. Likewise assuming $D_{bath,2}>D_{bath,1}$, then

$$D_{50}\left(D_{bath,1}\right)=D_{50}\left(D_{bath,2}\right)\Rightarrow \text{no out-of-field effect from } D_{bath}; \tag{6.2a}$$

$$D_{50}\left(D_{bath,1}\right)>D_{50}\left(D_{bath,2}\right)\Rightarrow \text{enhanced out-of-field effect}; \tag{6.2b}$$

$$D_{50}\left(D_{bath,1}\right)<D_{50}\left(D_{bath,2}\right)\Rightarrow \text{protective out-of-field effect}. \tag{6.2c}$$

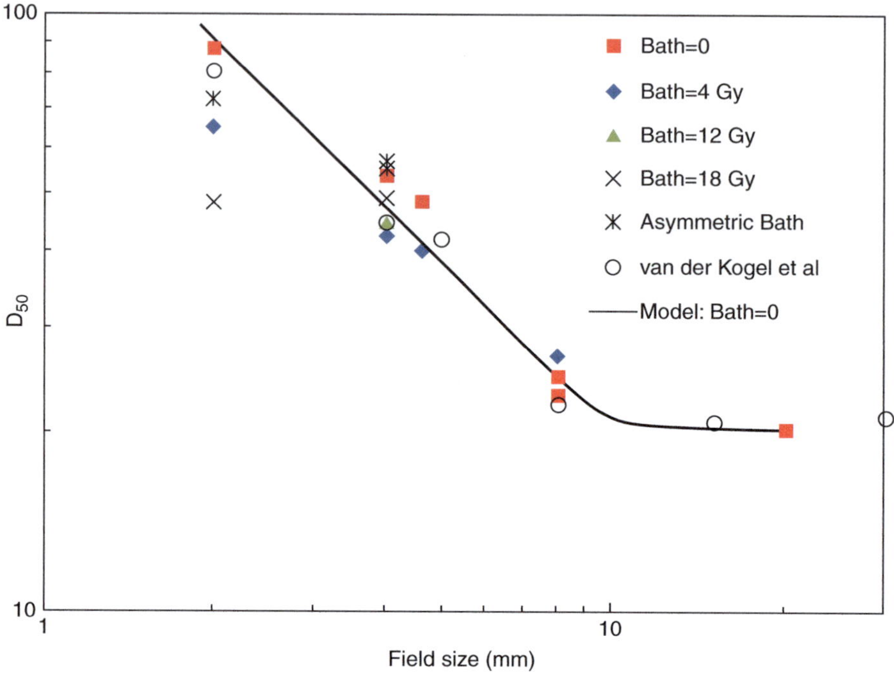

Fig. 6.5 D_{50} values of the bath-and-shower technique plotted along with D_{50} from van der Kogel volume reports on a log-log scale. The solid curve represents the model given in the text. Also noted in the text are pairs of points for which the out-of-field effects are protective, contradicting the assumptions of the bath-and-shower studies. The Philippens et al. (2009) data are shown at 4.6 mm.

where D_{50} doses in (6.1) and (6.2) refer to total doses, i.e., bath plus shower doses. The magnitude of the effect can be taken as D_{50} ($D_{bath,1}$)-D_{50} ($D_{bath,2}$), and reflects the authors' qualitative discussion of the bath-and-shower effect only if this difference is positive. Again, all D_{50}'s are the total $D_{bath} + D_{shower}$.

The data from Table 6.1 are plotted in Fig. 6.5, where the data are presented using D_{50} of the shower dose *plus* the bath dose. This total dose is the dose primarily, and possibly totally responsible for paresis. It gives a clearer understanding of the relationship between the shower dose, bath dose, and field size.

Table 6.1 D_{50} values for "bath and shower" experiments

Bijl et al	Field size, mm (shower)	Bath dose, Gy	D_{50}, Gy (shower only)	D_{50} max dose (Shower + Bath)	Toxicity direction (should increase downward)
2002	2	0	87.8	87.8	
2006	2	4[a]	68.6	72.6	
2006	2	4	61.2	65.2	
2003	4	18[a]	38.4	56.4	
2003	4	18[a]	37.2	55.2	
2002	4	0	53.7	53.7	
2003	4	18	31.3	49.3	
2006	2	18	30.9	48.9	
2003	4	12	33.4	45.4	
2003	4	4	39	43	
2006	8	4	23.1	27.1	
2002	8	0	24.9	24.9	
2006	8	0	23.2	23.2	
2002	20	0	20.4	20.4	

Data of D_{50} values for three Bijl et al. studies covering the bath-and-shower experiments. The data are sorted by decreasing values of D_{50} total dose (bath plus shower), thus the relative toxicity (toxicity/Gy) decreases from top to bottom. Green arrows indicate pairs of regimens that do not conflict with the possibility of a bath-and-shower effect, i.e., out-of-field effect. Red arrows indicated a conflict with the bath-and-shower effect as indicated by the D_{50} value changing in the wrong direction for the two regimens linked by the arrows, for example shower+bath having a higher D_{50} than shower alone. The 12 leftmost arrows connect to and from symmetric baths while the five rightmost arrows (separated by the dashed line) connect to or from an asymmetric bath techniques.
[a]Signifies asymmetric bath

Although, there is general agreement between data taken years apart, in different institutions, and with different radiation beams, some internal inconsistencies can be seen. To understand these inconsistencies it is first necessary to list some of the logical constraints on the behavior of the data. These constraints apply no matter what biomathematical model of the volume effect one might try to deploy against the data. "D_{50}" shall refer to the value for the total dose, bath plus shower. Variables are bath dose, shower dose, and field size. These constrains are given here.

Logical constraints for bath-and-shower technique:

1. D_{50}'s increase monotonically with decreasing field size (shower) assuming other variables are held constant.
2. (D_{50} for $D_{bath} = 0$) $\geq$ (D_{50} for asymmetric bath) $\geq$ (D_{50} for symmetric bath) for a given field size.
3. If a bath effect exists, D_{50}'s decrease monotonically with increasing D_{bath} because bath doses are more toxic than the same dose delivered with a smaller field size.

The problematic features of Fig. 6.5 are as follows.

1. The 4-mm data violate constraint 3 since the asymmetric bath D_{50} is higher than the D_{50} for $D_{bath} = 0$, as if the asymmetric bath had a protective effect on the shower region.
2. The bath data for the 4-mm field size additionally violate constraint 3 since D_{50} for $D_{bath} = 18$ Gy $> D_{50}$ for $D_{bath} = 12$ Gy $> D_{50}$ for $D_{bath} = 4$ Gy when that order should be reversed. Please note that this order *is* reversed if the shower D_{50} alone is considered, but that is not the dose that results in myelopathy. It is the *total* dose.
3. The data for $D_{bath} = 18$ Gy violate constraint 1 since D_{50} for 2 mm field size $< D_{50}$ for 4 mm.
4. The 8-mm data violate constraint 3 since the D_{50} for $D_{bath} = 0$ is lower than the D_{50} for the 4 Gy bath, as if the bath had a protective effect. The difference is small and will be assumed to be due to stochastic variation due to small sample sizes.

Ultimately the authors "hypothesize[d] that: (1) the adjacent low dose (bath dose) has a negative impact on the regenerative capacity in the high-dose field (shower), and (2) the bath dose induces indirect effects that prevent the regeneration of tissue, which leads to the development of white-matter necrosis at lower doses" (Bijl et al. 2006). Further, "the field length effects could *partly* be related to a limited migration of functional cells or precursors from the edges of small-irradiated field lengths" (Emphasis added) (Bijl et al. 2002). In 2003, they acknowledge that, "based on the classical pathogenesis theory of white matter necrosis with vascular or parenchymal cells as targets, the present large decrease of the ED_{50} values cannot be explained," nonetheless they suggest that "Migration of progenitor cells to depleted segments could be the explanation for the increase of the ED_{50} values when exposing segments shorter than 8 mm" (Bijl et al. 2003). Also they hypothesize the cytokine action may be responsible for an "out-of-field effect."

In 2005, the authors conclude that white matter necrosis is not a result solely of vascular endothelial damage and again suggest that "migration of progenitor cells is possibly involved in the restoration of radiation-induced white matter damage" (Bijl et al. 2005). The final conclusion is "contribution from migrated progenitor cells to the repair of radiation-induced damage in the high-dose segment may only partly explain the bath effect" and also that "interference with stem cell migration is *not* the most likely mechanism of a bath effect." (Emphasis added.) This leaves only cytokines and/or growth factors as the agents (mentioned thus far) that could be responsible in some way for the dramatic reduction in D_{50} with very small field lengths.

The clinical implication of these data is that for very small hotspots surrounded by a low-dose region, the traditional understanding of the sparing effect of small volume irradiation may not hold true. Whether or not this effect applies in other tissues, CNS applications that come immediately to mind are small SRS volumes that are surrounded by low doses and the treatment of oligometastases in the brain with multiple arcs that produce low-dose exit regions. But do the data convincingly demonstrate that there *is* a bath-and-shower effect at all?

It is instructive to take data for a given field size in pairs. Again, logic requires that the D_{50} (in total dose) for a treatment cannot increase when a bath dose is applied or is increased. If there is a bath-and-shower effect, that D_{50} would decrease. Taking all pairs of points for which the (shower) field size is the same, but the bath dose is different, we find that in 9 of the 17 pairs, the D_{50} does decrease, but in 7 of the 17 pairs it increases. See Table 6.1, where all arrows start and end in cells that have the same field size. If an effect only shows up 59% of the time, its existence is questionable. This serious problem may have arisen in part due to the stochastic variations in the data that results from so few animals in the dose groups.

Pushing the interpretation of the bath-and-shower effect as far as possible, one would still have to conclude that there is no evidence for its existence at the 8-mm field size since the bath dose would have to have a protective effect. For the 4-mm field size, a bath dose adds toxicity for the 4-, 12-, and 18-Gy bath doses. However, the opposite is true for the 18-Gy asymmetric bath, both cephalad and caudad positions. Furthermore, the added toxicity of the bath doses occurs in the wrong order, that is, the 4-Gy bath is the most toxic and the 18-Gy is the least toxic. Of the ten pairs of bath doses for the 4-mm field size, five are the wrong way round for an out-of-field effect. Thus, only the 2-mm data are not in conflict with the bath-and-shower effect. However, even though the authors have referred to that shower dose as homogeneous irradiation, it can be seen from Fig. 6.1 that the 2-mm field is decidedly nonuniform. Adding a uniform dose of 18 Gy across the profile of the 2-mm beam delivering 30.9 Gy results in a beam with a full width at half maximum of about 3.4 mm. This effect nearly eliminates the bath-and-shower effect for the 2-mm beam.

We should be clear regarding what is meant by the statement that the bath-and-shower effect does not exist within the context of the data discussed thus far. It means that the irradiation of areas adjacent to very small volumes of spinal cord has no discernible effect on the dose response within the smaller field. By "discernible effect" we mean one that can be consistently detected with statistical significance.

We leave the bath-and-shower experiments with the conclusion that there is no effect shown by the 8-mm data. For the 4-mm field size, half the data are in conflict with the effect and half are in agreement, so there is no consistent process behind the 4-mm data that is discernible. What role stochastic variation may play in experiments with such sparse data is uncertain. Finally for the 2-mm shower, the majority of the effect is consistent with the increase in effective field size when the bath dose is added to a shower dose with a significantly nonuniform profile.

We now consider the mathematical model that is described by the solid line in Figs. 6.3 and 6.5. The critical element model was derived by assuming that the probability of not creating a lesion in an irradiated volume was the product of not creating a lesion in any subvolume (Schultheiss et al. 1983). (This is simply an application of the law of total probability.) However, this assumption is dependent on a lesion being vanishingly small compared to the size of the subvolume. To accommodate the finite size of the lesion, we must determine the probability of creating a lesion (the critical element model) and that the lesion is a certain size. It is often seen in histology samples of irradiated spinal cords that a lesion exists that was too small to elicit a clinical response. Thus from Bayes' theorem, the probability of paralysis in the rat is

$$P(\text{creating a lesion AND the lesion is sufficiently large}) = P(\text{creating a lesion}) \times$$
$$P(\text{lesion is sufficiently large} \mid \text{a lesion was created}).$$

The probability of creating a lesion as described by the critical element model is

$$P(D,v) = 1 - \left[1 - P(D,1)\right]^{v} \tag{6.3}$$

where $P(D,v)$ is the dose-response function for the organ when the relative volume v is homogeneously irradiated to dose D. When the organ is inhomogeneously irradiated, Eq. (6.3) becomes

$$P(D,v) = 1 - \prod_{i}\left(1 - P(D_i,1)\right)^{\Delta v_i} \tag{6.4}$$

where Δv_i is the volume element irradiated to dose D_i. The ordered pairs $(\Delta v_i, D_i)$ comprise the differential dose-volume histogram.

The probability that the lesion is sufficiently large is

$$P(m_c \geq m) = \sum_{i=0}^{m_c} B\left(m_c - i, n\frac{\text{fs}}{20}, p_{s+b}\right) dB\left(i, n\frac{20 - \text{fs}}{20}, p_b\right) \tag{6.5}$$

where m is the minimum size of the lesion as measured in unknown tissue injury units, n is the number of those units irradiated to dose d, p is the probability of sterilizing a single unit, fs is the field length, $B(\cdot)$ is the cumulative probability distribution for the binomial distribution, and $dB(\cdot)$ is the differential distribution or simply the combinatorial factor. p may be given by many forms, but we have chosen

$$p = \exp\left(-\alpha d - \beta d^2\right) \tag{6.6}$$

where p_{s+b} and p_b in Eq. (6.5) correspond to the sterilization probability for $d_s + d_b$ and d_b, the shower and bath doses respectively. Thus Eq. (6.5) gives the probability that i TIU's are sterilized in the bath region and $m_c - i$ or more are sterilized in the shower region. The probability of disabling m out of n units is essentially the critical volume model, slightly modified for the special case of the bath-and-shower field.

In Fig. 6.6a, we see this model fitted to the 0-bath data. The model parameters are given in Table 6.2; the fit to the data is good with $p(\chi^2) \sim 0.6$ when two outliers for the 4-mm data are excluded. In Fig. 6.5, it can be seen that the 8- and 4-mm data for non-zero bath doses are distributed about the line, but all the 2-mm bath doses fall below the line, giving some evidence for the existence of the bath-and-shower effect for field sizes of 2 mm. For completeness, Figure 6.6b shows the Hopewell et al. data fitted to the same model (Hopewell et al. 1987).

In 2013, Laissue published the results of x-ray microbeam radiation on the Fischer rat spinal cord (Laissue et al. 2013). Two experiments were performed—one involved a single open beam of 1.35 mm length and the other deployed 52 beams of approximately 35 µm width and 210 µm center-to-center separation. The single beam D_{50} was reported as 130.1 Gy. Dose groups were 57, 114, 146, 182, 227, and 454 Gy. No rat developed paralysis below 146 Gy and all rats at that dose and higher developed paralysis. The rats in the two lowest dose groups were killed

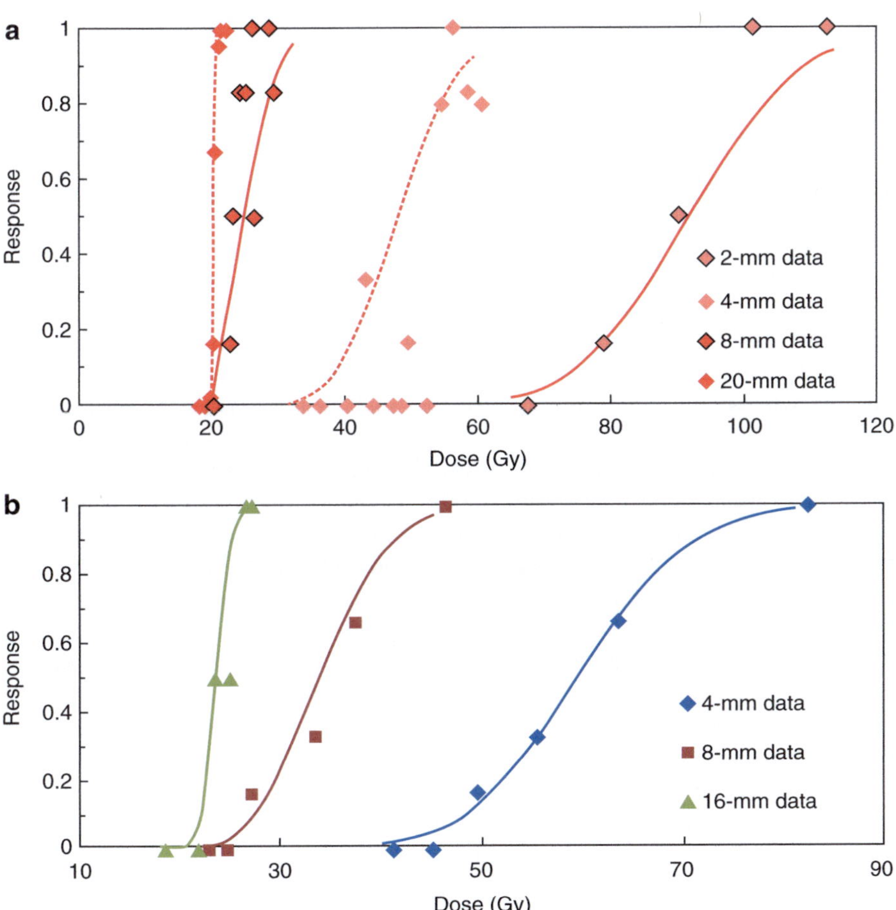

Fig. 6.6 The dose-response model defined by Eqs. (6.3)–(6.6). (**a**) is the model applied to data from Bijl et al. (2002) and (**b**) is the model applied to Hopewell et al. (1987). Interestingly, the Philippens et al. (2009) data follow the Bijl et al. 4-mm dose response very closely.

Table 6.2 Parameter values for small field data for the model defined by Eqs. (6.3)–(6.6)

	Bijl et al. (2002)	Hopewell et al. (1987)
D_{50}	20.3	23.6
k	118	34.6
m_c	33.6	25.5
N	2186	2309
α	1.52×10^{-3}	6.79×10^{-4}
β	3.68×10^{-6}	5.07×10^{-6}

Table 6.3 Philippens et al. bath-and-shower results with shower given at various times following the bath dose. The D_{50} values include the 4-Gy bath dose

Schedule	D_{50} (Gy)
8 min	40.8
3 h	44.4
12 h	44.8
24 h	51.9
No bath	48.7

at day 230. Latencies for paresis were 2 to 15 days in the animals irradiated to 146 Gy or higher. Since the dose-response data were completely separated, no MLE of this parameter would have been possible. Four animals per dose group were used.

Figure 6.3 shows the Bijl et al. zero-bath data, the van der Kogel D_{50} values, the Hopewell et al. data, and the D_{50} from the Laissue et al. The data points appear to be in good agreement with the model proposed above in Eqs. (6.4)–(6.6). As shown in Fig. 6.5, when plotted on a log-log scale, the small field size D_{50}'s fall on a straight line. In all cases, the D_{50} values are those reported by the authors. This appears to be compelling evidence in favor of the critical volume model for small field sizes. As noted in Chap. 9, the critical volume model is a nonconsecutive k-out-of-N model use in reliability theory. Furthermore as noted above, the Bijl et al. data are well fitted by the model, but the other studies are represented in the graph only by their D_{50}'s, which is not necessarily an indication that the response data fit the model.

To continue the examination of the result of the Philippens et al. paper from 2009, we examine the result as shown in Table 6.3. This gives the D_{50} for bath plus shower doses at variation times. The bath dose of 4 Gy was given at $t = 0$ (and the shower dose followed at the times shown with no more than six animal per dose point. A consistent and reasonable feature of these data is that as the time between the bath and the shower doses increases, the effect of the bath dose diminishes. A somewhat inconsistent feature of the data is that 24 h after a bath dose of 4 Gy the D_{50} is indistinguishable from the D_{50} for zero bath dose, but the original and the re-analysis both indicate some remaining effect from the bath dose, but not more than would be expected from any 2-fraction regimen. Also, the bath effect is the same at 4 and 12 h. One can easily ascribe this behavior to variability and poor resolution resulting from too few animals per dose point. Although the authors conclude that the LQ model with incomplete repair does not fit the data, this is not strong evidence against the LQ model with IR. However, this experiment is like none other in the literature. There are a number of examples of split dose experiments, but for those, the split is generally greater than 24 h. Ruifrok et al. (1993) and Ang et al. (1984) explored the repair of sublethal damage kinetics using 2-fraction dose-response experiments. Ruifrok et al. used young animals (6 weeks old) and both experiments suffered from sparse data, but Ruifrok et al. had only a few separated dose-response curves. In both cases, the D_{50} increased over the 24 h period. However, the Philippens' 09 paper can be understood as an experiment using a small priming dose since the bath dose was very small relative to the D_{50} values. This is an area where more data might yield surprising results.

The authors of this paper also state that a modified LQ model with incomplete repair gives an adequate fit to the data as measured by the variance. As so many other authors do, they assume a value for the α/β ratio, but then they allow β to vary. However, re-analysis shows that including β as a covariate results in its ML estimate being negative, although with not much improvement in the fit. Nonetheless, one cannot claim that the LQ model fits if the a/b ratio is negative. Furthermore, the reason their model fitted the data was that they included too many extreme values in the degrees of freedom. Adjusting the degrees of freedom for the excessive number of extreme responses results in a poor fit.

Both clinical and histological endpoints were reported for this paper. Only the histological endpoint was used in the dose-response analysis reported above. However, a good fit to the clinical data was seen on re-analysis of the clinical data using the critical volume model of Eqs. (6.5) and (6.6). Thus in these data the bath effect is explained by the critical volume model. See Fig. 6.7. Data from times of 3 and 12 h were combined, and times after bath dose were modeled using a single fraction to represent unrepaired dose. $p\ (\chi^2) \sim 0.7$

In 2004, the same group of authors from Philippens et al. (2009) published a volume effect study on the TL spinal cord using Wistar rats irradiated with a 4 MV beam with field lengths of 5 to 40 mm centered at T12-L2 (Philippens et al. 2004). Both clinical and histological endpoints were recorded. They also encountered difficulty with the set-up on the linac that resulted in the 5-mm and 15-mm dose groups treating higher spinal cord levels than intended. Conversely in all but four rats in the 10-mm and 40-mm region, the lesions were in the nerve roots.

Like so many papers written after 2000, this work is seriously underpowered, using only six animals per dose point. It is difficult to criticize any particular author or group for this transgression since it seems to be endemic. However, it is ironic to see the detailed and complex mathematics that go into modeling and analysis efforts in the face of ignoring such a basic concept as allocating sufficient subjects per group to justify and accredit the statistical analysis.

It is of considerable concern that different treatment groups displayed different endpoint histologically in this study. It is well documented that there is a spectrum of lesions that cause radiation myelopathy, and this is the reason for using a clinical endpoint. However, *nerve root degeneration is not radiation myelopathy*. These

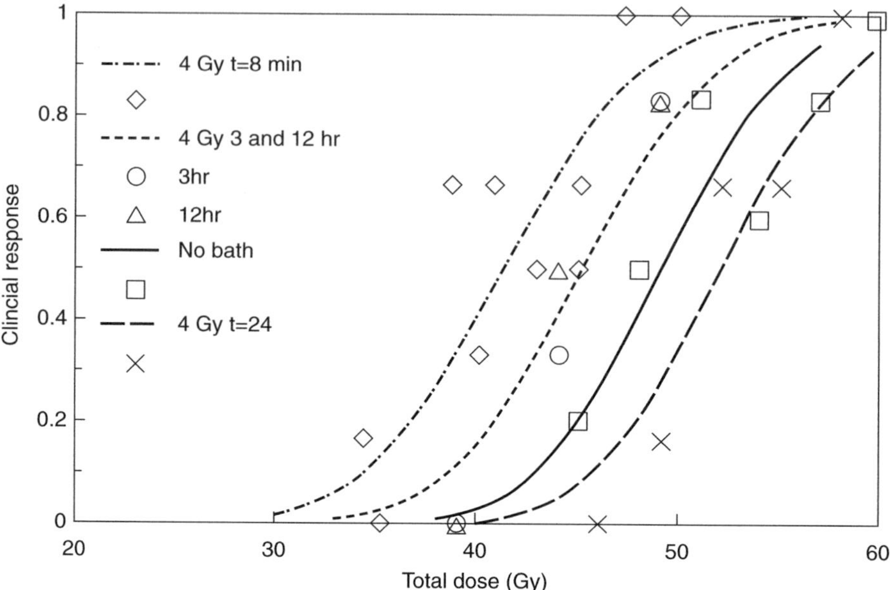

Fig. 6.7 Clinical response in Philippens et al. (2009) bath-and-shower experiment. Bath dose given at $t = 0$ and shower doses at various times after. Data are well described by the critical volume model.

authors go into some detail in describing the different clinical manifestation of this radiculopathy as compared to myelopathy. Combining the response of different tissues (nerve roots versus spinal cord parenchyma) as a single outcome cannot be justified when biostatistical models apply only to one.

In both Philippens et al. (2004, 2009), the model assessment method published by van Luijk et al. supplemented the more conventional variance method (Van Luijk et al. 2003). In this method, parameters estimates are obtained by MLE and a likelihood ratio is calculated. Call it Δ_0. The likelihood ratio is asymptotically distributed as χ^2 with the degrees of freedom equal to the number of covariate patterns minus the number of estimated parameters. Rather than simply using this to determine a p-value for the fit, van Luijk et al. suggest an additional procedure be performed. Using the model with the estimated parameters values, 1000 outcomes are generated by using random generation of binomial data for each covariate pattern based on the probability estimate for that covariate pattern and the number of subjects for that point. For each of the 1000 random experimental histories, a likelihood ratio is calculated. The p-value is taken to be the fraction of the likelihood ratios with higher values than Δ_0.

This intriguing method generates the distribution of (squared) residuals for the specific experimental design (covariate patterns and number of subjects) for a given generating function, i.e., the fitted dose-response function. It is beyond the scope of this volume to critique this methodology. However, if this method is valid, it is brilliant in its simplicity, but somewhat difficult to implement. It obviates this issue of sparse data by generating a distribution designed specifically for those data rather than using a distribution that is only asymptotically correct. A specific criticism may be that 1000 histories is generally too few for a Monte Carlo method such as this. Some preliminary Monte Carlo modeling done with 10,000 histories produced promising results (not reported for this work). This method must be tested against others designed to handle sparse data. Finally with well-designed experiments, the conventional methods should work satisfactorily and without resorting to MC methods.

In 2005, van Luijk published dose-response analyses for the 2002 and 2003 Bijl studies, testing 14 models for their fit to the data. The conclusion of the paper was that no extant model of dose-volume effects fit the rat spinal cord volume effects data, and it seemed that the authors implied that this is a general conclusion, not limited specifically to the Bijl data. The authors attribute the models' failure to fit the data to non-local effects involved in damage repair. Results in the preceding paragraphs show that a model that extends the critical element model with the critical volume component fits the no-bath data, and that much of the bath-and-shower data are not internally consistent. However, it should be noted that van Luijk did not assume a uniform dose over the irradiated volume as was done above.

Looking at the volume effect in a different way, Bijl et al. examined the differences in radiosensitivity in the rat spinal cord across its cross section (Bijl et al. 2005). Again using proton beams, they irradiated the central versus the lateral regions of the spinal cord and used two penumbrae for irradiating the lateral portions. The 80–20% dose falloff was 1.1 mm for the wider penumbra and 0.8 mm for the tighter penumbra, as measured using a dosimetry system with 0.22 mm resolution. The beam configuration that achieved these beams was as follows. For the wide penumbra, center of the spinal cord was carefully aligned with the edge of the field defining collimator, thus half the spinal cord (lengthwise) was within the geometric beam projection. The collimator was 15 cm from the spinal cord. For the narrow penumbra, an outboard collimator was used that was positioned down-beam from the field defining collimator something like trimmers for cobalt machines. This edge defining collimator was on the central axis of the beam at a distance of 7 cm from the spinal cord. The investigators also used an open beam 2-mm wide centered on the spinal cord. Since the rat spinal cord is about 3.5 mm in width, the lateral funiculi and lateroposterior funiculi received less than 50% of the dose.

The investigators observed a higher D_{50} for the narrow- versus (33.4 Gy) the wide-penumbra beam (28.9 Gy). No more than 6 and sometimes as few as three animals were used per dose group. The dose-response data for the narrow -penumbra experiment were quasi-separated, making a MLE of D_{50} impossible. The authors still concluded that there was a significant difference between the two groups. To assess this assertion, the data were re-analyzed together using probit analysis as they did, but including a DMF for the wider beam. It was necessary to multiply the wide-beam data by a DMF of 1.15, which was significantly different from 1 and a χ^2 statistic for the model indicating a good fit ($p \sim 0.2$). Thus the conclusions regarding these data were substantiated. The wide beam did have greater toxicity. Moreover, upon careful consideration of their Fig. 2, one can tell that the narrower beam contains a larger volume of white matter at the higher doses. This is easily understood by considering a beam whose left-to-right profile is a step function at the center of the cord. As the gradient becomes more gradual, the volume of cord above 50% decreases and the volume below 50% increases.

The treatment with a central beam resulted in a much larger increase in the D_{50} (compared to a 20-mm uniform field) than the lateral side irradiation. For the central beam, the D_{50} was 71.9 Gy, so a difference between this and the grazing beams was not in question. The authors specifically state that the difference could not be a result of a volume effect because of the shape of the DVH seen in their Fig. 6. However, a DVH can be difficult to interpret, and in this case, the important differences occur at the higher doses, where the DVH's blend confusingly together. In fact, the typical cumulative DVH as shown here does not give the volume that was treated to a specific dose; rather it gives the volume that was treated to at least that dose. To get a useful idea of how much volume receives a given dose, one must use a differential dose-volume histogram. However, it would appear that the high dose volume for the central beam is not much larger than for the grazing beams.

It is clear from their Fig. 2 that the lateral funiculi are spared by the grazing beams. All else being equal, this implies that these funiculi are more sensitive to radiation than the posterior or anterior columns. van der Kogel has found that spinal cord radiation lesions appear preferentially in the lateral tracts. Of course this could simply reflect the greater volume of lateral as opposed to dorsal tracts. These experiments, which required extreme care and precision (van Luijk et al. 2001), support the idea that there is an intrinsic difference in the radiosensitivity of the lateral versus the dorsal tracts in the rat spinal cord.

The authors conclude that "the regional difference of radiosensitivity within the white matter is not only the result of vascular damage because there is no regional difference in vascular density or blood flow within the white matter." They discuss Nieder et al. experiment with IGF-1 and bFGF, discussed earlier in Chap. 5 (Nieder et al. 2002). However, they do not relate it to any pathogenesis that could yield a regional difference in sensitivity. Then they once again evoke OPC migration as a putative mechanism that could cause a differential effect. They argue that migration of OPC's into the dorsal tracts is "not hampered by anatomical obstacles," but migration of OPC's from contralateral tracts "is partially hampered by the anterior median fissure and posterior median septum." Although the authors focus solely on OPC migration, presumably their arguments could apply equally well to endothelial cell migration. The idea that OPC migration influences radiation response in the spinal cord was largely abandoned by this group eventually.

A 2007 by Philippens et al. paper had a similar title to the 2003 paper by Bijl et al., but they are quite different studies (Philippens et al. 2007). Both refer to inhomogeneous irradiation in their titles, but in the Bijl case this refers to the bath-and-shower technique whereas in the Philippens case it refers to irradiation with Ir-192 HDR treatment. However, the latter paper is not a brachytherapy study. Treatment was delivered with 1, 2, or 6 catheters, with the first two configurations used to create inhomogeneous dose distribution and the last to create a uniform distribution.

The choice of dose delivery system here seems unusual, especially since brachytherapy treatment was not a subject of investigation. The choice to use HDR to create a uniform dose is especially curious given the effort required for each individual treatment. No dose distributions were presented. It would have been useful to show the uniform distribution created by the 6-catheter technique so that an understanding could be gained regarding how the reference dose related to the overall dose distribution to the spinal cord.

Many dose-volume response models were fitted to the data and a number were found to provide satisfactory representations of the outcomes. This serves to prove that when one of the more standard radiation dose-response models fits the data, several others will also. An important finding from this study was "a higher radiosensitivity was observed for the lumbar nerve roots than for the thoracic WM."

References

Ang KK, van der Kogel AJ, Van Dam J, van der Schueren E. The kinetics of repair of sublethal damage in the rat cervical spinal cord during fractionated irradiations. Radiother Oncol. 1984;1(3):247–53. https://doi.org/10.1016/S0167-8140(84)80007-9.

Bijl HP, Van Luijk P, Coppes RP, Schippers JM, Konings AWT, van der Kogel AJ. Dose-volume effects in the rat cervical spinal cord after proton irradiation. Int J Radiat Oncol Biol Phys. 2002;52(1):205–11. https://doi.org/10.1016/S0360-3016(01)02687-6.

Bijl HP, Van Luijk P, Coppes RP, Schippers JM, Konings AWT, van der Kogel AJ. Unexpected changes of rat cervical spinal cord tolerance caused by inhomogeneous dose distributions. Int J Radiat Oncol Biol Phys. 2003;57(1):274–81. https://doi.org/10.1016/S0360-3016(03)00529-7.

Bijl HP, Van Luijk P, Coppes RP, Schippers JM, Konings AWT, van der Kogel AJ. Regional differences in radiosensitivity across the rat cervical spinal cord. Int J Radiat Oncol Biol Phys. 2005;61(2):543–51. https://doi.org/10.1016/j.ijrobp.2004.10.018.

Bijl HP, Van Luijk P, Coppes RP, Schippers JM, Konings AWT, van der Kogel AJ. Influence of adjacent low-dose fields on tolerance to high doses of protons in rat cervical spinal cord. Int J Radiat Oncol Biol Phys. 2006;64(4):1204–10. https://doi.org/10.1016/j.ijrobp.2005.06.046.

Cox DR. A simple example of a comparison involving quantal data. Biometrika. 1966;53:215–20.

Hinks GL, Chari DM, O'Leary MT, Zhao C, Keirstead HS, Blakemore WF, Franklin RJM. Depletion of endogenous oligodendrocyte progenitors rather than increased availability of survival factors is a likely explanation for enhanced survival of transplanted oligodendrocyte progenitors in X-irradiated compared to normal CNS. Neuropathol Appl Neurobiol. 2001;27(1):59–67. https://doi.org/10.1046/j.0305-1846.2001.00303.x.

Hopewell JW, Morris AD, Dixon-Brown A. The influence of field size on the late tolerance of the rat spinal cord to single doses of X rays. Br J Radiol. 1987;60(719):1099–108. https://doi.org/10.1259/0007-1285-60-719-1099.

Jackson A, Kutcher GJ. Probability of radiation-induced complications for normal tissues with parallel architecture subject to non-uniform irradiation. Med Phys. 1993;20(3):613–25. https://doi.org/10.1118/1.597056.

Laissue JA, Bartzsch S, Blattmann H, Brauer-Krisch E, Bravin A, Dallery D, Djonov V, Hanson AL, Hopewell JW, Kaser-Hotz B, Keyrilainen J, Laissue PP, Miura M, Serduc R, Siegbahn AE, Slatkin DN. Response of the rat spinal cord to X-ray microbeams. Radiother Oncol. 2013;106(1):106–11. https://doi.org/10.1016/j.radonc.2012.12.007.

Nieder C, Price RE, Rivera B, Andratschke N, Ang KK. Experimental data for insulin-like growth factor-1 (IGF-1) and basic fibroblast growth factor (bFGF) in prevention of radiation myelopathy. Strahlenther Onkol. 2002;178(3):147–52. https://doi.org/10.1007/s00066-002-0897-8.

Philippens MEP, Pop LAM, Visser AG, Schellekens SAMW, van der Kogel AJ. Dose-volume effects in rat thoracolumbar spinal cord: an evaluation of NTCP models. Int J Radiat Oncol Biol Phys. 2004;60(2):578–90. https://doi.org/10.1016/j.ijrobp.2004.05.029.

Philippens M, Pop LAM, Visser AG, van der Kogel AJ. Dose-volume effects in rat thoracolumbar spinal cord: the effects of nonuniform dose distribution. Int J Radiat Oncol Biol Phys. 2007;69(1):204–13. https://doi.org/10.1016/j.ijrobp.2007.05.027.

Philippens MEP, Pop LAM, Visser AG, Peeters WJM, van der Kogel AJ. Bath and shower effect in spinal cord: the effect of time interval. Int J Radiat Oncol Biol Phys. 2009;73(2):514–22. https://doi.org/10.1016/j.ijrobp.2008.09.028.

Ruifrok ACC, Kleiboer BJ, van der Kogel AJ. Repair kinetics of radiation damage in the developing rat cervical spinal cord. Int J Radiat Biol. 1993;63(4):501–8. https://doi.org/10.1080/09553009314550661.

Schultheiss TE, Orton CG, Peck RA. Models in radiotherapy: volume effects. Med Phys. 1983;10(4):410–5. https://doi.org/10.1118/1.595312.

van der Kogel AJ Effect of volume and localisation on rat spinal cord tolerance. In: Fielden EM, Fowler JF, Hendry JH, Scott D (eds) Proceedings of the 8th international congress of radiation research, London, 1987. Taylor and Francis, London 352.

van der Kogel AJ. Central nervous system radiation injury in small animal models. In: Gutin PH, Leibel SA, Sheline GE, editors. Radiation injury to the nervous system. New York: Raven Press; 1991.

van der Kogel AJ. Dose-volume effects in the spinal cord. Radiother Oncol. 1993;29(2):105–9. https://doi.org/10.1016/0167-8140(93)90234-Y.

van Luijk P, Bijl HP, Coppes RP, van der Kogel AJ, Konings AWT, Pikkemaat JA, Schippers JM. Techniques for precision irradiation of the lateral half of the rat cervical spinal cord using 150 MeV protons. Phys Med Biol. 2001;46(11):2857–71. https://doi.org/10.1088/0031-9155/46/11/307.

van Luijk P, Delvigne TC, Schilstra C, Schippers JM. Estimation of parameters of dose–volume models and their confidence limits. Phys Med Biol. 2003;48(13):1863.

van Luijk P, Bijl HP, Konings AWT, van der Kogel AJ, Schippers JM. Data on dose-volume effects in the rat spinal cord do not support existing NTCP models. Int J Radiat Oncol Biol Phys. 2005;61(3):892–900. https://doi.org/10.1016/j.ijrobp.2004.10.035.

7.1   Retreatment

7.1.1   The Dose-Repaired/Recovered Model

Scientists investigating retreatment effects have generally sought a method to describe how much additional dose can be delivered at a time, t, after the initial dose has been delivered, where t is considered long compared to a course of treatment. Of course this depends on how much of the damage from the initial course is repaired by time t. Thus the dose that can be delivered and the damage that is repaired by time t become conflated. A few have argued, as it is here, that dose is not a good surrogate for radiation damage. (This issue is addressed in more detail in Chap. 9.) If it were possible to capture fully the complete dose regimen in a single biological dose metameter, then there would be a one-to-one correspondence between this biological dose and damage. Failing that, it is tempting and potentially edifying to express the amount or fraction of damage repaired between treatments in terms of dose. It is nonetheless an incomplete description of damage.

Thus, scientists investigating retreatments will gauge the repair at time t by how much dose must be given in a second treatment to achieve an outcome uniquely relatable to a specific dose given in a single treatment course. As described in Chap. 5, for a dose metameter to be additive, it must be linear in N, the number of fractions. In addition to the TDF, this is satisfied by the ERD given by

$$\mathrm{ERD} = D\left(1 + d / \alpha / \beta\right). \tag{7.1}$$

(This additivity is also true for D_d, the isoeffect dose given in constant fraction size d, e.g., $D_{2\,\mathrm{Gy}}$.) Thus for their work in retreatment, Wong et al. (Wong and Hao 1997) define the fraction of initial radiation damage recovered as

$$f = \left(\mathrm{Retreatment\ ERD\ at\ time}\ t - \mathrm{retreatment\ ERD\ at\ day\ 4}\right)] / \mathrm{initial\ ERD}, \tag{7.2}$$

T. Schultheiss, *Radiation Myelopathy*,
https://doi.org/10.1007/978-3-030-94658-6_7

where both ERD's in the numerator refer to the ERD_{50}, the ERD for a 50% response rate. Note that in the numerator, it is not important whether the ERD's contain the ERD for the initial dose or represent only the retreatment ERT since the metameter is linear and the initial dose ERD's would cancel each other. So in this consideration, the repair or recovery is gauged relative to the initial dose, but the amount of recovery is determined by the change in ERD_{50}.

Ruifrok et al. (1992a, c) describe the concept of "extra effect dose in % of TE [total effect]." The definition of TE is $TE = D\ (\alpha/\beta + d)$, which is simply the ERD multiplied by the α/β ratio. Although their explanation of TE and the fraction of dose recovered is somewhat less direct, it ultimately is the mathematical equivalent. The TE concept is from Thames and Hendry (Thames and Hendry 1987), and it is ultimately derived from Barendsen (Barendsen 1982).

The typical retreatment experiment that will be discussed below is characterized by an initial single or fractionated dose followed weeks or months later by a dose-response experiment using single or fractionated doses. By including the ERD surrogate for fraction of damage repair in the systemic component, z, of the GLM (Eq. 3.8), the fraction of damage repaired (as gauged by its surrogate ERD) can be directly estimated for a given initial dose and retreatment time using logistic regression. Thus,

$$z = \beta_o - \alpha \cdot \left(D_{\text{init}} + D_{\text{rerx}} \right) - \beta \cdot \left(D_{\text{init}}^2 / N_{\text{init}} + D_{\text{rerx}}^2 / N_{\text{rerx}} \right)$$
$$+ f \cdot \alpha \cdot D_{\text{init}} + f \cdot \beta \cdot D_{\text{init}}^2 / N_{\text{init}} \tag{7.3}$$

where, using the formalism of Eq. (3.8), $\beta_1 = \alpha$ and $\beta_2 = \beta$. D_{init} and D_{rerx} are the total initial and retreatment doses given in N_{init} and N_{rerx} fractions, respectively, and f is the recovered or repaired fraction of the initial ERD given in Eq. (7.2). Note f is generally considered to be dose and time dependent. This dependence of course requires its own functional specification resulting in a nonlinear logistic regression.

In 1980, White and Hornsey investigated slow repair in a split-dose experiment where two equal doses were given 1, 15, 60, and 120 days apart with dose-response curves being obtained at each time interval (White and Hornsey 1980). This is not a typical retreatment experiment where the initial dose is constant and a dose-response curve is elicited after a number of weeks. Therefore, the repair fraction does not have the same interpretation since the initial dose varies along with the retreatment dose. The data all consisted of two-fraction experiments, with eight to ten animals irradiated with equal doses separated by the above times. The original analysis was based entirely on D_{50} values, a technique criticized in this volume for not efficiently utilizing all the data. Furthermore the D_{50}'s were estimated by eye. The re-analysis here is nearly identical to the retreatment analysis to follow except that the logistic link function was used rather than the log–log because the latter failed to fit the data. It should be noted that these analytical methods were not well known in the radiation community at the time of publication.

The results of the re-analysis are given in Table 7.1 and Fig. 7.1. The figure shows (1) no apparent recovery delay and (2) recovery levels that are higher than in

Table 7.1 Model parameter for Eqs. (7.3) and (7.4) for the re-analysis of the data from White and Hornsey (1978). Results are depicted in Fig. 7.1

β_0	14.9
α	1.45×10^{-4} Gy^{-1}
β	0.0276 Gy^{-2}
α/β	0.00524 Gy
t_0	138.8 days
$p\ (\chi^2)$	0.44
D_{50} 1 day	32.81 Gy
D_{50} 15 days	33.9 Gy
D_{50} 60 days	36.4 Gy
D_{50} 120 days	38.8 Gy

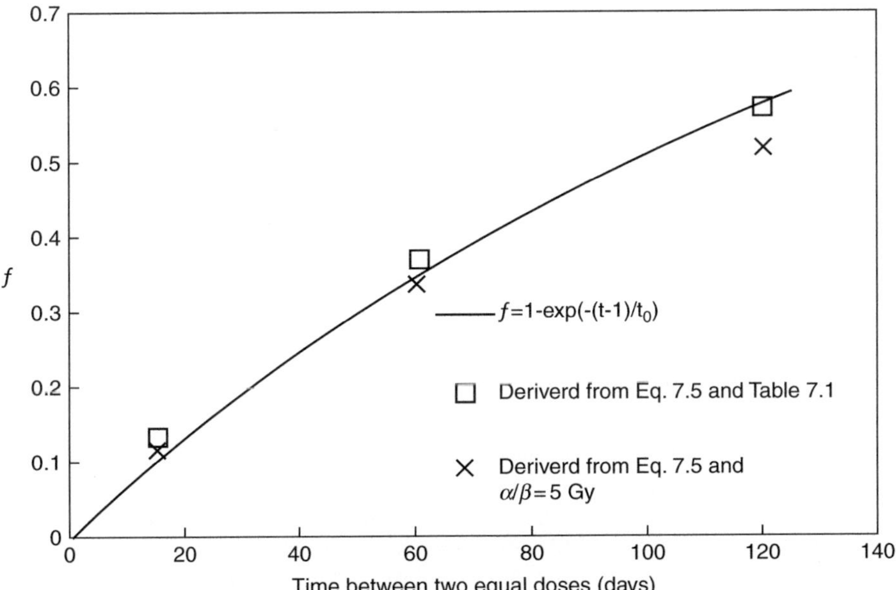

Fig. 7.1 Repair fraction derived from split-dose experiment consisting of dose-response curves from two equal fractions (White and Hornsey 1978). The data points are derived from D_{50}'s from re-analysis and Eq. (7.5). The line reflects the value of t_0 from the logistic regression.

the rodent experiments of Wong et al. and Ruifrok et al. described below. However, the shape of the recovery curve indicates that it follows mono-exponential saturation dynamics. In the dose-response model, it was assumed that

$$f = 1 - \exp\left[-(t-1)/t_0\right],\tag{7.4}$$

without any dependence on initial dose.

An alternative method for determining the values of f at the different time points uses the values of D_{50} calculated for the dose-response function for those times. Of course the D_{50}'s must be converted into ERD_{50}'s because this is the metameter that is a surrogate for the radiation damage that is being repaired. We follow Wong et al. for the calculation of percent of initial radiation damage recovered, and for D_{init} we use $\frac{1}{2}D_{50}$ since that is the initial dose of two equal fractions. From this it is easy to show that

$$f = 2 \cdot \left[1 - ERD_{50}\left(t = 1 \right) / ERD_{50}\left(t \right) \right] \tag{7.5}$$

where t is the time that the second of two equal fractions is delivered. We take the convention that the first radiation dose on day $t = 1$. Note that neither this method, nor the modeling of the dose response with f included in the systematic component for the GLM, z (Eq. 7.3), specifically account for the potential dependence of f on the initial dose. Figure 7.1 shows the values of f using Eq. (7.4) and t_0 from Table 7.1. As can be seen from Table 7.1, the α/β ratio is essentially zero for these data, although this design is not efficient for assessing α/β. Also shown in Fig. 7.1 are the results from Eq. (7.5) if $\alpha/\beta = 5$ Gy is assumed. One can see that in this formulation, f is very insensitive to the value of α/β, which is what would be expected.

In the same study, White and Hornsey performed a similar experiment in which lumbar cord irradiation was used. However, the dose-response curves were not shown, only the results of their analysis. In this case, there seemed to be delay of recovery and an overall similarity with the cervical spine results. As a reminder, because the radiation was applied to the lumbar region, it is likely that there were no spinal cord lesions, only nerve root damage. The lumbar experiment was terminated early due to comorbid disease.

A 1993 paper by Wong et al. addressed the influence of initial dose on long-term repair (Wong et al. 1993b). A series of fractionated treatments of 2.15 Gy per fraction was given, followed by a 20-week break and then retreatment with graded single doses. The initial treatments were 0, 10, 20, 30, and 36 fractions of 2.15 Gy. These doses were believed to represent 0, 25, 50, 75, 90% of the D_{50} based on a value of $D_{50} = 86.1$ Gy for 40 fractions obtained from Wong et al. (1992). This D_{50} value was determined by interpolation using the 0/9 and 8/9 responses at 83.6 and 88.2 Gy, respectively, from the 1992 paper. One emphasis of the 1993 paper was comparing the amount of damage repaired to the damage resulting from D_{50} using ERD as a surrogate for damage. Unfortunately, the dose-response model from the 1992 publication did not fit the data (described in Chap. 5) and the 40-fraction data were quasi-separated. Therefore the D_{50} could not in fact be estimated using logistic regression. However, a simple change to the analysis can solve this problem. Rather than import the D_{50} directly from the 1992 paper, the re-analysis here simply adds the 40-fraction data from the 1992 paper to the data in the 1993 paper and pools all the data for a maximum likelihood analysis. After all, using the original data seems preferable to using a number (D_{50}) extracted from the original data. This is an interesting case where a model may fit the data, but the fitted parameter values make the model unrealistic.

Obviously the initial 2.15-Gy treatments represent increasingly greater subclinical damage due to higher doses. The authors conclude that in addition to a monotonic shift of the dose-response curves depending on the initial dose, the amount of radiation injury recovered was influenced by the initial dose. They did not elaborate on the nature of that influence. They did determine the percent of the ERD_{50} that the retreatment represented as a function of the percent of ERD_{50} represented by the initial treatment. They used this exercise as a preliminary calculation to the percent of the initial damage (as described by the ERD) that was repaired. The results of their calculation was that the amount of damage from the 10, 20, 30, and 36 fractions that was repaired was 26, 41, 43, and 35%, respectively. An α/β value of 3 Gy was assumed in all their calculations.

Unfortunately, their multi-step process of determining the percentage of damage repaired after 20 weeks was slightly flawed. The process itself involved determining the D_{50}'s for all the single dose experiments, both the retreatment and single course treatment. They took the single course treatment as 100% of the ERD, but they also took 40×2.15 Gy as 100%. These two values are inconsistent if $\alpha/\beta = 3$ Gy, as they assumed. Using all of the authors' assumptions except that 40 fractions of 2.15 Gy is 100% of ERD, one can calculate the percentage of damage repaired using

$$\mathrm{ERD}_{50\,\mathrm{rerx}} + (1 - f) \cdot \mathrm{ERD}_{50\,\mathrm{init}} = \mathrm{ERD}_{50\,N=1} \tag{7.6}$$

where $\mathrm{ERD}_{50\,\mathrm{rerx}}$ is the ERD of the single dose D_{50} for retreatment and $\mathrm{ERD}_{50\,\mathrm{init}}$ is the ERD of the initial fractionated dose. From this equation, the amount of damage repaired from the 10, 20, 30, and 36 fractions was 29.6, 43.8, 45.8, and 38.9%, respectively. Thus, it is not necessary to know the D_{50} for 2.15 Gy nor the percentage of the ERD_{50} that each initial fractionated treatment represents. It seems however, that the percentage of damage repair is not a monotonic function of initial dose. See Fig. 7.2.

It turns out that the data of this paper are analysis friendly in that a number of different scenarios (or models) fit the data. Although the authors state that they used

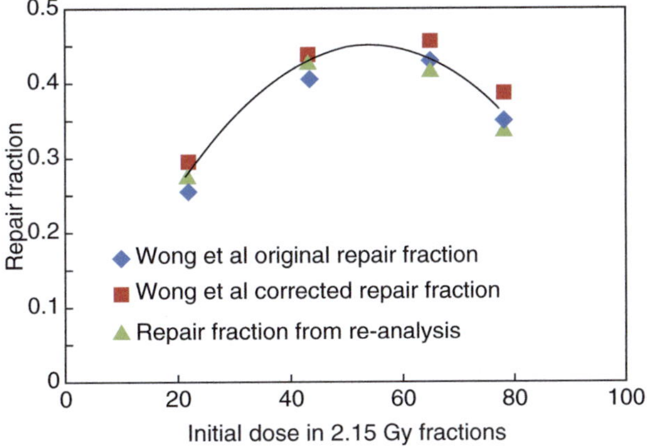

Fig. 7.2 The repair fraction as a function of the initial dose.

logistic regression, implying the use of the logit link function, for the analysis here the log–log link function was used. The systematic component is given by

$$z = \beta_o - \alpha \cdot \left(D_{init} + D_{rerx}\right) - \beta \cdot \left(D_{init}^2 / N_{init} + D_{rerx}^2 / N_{rerx}\right) \tag{7.7}$$

or Eq. (7.3) with Eq. (7.7) representing the case of no long-term repair and Eq. (7.3) the case where a fraction f of the initial ERD is repaired. D_{rerx} represents the single doses used the retreatment dose-response experiments. Using Eq. (7.7), a good fit to the 1993 data is achieved ($P\ (\chi^2) \sim 0.4$). This indicates that the effect of retreatment can be captured in this experiment solely by adjusting the values of α and β, without specifically addressing the intertreatment interval or repair. The maximum likelihood estimate of the α/β value was 0.56 Gy. If a single value of f for all doses is included in the analysis, it is driven to 0. However, the problem with this fit is that it predicts a D_{50} of about 130 Gy for the single course 2.15 regimen, which is of course unrealistically large.

In order to get the proper influence of the single course treatments near 2 Gy per fraction, the data for the 40 fraction experiment (1992) were added to the analysis. Only the three points in Table 7.2 were added since all other data points outside this range resulted in either 0% or 100% response; therefore they added degrees of freedom only but not information to the analysis.

With these data added to the analysis, Eq. (7.7) no longer provides an adequate fit to the model. Equation (7.3) must be used, but it requires only a single value for the repair fraction. Fitting this model to the data yields maximum likelihood estimates of $f = 0.38$ and $\alpha/\beta = 2.53$ Gy [$p\ (\chi^2) = 0.34$]. Assigning a different value of f to each initial ERD was investigated with the result that $f_{N = 10} = 0.28$, $f_{N = 20} = 0.43$, $f_{N = 30} = 0.42$, $f_{N = 36} = 0.34$, and $\alpha/\beta = 2.54$ Gy [$p\ (\chi^2) = 0.64$]. Thus, a better fit by the Pearson χ^2 statistic was obtained for the model with the 3 additional values for the repair fraction. The AIC statistic supported this finding. The values are in excellent agreement with those in the paper, which were obtained using a completely different method and an assumed value of α/β of 3 Gy, once again supporting the idea that the repair fraction is not a function of α/β. The value of α/β found on re-analysis should be taken very cautiously since the only variation in dose per fraction was in the final single dose treatments, all other treatments being given with 2.15 Gy per fraction.

Although re-analysis supports the findings regarding the repair fraction, the methods used were quite different. The re-analysis has the advantage that all the data were used as opposed to collapsing each dose-response curve into a single D_{50} value and then using these values to assess the effects of retreatment. It is important to stress once again that the formerly popular technique of first distilling a

Table 7.2 Data from Wong et al. (1992) added to analysis of Wong et al. (1993b)

Dose (Gy)	Dose per fraction (Gy)	# of animals	# of responders
83.6	2.09	9	0
88.2	2.21	9	8
91.2	2.28	9	9

dose-response curve into a single value, D_{50}, was ill-advised and unsupportable if the data exhibit quasi- or complete separation as was so often the case. However, once the field started deploying more sophisticated statistical techniques, this isoeffect strategy waned. This particular case shows that the step of collapsing dose-response curves into D_{50} values will yield accurate results if the model deployed is supported by the data. However, if this is not the case, the collapsing technique can make it appear that the model and the data agree.

In the paper, Wong et al. assert that after initial treatment of 25% to 90% of D_{50}, the retreatment D_{50} decreased from 81% to 42%. Of course, with larger initial treatments one expects smaller retreatment doses to be required to achieve the same level of effect. Although they do not explicitly state it, the implication seems to be that increasing the initial damage reduces the amount of recovery as a percentage of the initial damage. Their actual words are that there is "an influence of the initial injury on the amount of radiation injury recovered." They were correct not to claim that the percent of amount damage initial damage reduced the percent of damage repair, since the behavior of the repair fraction was not that simple in their data. In fact, there seems to be a quadratic dependence of the repair fraction on initial dose as shown in Fig. 7.2.

If one examines the "absolute" damage repaired as measured by dose (or ERD) rather than the repair fraction, a different picture emerges. This is shown in Fig. 7.3.

In this figure, the ordinate shows the initial dose multiplied by the repair fraction (determined by the re-analysis above) versus the initial dose. Broadly speaking, there appears to be no dose repair below about 10 Gy and no additional repair above about 60 Gy for the treatment regimens used. Whether the shape of this curve is indicative of the actual repair process is speculative. Since the repair fraction changes only modestly, but the initial dose increases linearly, this effect could simply be an artifact.

Wong et al. were correct regarding the need to assess the amount of repair as a function of initial dose. Although the treatment of initial dose as a surrogate for initial damage is not founded in biological theory, it is the only metric available.

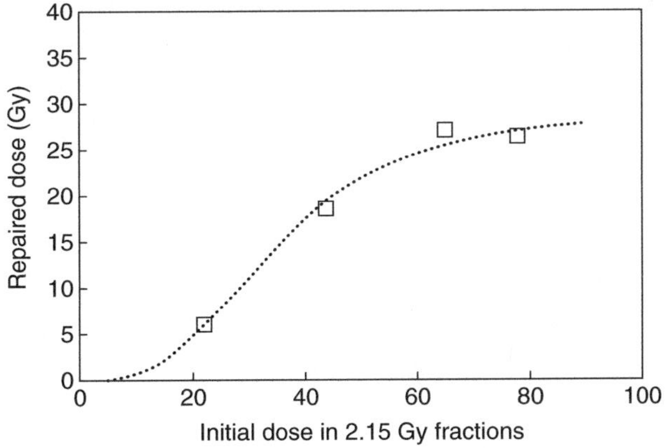

Fig. 7.3 Repaired dose as a function of total initial dose, obtained by multiplying the initial dose by the repair fraction, f.

Also in 1993, the Wong group published a retreatment paper, again using a 20-week intertreatment interval, but the retreatment doses were fractionated (Wong et al. 1993a). Two sets of experiments were performed. In the both sets, the initial treatment was 3×9 Gy, representing 3 fractions of the 4-fraction D_{50}. In the first set of experiments, the retreatment doses were given in 1, 2, 5, 10, and 20 fractions. In the second set of experiments the retreatment doses were given in 1, 2, 4, 10, and 20 fractions that followed an initial top-up dose of 12.8 Gy. In the second set of experiments, the authors acknowledged that the 10-fraction data were aberrant and should not be included in the ultimate analysis. Since all initial treatments were the same, this experiment cannot and was not designed to address the dependence of retreatment tolerance on the size of the initial dose or initial damage.

For this study, all data were analyzed together rather than relying on D_{50} values to represent the whole dose response. (This was essential since 4 of the 5 dose-response experiments for the first set of data were quasi- or completely separated and no MLE estimates could have been obtained.) Their analysis showed that for the first set of data, without the top-up dose, the α/β was 3.07 Gy. For the second set of data, the α/β value was 4.65 Gy if the 10-fraction data were included and 3.34 if that group was left out. However, both analyses of the second set of data were adequately fitted by the LQ model with p (χ^2) = 0.23 and 0.82 for the data with and without the 10-fraction group. Since no assessment of fit was given for the first set of data, one can assume that the LQ model did not fit those data adequately.

Re-analysis was able to duplicate the results exactly. The p (χ^2) for the first set of experiments was found to be 0.09, which does indeed indicate a poor fit. It was found that the fitting of the second set of data included redundant values of 0% and 100% that we showed in Chap. 3 should be disregarded. When one uses the trimmed degrees of freedom, we find p (χ^2) = 0.65, still an excellent fit, and α/β = 3.27 Gy.

A quandary arises when all the data are combined. Re-analysis produces a value of α/β = 2.72, but p (χ^2) = 0.09, showing a poor fit to the data. Basically, this resulted because the combined analysis gave parameter values very close to the original, but different enough to make each set of data fit less well. The authors also pooled both sets of data, obtaining α/β = 3.1 Gy, but they did not comment on the goodness-of-fit statistic.

The authors' main conclusion is that since the α/β ratio for the 1992 data without a top-up dose and the data presented in this 1993 publication were similar, the fractionation sensitivity of the rat spinal cord is the same for retreatment as for *de novo* treatments. However, since neither the data from the 1992 paper nor the data from this paper were well fitted by the LQ model, *this conclusion is not supported by the data*. One can only say that a portion of the data in this study were well fitted by the LQ model with α/β = 3.3 Gy.

The study of long-term repair (or recovery) continued in the Wong and Hao (1997) paper that used three different initial dose regimens and seven different retreatment times. The initial dose regimens were 2×9 Gy, 3×9 Gy, and 3×10.25 Gy (Wong and Hao 1997). The retreatment times were 4 days, 6, 8, 12, 20, 29, 40, 52 weeks. Retreatment doses were given as single fractions. The data consisted of 83 point and 19 dose-response curves, but some having only two or three dose points. Most of the dose-time points were assigned eight or nine animals;

6 of the 19 dose-response curves were quasi- or completely separated. Slightly less than half of the responses were 0% or 100%.

The objectives of this ambitious study were two-fold: to assess the dynamics of damage recovery after an initial insult and to determine how the initial dose impacted the recovery. The fractionation behavior of retreatment had been studied in the 1993 papers and by others, allowing the investigators to simplify this study by using single doses for retreatment.

By assuming an α/β ratio of 3 Gy and calculating D_{50} values obtained by "simple logistic regression," which probably means 2-parameter linear logistic regression, the authors compared the ERD_{50} for the initial treatment plus retreatment at 4 days (the controls) to the effective ERD_{50} doses at the various time points. From this they determined the percentage of the initial ERD that remained in effect at the time of retreatment using Eq. (7.2). Their Table 5 is reproduced here in Table 7.3, with one apparent transcription error corrected (3×9 Gy, 6 weeks). The statistical analysis involved in producing this table was the determination of the retreatment ERD_{50}'s. This effort was not re-analyzed here because of the separation problem in the dose-response curves and because collapsing data to D_{50} values is not as informative and a maximum likelihood analysis of the pooled data. The repair fraction *was* independently estimated here. See below.

The values from Table 7.3 are plotted in Fig. 7.4. Although there are some differences in the curves for the different initial treatment doses, these differences are

Table 7.3 The repair fraction (%) at various times following three different initial dose regimens

Weeks after	2×9 Gy	3×9 Gy	3×10.25 Gy
0.14	0%	0%	0%
6		−0.56%	
8	10.5%	−1.4%	1.1%
12	19.8%	15.9%	11.1%
20	35.1%	32.7%	22.9%
28	33.3%	39.3%	33.0%
40		40.6%	40.4%
52		48.6%	

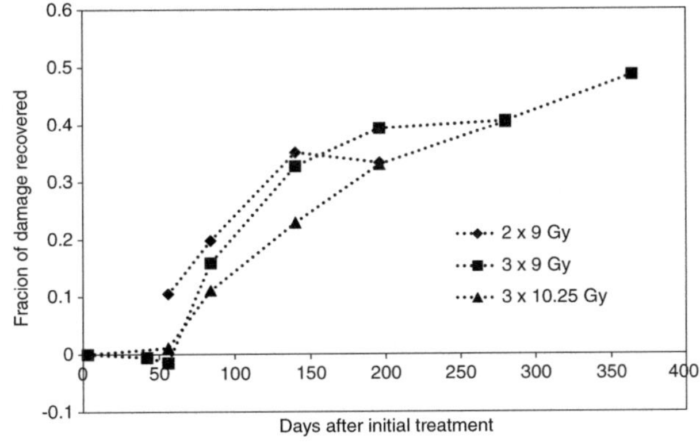

Fig. 7.4 Data on repair of initial dose versus time after retreatment from Wong and Hao (1997). Data points were derived using D_{50} values determined at each dose and time point.

modest and do not indicate that the repair depends strongly on the retreatment dose. However, the temporal dependence of repair is dramatic and seemingly ongoing at 1 year.

In an effort to describe the effect of retreatment using a single dose-response model, the authors tested 6 different functional forms of the log odds ratio. For the simple LQ model, this function simply includes a constant, a dose term, and a dose *times* dose per fraction term. To this, the authors added various functions of time to retreatment, including linear, quadratic, Gompertzian, and a dose times time term. None of these models fit the data. For the last two models, they used a quadratic time function times the initial dose with the α and β values being either constant or different for the initial and retreatment course. These models contained 4 or 6 parameters depending upon whether the values of α and β were assumed to be the same or different.

The authors reported both the χ^2 values and their p-values for these models, so it is simple to infer the number of degrees of freedom for these two models. For the model with *different* initial and retreatment α and β values, the χ^2 was 81.8 and p $(\chi^2) = 0.30$, thus d.f. = 76. Since this is a 6 parameter model and there are 83 data points, therefore one data point was probably treated as an outlier. Similarly, the number of degrees of freedom for the 4-parameter model with α and β constant for the initial and retreatment were 78. The α/β ratios were 6.63 Gy and 4.47 Gy for the initial and retreatment, and it was 4.21 Gy when a constant value was modeled.

Unfortunately, there were a number of extreme values of response that did not contribute to the fit and therefore should not be counted in the degrees of freedom; see supra, Chap. 3. The number of these superfluous extreme values was 13. Reducing the degrees of freedom by 13 results in a poor fit to the data with p $(\chi^2) \sim 0.06$.

By examining the shape of the curves in Fig. 7.4, one can see that the shape approximates the curve for mono-exponential saturation, albeit offset from zero. Using this function, a re-analysis was done where the value of the fraction of the initial dose repaired by the time of retreatment is modeled by

$$f = 0 \qquad t < t_d$$

and

$$f = C\left(D_{\text{init}}\right) \cdot \left[1 - \exp\left(-\left(t - t_d\right)/t_0\right)\right] \qquad t > t_d \qquad (7.8)$$

where t is the time of retreatment in days, t_d is the elapsed time after treatment before recovery starts, $C\left(D_{\text{init}}\right)$ and t_0 are constants. $C\left(D\right)$ represents the maximum recovered dose as a potential function of the initial dose. The systematic component of the GLM then becomes

$$z = \beta_0 + \left(1 - f\right) \cdot \alpha \cdot \text{ERD}\left(0\right) + \alpha \cdot \text{ERD}\left(t\right), \qquad (7.9)$$

which is equivalent to Eq. (7.3). ERD (t) is the retreatment ERD at time t.

Applying the log–log link function, the ML estimates were obtained for β_o, α, β, t_d, and t_0. The term $C\left(D_{\text{init}}\right)$ was estimated for each initial dose schedule, D_{init}. The ML parameter estimates are given in Table 7.4.

Table 7.4   Maximum likelihood estimates of the model parameters of Eqs. (7.8) and (7.9)

$\alpha = 0.232 \ \mathrm{Gy}^{-1}$	$C \ (2 \times 8 \ \mathrm{Gy}) = 0.325$
$\beta = 0.040 \ \mathrm{Gy}^{-2}$	$C \ (3 \times 8 \ \mathrm{Gy}) = 0.425$
$\beta_o = 21.9$	$C \ (3 \times 10.25 \ \mathrm{Gy}) = 0.381$
$\alpha/\beta = 5.8 \ \mathrm{Gy}$	$p \ (\chi^2) = 0.46$
$t_d = 54.7 \ \mathrm{days}$	
$t_0 = 89.9 \ \mathrm{weeks}$	

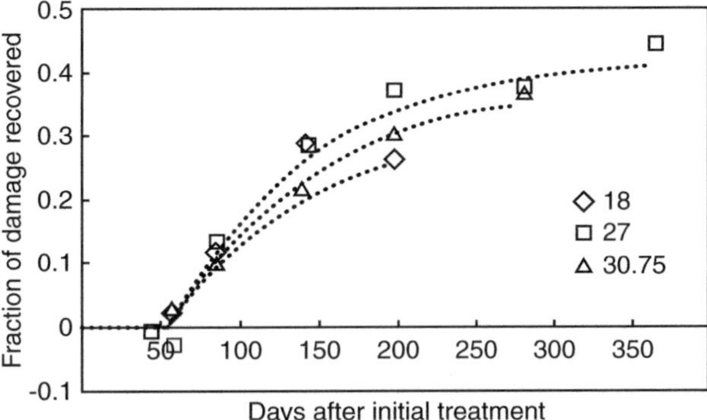

Fig. 7.5   Repair fraction of initial dose versus time after retreatment. Data re-analyzed from Wong and Hao (1997). Data points were obtained by determining individual values of f in Eq. (7.9) in a pooled analysis of all data. Curves represent the graphical result of the analysis of the pooled data using the estimates of the model parameters given in the text.

Thus a good fit was achieved for the entire data set analyzed simultaneously. To obtain individual estimates of the value of f for a given initial dose schedule and time to retreatment, an additional pooled analysis was performed where each f was estimated individually. These individual values are shown in Fig. 7.5 along with the curves for the initial dose schedules from Eqs. (7.8) and (7.9) and the parameter values above. Thus these data indicate that there is a delay of about 8 weeks before recovery from the initial damage is seen. This delay time appears not to depend on the initial doses used in this experiment. The dynamics of recovery are also independent of initial dose, and *the amount of recovery might depend on initial dose, but not in a monotonic manner for these data.* Using a single value for $C \ (D_{\mathrm{init}})$ did not produce a good fit. The average value of f at 6 months is 0.286. Recovery is 95% complete at about 46 weeks, essentially 1 year.

Since the repair fraction can be calculated from Wong and Hao (1997) as a function of dose, we can compare the result to Wong et al. (1993b). This comparison is shown in Fig. 7.6 where the behavior is seen to be very similar, but the repair fraction is lower. Nonetheless, there is no evidence here that the repair fraction is a monotonic function of initial dose.

In 1992, Ruifrok et al. published a paper that also addresses the retreatment fractionation sensitivity (Ruifrok et al. 1992a). The initial dose was 15 Gy in a single fraction. Retreatment dose-response curves were obtained for single fractions, 6-, 4-, and 3-Gray fractions. Retreatments were done at 1 day and 6 months. The original intention was to give the 3-Gray fractions twice daily at 10-hr intervals. However, incomplete repair was an obvious problem once the data were in hand so this schedule was dropped. There were 3-Gy fraction data for daily fractionation and 1-day

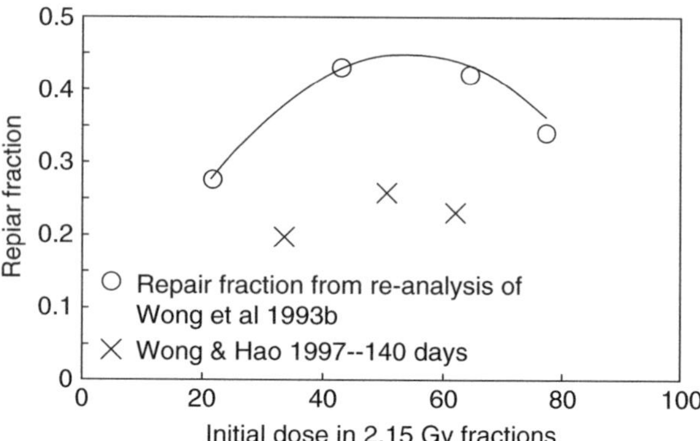

Fig. 7.6 Repair fraction at 140 days from Wong and Hao (1997) using α/β ratio from Table 7.4. The shape is the same as found earlier by Wong et al. (1993b), but with a smaller magnitude at 140 days.

retreatment interval. No dose-response curves were separated; generally 5 or 6 animals were assigned to a dose group.

The analysis consisted of individual dose-response analysis for each fractionation schedule, separate dose-response analysis for the 1-day and 6-month data, and an assessment of the repair fraction. The 1-day analysis yielded an α/β ratio of 2.3 Gy and a 6-month value of 1.9 Gy. These were not statistically different. The individual D_{50} values and the α/β ratios were reproduced by re-analysis here to within one decimal. Furthermore, the model assessment indicated a good fit to the data, although this was not done in the paper. However, the authors state that the amount of dose effectively remaining at the 6-month retreatment time was 45%, but they do not indicate how this was calculated. A re-analysis was performed here that included all data simultaneously, including a factor for repair. The re-analysis deployed the ERD concept since it is the metameter that is additive, in the partial tolerance sense. A single value of the repair fraction, f, was found to have the value of 0.291. The value at 6 months elicited above from the data of Wong and Hao (1997) was 0.286. In all likelihood, this surprising agreement was serendipity, but impressive nonetheless. The α/β value from the analysis of the combined data was 2.1 Gy.

Pigs Medin et al. compared retreatment with hemi-lateral fields 1 year after uniform treatment with 30 Gy in 10 fractions to *de novo* results with hemi-lateral fields (Medin et al. 2012). The hemi-lateral fields were designed as an approximation to SRS treatment. The authors found nearly complete recovery (96%) of the initial 30 Gy in 10 fractions with an excellent fit of the model to the data. From information on the pathogenesis in this pig strain published later, these data may have included or even dominated by nerve root lesions.

Rhesus Monkeys The M.D. Anderson group performed retreatment experiments using rhesus monkeys (Ang et al. 1993, 2001). With the dose-response data in hand, they designed experiments using 44 Gy in 2.2 Gy fractions followed by retreatments at 1, 2, and 3 years and then following the animals for an additional 2 years. The design and the results of the experiments are given in Table 7.5.

Table 7.5 Experimental design and results of rhesus monkey treated using 2.2-Gy fractions

D_{init} (Gy)	D_2 (Gy)	1 years	2 years	3 years
44	57.2	2/14	2/21	
44	66		1/6	1/12
44	39.6		0/4	
44	48.4		0/4	
70.4	13.2		1/4	
70.4	22		1/4	
70.4	30.8		0/4	

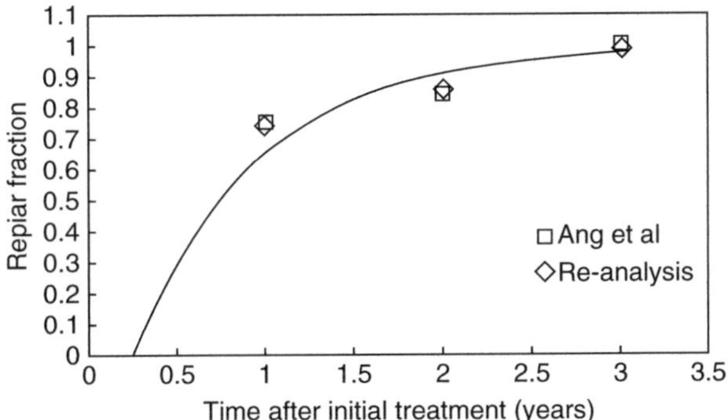

Fig. 7.7 Repair fraction in rhesus monkeys treated 1–3 years after 44 Gy. Repair fraction is calculated using Eq. (7.2) with ERD replaced by dose since the fractions size is constant at 2.2 Gy.

The last five lines of Table 7.5 comprise the preliminary data while the main experiment is given on the first two lines. Ang et al. (2001) give the repaired fractions for 1, 2, and 3 years after 44 Gy as 76%, 85%, and 101%, respectively, assuming parallel probit dose-response functions. Since all previous studies by this group had used the log-linear logistic dose-response function, the repairs fractions were recalculated here under the same assumptions with result within about 1% of the original authors (75%, 86%, and 99.5%). These results are graphed in Fig. 7.7, along with a line showing first order repair kinetics. *This line is drawn for illustration purposes only and is not the result of statistical analysis.* It depicts repair starting at 3 months. However, because of the linear nature of the data (actually slightly concave upward), a fitted curve with first order kinetics would be forced through the origin, assuming delay time from initial treatment was forced to be non-negative. Thus these data show a substantial long-term repair fraction, but they do not seem to conform with the conventional wisdom that long-term repair starts weeks or months following initial treatment.

Another issue with these data arises from the 70.4-Gy initial treatment groups. These animals were part of the original dose-response experiment and did not show symptoms within 24 months. Therefore they were retreated. Repair fractions can be estimated using the same assumptions as above, but the model that assumes the same slope for retreatment is soundly rejected. Obviously from an examination of the data, one can see that the retreatment dose response is essentially flat. The problem is compounded (and perhaps caused) by having so few animals per dose group. In retrospect, it would have been better to have two groups of six. Only two alternatives seem likely to explain the conflict between 44 Gy initial dose and 70.4 Gy initial dose. First, because there was such a high initial dose and certainly very close to the tolerance for animals RM-free at 2 years, it may be that long-term repair was altered in some way as to flatten the dose-response curve. Alternatively, the symptoms that were exhibited in this part of the experiment may indicate that the animals' tolerance was exceeded by the first dose, but the latency for these animals was greater than 2 years. Latencies longer than 2 years are common in humans. A two-year limit was selected for the rhesus monkeys primarily because of the expense of housing the animals versus the benefit of holding them for an additional year or more.

7.1.2 The Volume-Repaired Model of Retreatment Response

A major conceptual problem regarding the modeling of retreatment responses is the equating of damage repaired with dose repaired or recovered or forgotten. To circumvent this issue, we can model damage as volume of damaged targets. Then during the time interval between the first treatment and the retreatment, the simplest model of repair would be that a certain fraction of these targets is completely repaired and the complementary fraction of these targets experience no repair. If D_1 and D_2 are the first and second treatment doses, then the repaired volume will see only dose D_2, the initial damage having been completely repaired, and the unrepaired volume will experience $D_1 + D_2$. Thus damage is repaired, not dose, and the retreatment problem is now expressed as a volume effects model. Applying this model to the rhesus monkey data produces the results shown in Fig. 7.8. It is now possible to pool all the rhesus monkey data, including the volume effects data, in a single analysis. Furthermore, for the first time, the preliminary data on retreatment using the problematic data from the initial dose of 70.4 Gy are included in a retreatment analysis.

Mathematically, this model can be written as Eq. (7.10)

$$P(D_1, D_2) = 1 - \left(1 - P(D_1 + D_2)\right)^{1-v_r} \left(1 - P(D_2)\right)^{v_r} \tag{7.10}$$

where D_1 is the initial dose, D_2 is the retreatment dose, v_r is the constant that represents the relative volume that is completely repaired and $1-v_r$ represents the volume in which there is no repair. The dose-response function is given by $P(D)$. The equation is easily understood as volume v_r only experiencing dose D_2, while volume $1-v_r$ experiences dose $D_1 + D_2$. The parameters of this model are given in Table 7.6, where a logistic dose-response function in the form of Eq. (3.6) was used.

The model shown in Fig. 7.6 fits the data well ($p(\chi^2) \sim 0.9$), but the excellent fit is augmented by the good fit to the *de novo* data. Nonetheless, the Pearson residuals for the retreatment data are all less than 1.1 and all appear to be lower than in the Ang et al. paper (Ang et al. 2001, 1993). Of course the data for retreatment after

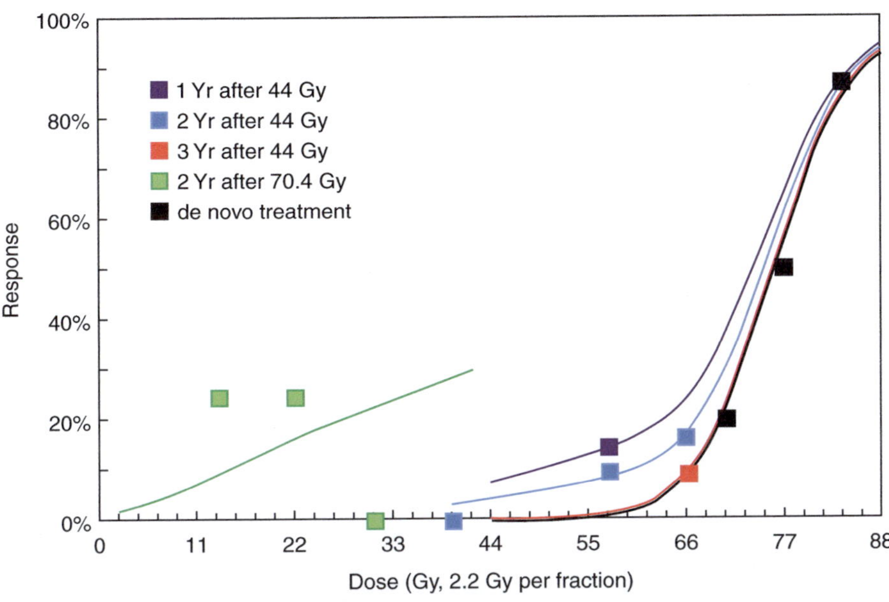

Fig. 7.8 New retreatment model where the repair is modeled as fraction of treated volume where damage is fully repaired plus the complementary fraction where no damage is repaired. This graph shows the *de novo* data in black. The abscissa is the retreatment dose, except for the *de novo* response data.

Table 7.6 Maximum likelihood estimates of the parameters in the volume-repaired model of retreatment applied to the rhesus monkey data (2.2 Gy per fraction). The numbers in parentheses are 95% confidence limits

D_{50}	75.1 Gy
k	17.4
Years after 44 Gy	v_r
1	0.971 (0.880,0.997)
2	0.985 (0.944,1)
3	0.999 (0.943,1)
2 Years after 70.4 Gy	0.948 (0.790,0.992)

70.4 Gy do not even behave like dose-response data. Therefore a model that is not rejected by these data would have to have a very low slope, which is a characteristic of the low-dose portion of the volume-repaired model.

7.1.3 Human Retreatment

In a study of latent periods, Schultheiss et al. reported that the latent periods following a second retreatment were shorter than those whose treatment was given in a single course (Schultheiss et al. 1984). This finding was corroborated in single institutional data from Wong et al. (1994). This observation probably does not have particular significance to the histopathology or the radiobiology of retreatment (Hubbard and Hopewell 1978), but there is a clinical implication made by Schultheiss et al. That is, one should not anticipate that a short life expectancy when retreating the spinal cord would mean that the patient would probably not survive the latent period. Contradicting this finding is the fact that in the retreatment of rhesus monkeys, the latent periods after the retreatment following the initial treatment of 44 Gy were significantly longer than the latent periods for the *de novo* treatment with 70.4–83.6 Gy ($p < 0.002$) or the *de novo* treatment with only the 70.4 and 77 Gy groups ($p < 0.05$).

The Wong et al. paper describes 24 cases of radiation myelopathy in patients treated with a single course and 11 patients with RM following retreatment. There were no real surprises in the retreatment data, now that we have dose-response curves for both cervical and thoracic levels of the spinal cord. Using the α/β ratio from the dose-response analysis presented in Chap. 4, the effective dose in 2-Gy fractions, $D_{2\,Gy}$, can be calculated for each course of treatment. For the course of treatment with the larger dose, the *smallest* value of $D_{2\,Gy}$ was 52.5 Gy. This is associated with a probability of RM of 0.4%. Given that these patients were among those treated at Princess Margaret Hospital between 1955 and 1985, it seems reasonable to see 11 RM cases in retreated patients or a total of 35 RM cases out of what was likely on the order of tens of thousands courses of radiation treatments during this period.

Nieder et al. in two publications used these and other data to construct an algorithm for risk assessment in retreatment of the human spinal cord. The risk level was based on the total ERD of the combined treatments, plus a consideration based on the retreatment interval and the maximum ERD of a single course (Nieder et al. 2005, 2018). These authors combine the Wong data on humans with published retreatment human data for which there were no RM cases. Inventing a risk metric seems to be a more sensible approach than trying to create a uniform cohort of unrelated cases and controls to subject to dose-response analysis. However, because the ERD they used deployed an α/β value of 2 Gy rather than a much lower value that is probably more appropriate to the human, there is the likelihood that the risk is underestimated for courses using very high doses per fraction and possibly overestimated for courses with relative low doses per fraction. For example, two fractions of 9 Gy followed just more than 6 months later by a single fraction of 8 Gy would be considered low risk, whereas 26×2 Gy followed at any time by 5×1.8 Gy would be gauged at high risk. Because in the main, their algorithm seems quite reasonable, it is reproduced here.

$$\text{Risk score} = \text{Int}\left[\left(\text{ERD}_{total} - 110.1\right)/10\right]$$
$$+\,4.5 \text{ if either ERD} \geq 102\,\text{Gy} \tag{7.11}$$
$$+\,4.5 \text{ if the time between treatments} \leq 6\ \text{mo}$$

where "Int (x)" is the greatest integer in x and the total ERD is capped at 200 Gy. The risk levels are then

$$0\text{--}3 \Rightarrow \text{low risk}$$
$$4\text{--}6 \Rightarrow \text{intermediate risk}$$
$$\geq 6 \Rightarrow \text{high risk}.$$

Low, intermediate, and high risks are not specifically defined, but the implication is that they relate to the probability range of RM. However, some may consider even 1% high risk of such a serious complication. Clearly the risk depends on the context. While this tool could be considered imprecise, it captures the elements of the major components of risk in retreating the spinal cord.

It should be emphasized that the risk score and its interpretation applies only to ERD's using $\alpha/\beta = 2$ Gy. Therefore this algorithm cannot be used if a different value of α/β is considered more suitable. However, if one converts a dose schedule to the equivalent dose in 2-Gy fractions, then one can apply the algorithm using one's own choice of α/β. To apply the algorithm in this case, the following equation must be used:

$$\text{Risk score} = \text{Int}\left[\left(\text{total } D_{2\,\text{Gy}} - 55.05\right)/5\right]$$
$$+ 4.5 \text{ if } D_{2\,\text{Gy}} \geq 51\,\text{Gy for either course} \tag{7.12}$$
$$+ 4.5 \text{ if the time between treatments} \leq 6 \text{ mo.}$$

Then if it is believed that the original algorithm puts too little weight on high doses per fraction, a smaller value of α/β should be used to convert the dose schedules to 2 Gy per fraction equivalents.

The Issue of Whether the Amount of Repair Diminishes with Increasing Dose Stewart and van der Kogel declaratively state that "retreatment tolerance is inversely related to initial dose" (Stewart and van der Kogel 1994). The logical alternative to this statement is that retreatment tolerance does not depend on initial dose. This could occur if 100% of the initial treatment is repaired, or if the unrepaired dose equivalent has a constant absolute value independent of the initial dose. Neither of these situations is plausible. However, what is typically meant by the statement is that the fraction of damage repaired decreases with increasing initial dose. As evidence, Stewart and van der Kogel cite five publications that constitute "all the rat data." van der Kogel is the first author on two papers. The first is a split-dose study. As stated at the beginning of this chapter, this type of study does not address the question of the repair fraction as a function of initial dose because it comprises dose-response data from two equal fractions separated by increasingly longer intervals. The second paper is from the proceeding of the 13th L.H. Gray Conference in 1986 (van der Kogel 1986). In a paper on target cell response, he reports preliminary results of retreatment after 5, 7, and 9 fractions using a fraction size of the 10-fraction D_{50}. Retreatments were given in the same fraction size and the retreatment D_{50} estimated. The data were reported solely in terms of D_{50}'s and in graphical form. To get from the reported metric to the fraction of the initial dose that was repaired between the treatments required some calculations, but the result showed that the fraction of damage repaired increased with time and *decreased* with initial dose. In a later publication (van der Kogel 1991), this study was attributed as "van der Kogel and Fawcett, unpublished." With a more complete description the results were unchanged, but it was also stated that the there were "adjustments in dose [so that] the final values were slightly different." Therefore, no further attempt was made to assess this study. Because of the way the results were reported, i.e., without supporting data, there can be no statistical re-evaluation or assessment whether the data support the conclusions or the underlying model. Nonetheless, this work is commonly cited as evidence of long-term repair diminishing with increasing initial dose.

The other three papers cited by Steward and van der Kogel were by Wong et al. and Ruifrok et al. One of the latter papers used immature rats in whom the retreatment effects are somewhat different, so this will be addressed later in this chapter (Ruifrok et al. 1992c). In the paper by Wong et al., it was shown in this chapter that the repair fraction increases and then decreases within a narrow range as the initial dose increases (Wong et al. 1993b). And the final paper by Ruifrok had only a single initial dose and dependence of the repair fraction on initial dose could not be assessed (Ruifrok et al. 1992a). *Thus none of the papers cited have data that offer statistical support to the idea that the repair fraction in adults depends on initial dose.*

Looking at retreatment studies in adult animals subsequent to the Stewart and van der Kogel paper, the rhesus monkey data cannot be used to support the assertion that f is a decreasing function of initial dose since most of the data have only 44 Gy as an initial dose, and one cannot eliminate late occurring myelopathies as being responsible for the responses retreatment after 70.4 Gy. The pig data from Medin et al. had a single D_{init} (Medin et al. 2012). The extensive study by Wong and Hao (1997) shows not a monotonic dependence of f on initial dose, but rather a concave downward dependence as in the 1992 data. The data, published in their entirety, represent a very similar experiment to van der Kogel's 1991 chapter published in summary form. The difference is that data from Wong and Hao show quite a different behavioral dependence of f on D_{init}. In summary, there are no statistically validated data that affirm the theory that the fraction of damage repaired after treatment of adult animals is linearly or monotonically dependent on the dose of the initial treatment. That is not to say that the theory is not true—only that as of this point there is no apparent statistical evidence for it.

Thus the conclusions that can be drawn from the combined efforts of the investigators exploring retreatment effect in the rat model are that recovery of damage starts at 6–8 weeks, increases probably according to first order kinetics approaching an asymptote, and is in the neighborhood of 30% at 6 months. The recovery, as measured by dose is about 45% at 1 year, but may be higher and is higher for immature animals. Whether this can be applied to the human is somewhat doubtful, given that retreatment times had to be kept relatively short to accommodate the rat's live expectancy, but the nonhuman primate data address this time issue more effectively. There is no evidence fractionation sensitivity is altered in the retreatment setting. Although it may be widely believed, there is no statistically compelling evidence in the rat data that the rate of recovery is dependent on the initial dose. Again, this is addressed somewhat in the nonhuman primate data.

7.2   RM in Immature Humans and Animals

Pig   van den Aardweg et al. performed a study in mature (average age 40 weeks) versus immature pigs (average age 18 weeks) (van den Aardweg et al. 1994). The field size was 10 cm and the radiation was delivered with C0–60 with a low dose rate at the cord of 21–65 cGy/min. The mature pigs also served as the large field size group in a volume study of RM discussed in Chap. 8. A finding that was not surprising in this study was that the spine of the immature pigs was significantly underdeveloped at ages 60–120 weeks compared to the animals irradiated as adults.

Permanent radiation myelopathy developed in the mature pigs between 7.5 and 16.5 weeks after irradiation. All of these pigs showed white matter necrosis with some vascular changes and mononuclear inflammatory reactions. The inflammatory reaction was observed outside of the radiation field as well as in association with the damaged white matter. The white matter changes were consistent with Type 3 changes as described in Schultheiss et al. (1988). An unusual finding was that in animals killed prior to 16.5 weeks, the intimal layer of the main ventral artery exhibited "leukocyte adhesion to the endothelium with subendothelial infiltration. In some cases the endothelium had separated from the internal elastic lumina and the space was filled with what appeared to be loose connective tissue." This was seen in 7 of 8 animals that developed RM prior to 16.5 weeks.

Two additional animals developed late paralysis 15 and 17 months after 27.5 and 26 Gy, respectively. These animals showed extensive glial scarring and "varying degrees of occlusion of the ventral artery due to myointimal proliferation." However, these changes were seen in all mature animals treated to these doses and killed at the end of the study, irrespective of their RM status.

Gray matter changes of petechial hemorrhages were seen in positive pigs and rarely in negative animals. Dilated blood vessels were also seen, rarely, in the gray matter.

A single immature animal developed permanent myelopathy similarly to the mature animals, 8.5 weeks after 25.5 Gy. Its only lesion was white matter necrosis. Animals killed between one and 2 years showed changes similar to the mature animals in the main ventral artery. A single pig from this group, treated to 24 Gy, showed signs of paralysis after 52 weeks. The pathological changes in this animal included an 80% occlusion of the main ventral artery, as in the mature pigs.

The dose-response curves for permanent myelopathy in the mature pigs and transient RM in the immature pigs are shown in Fig. 7.9. If the endpoints for the two groups were the same, the DMF for immature irradiation would be 1.035. However, considering that only one immature pig developed permanent RM, it obviously cannot be said that these data indicate a greater radiation sensitivity in the young animals and perhaps the opposite is true.

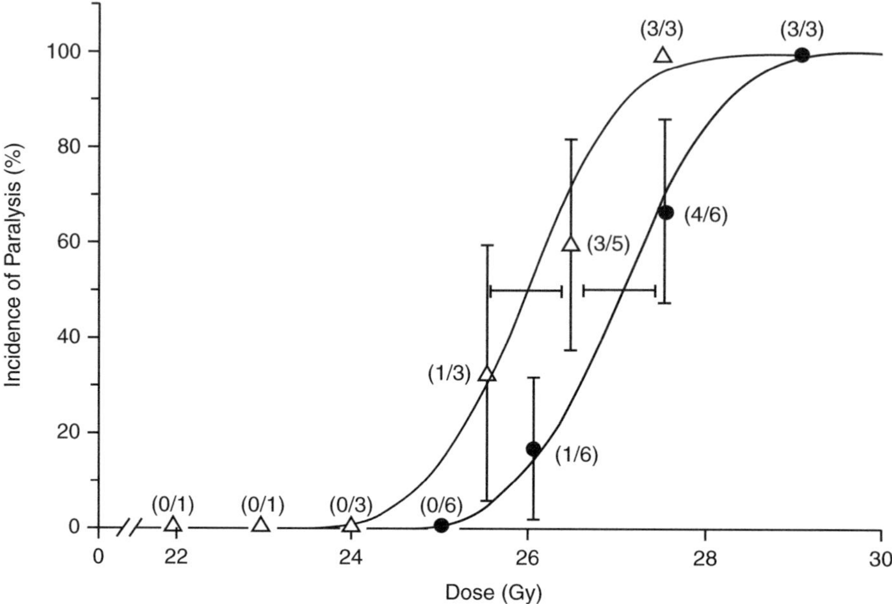

Fig. 7.9 Dose-related changes in the incidence of radiation-induced myelopathy in mature pigs (●) and transient radiation myelopathy (△) in immature pigs after the irradiation of a 10 cm length of the cervical spinal cord with single doses of 60Co γ-rays. Figures in parenthesis represent the number of responders out of the total number of animals at risk in each dose group. Reprinted from International Journal of Radiation Oncology*Biology*Physics, 29, G.J.M.J. van den Aardweg, J.W. Hopewell, E.M. Whitehouse, W. Calvo, A new model of radiation-induced myelopathy: A comparison of the response of mature and immature pigs, 763–770, Copyright (1994), with permission from Elsevier.

Rodents Ruifrok et al. published the dose response for immature rats at various ages and for various fractionation schedules for one-week-old rats (Ruifrok et al. 1992b). Mostly five, but in a few cases, six animals were used per dose group. Rats were treated at ages of 1, 2, 3, and 18 weeks. This resulted in quasi- or completely separated dose-response data for all but the 1-week-old rats. The authors suggested, without the benefit of statistical analysis, that only the 1-week-old group had a dose response different from the adult, 18-week-old rats. Upon analyzing the data by combining the 2-, 3-, and 18-week data into a single linear logistic dose-response analysis, we find an excellent fit to the data is obtained (p (χ^2) > 0.7). Thus it can be claimed that the data are consistent with a difference in dose response *only* for 1-week-old rat pups. However, the latent period is age-dependent, increasing with increasing age.

In an earlier publication, a similar experiment was performed using different fractionation schemes with $N = 1, 2, 3, 5$, and 10 fractions (Ruifrok et al. 1992b). The authors report an α/β ratio of 4.5 Gy for 1-week-old rats. Upon re-analysis, nearly the same results were obtained. As shown in Fig. 7.10, the original and the

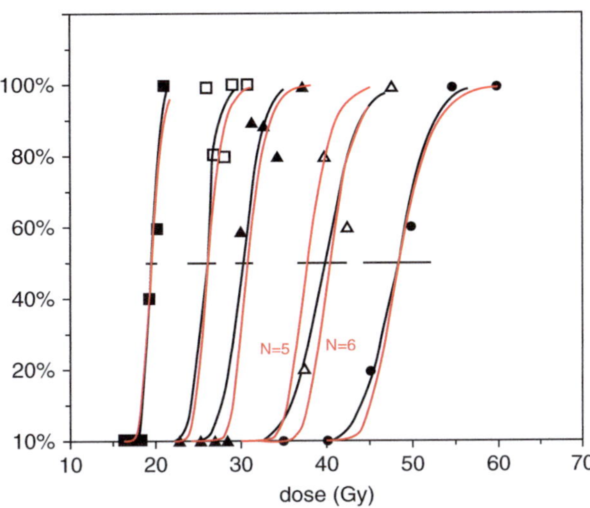

Fig. 7.10 Dose-response curves of 1-week-old Wistar rats in 1, 2, 3, 5, and 10 fractions. Reprinted from International Journal of Radiation Oncology*Biology*Physics, 24, Arnout C.C. Ruifrok,Bert J. Kleiboer,Albert J. van der Kogel, Radiation tolerance and fractionation sensitivity of the developing rat cervical spinal cord, 505–510, Copyright (1992), with permission from Elsevier.

results obtained here are quite close. However, the dose-response curve depicted in the original paper for the $N = 5$ data is in between the dose-response curves for $N = 5$ and $N = 6$ with the same parameter values. This is verified by the table of D_{50} values, where the authors' value of 39.9 Gy in fact corresponds to $N = 6$ and the true value for $N = 5$ is 37.2 Gy. It is probable that the dose-response curves for the original figure represent those that were fitted individually and not the analysis where the data were pooled. Taking this into account, and adjusting the degrees of freedom to account for the numerous sequential values of 0% or 100% response, we find that the LQ model does not fit the data and therefore no α/β value can be estimated from this experiment. Another interesting aspect of this paper is that the $N = 10$ data were treated twice daily with a 10-h interval. Analyzing the data without the $N = 10$ group yields a similar result for the $N < 10$ groups, but the predicted $N = 10$ dose-response function lies well to the right of the data, indicated that incomplete repair was not complete in 10 h.

The immature rats treated at 1 week of age had a latency on the order of 2 weeks. This increased with age. However, it seems that the response of immature rats cannot be compared directly to adults because of the volume issue. Does one treat the same absolute volume or the same spinal cord segments. It has been shown in Chap. 6 that for adult rats with field sizes less than 8 mm, the absolute volume is important.

Also in 1992, Ruifrok et al. published results of a re-irradiation experiment using 3-week-old weanling rats (Ruifrok et al. 1992c). The young rats were given initial doses of either 12 Gy or 14.9 Gy. They were re-irradiated at 1 day, 4, 12, or 24 weeks. In addition, the rats initially treated with 14.9 Gy included groups that were re-irradiated at 2 and 8 weeks.

All of the initial doses were given with 250 KV x-rays, as were the retreatments given at 1 day and 14 days. All other treatments were given with 4 MV x-rays. To calibrate the treatments to each other single dose dose-response data were collected using both photon energies. The authors state that they determined the RBE of the

Table 7.7 Parameter values for re-analysis of Ruifrok et al. (1992c) using Eqs. (7.8) and (7.9)

β_0	10.3
α	$1.99 \times 10^{-5}\ \mathrm{Gy}^{-1}$
β	$0.025\ \mathrm{Gy}^{-2}$
C (12 Gy)	0.78
t_0 (12 Gy)	30.4 days
C (14.9 Gy)	0.44
t_0 (14.9 Gy)	21.1 days

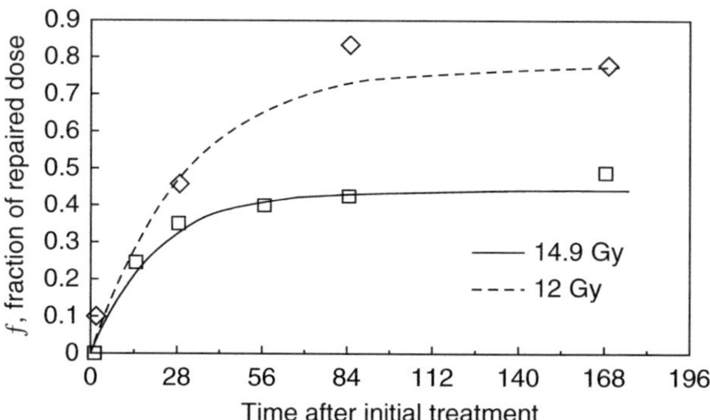

Fig. 7.11 Fraction of repaired dose (ERD) remaining after initial treatments of 12 or 14.9 Gy in 3-week-old rats.

250 KV beam to be 1.1 by taking the ratio of the D_{50} values for the two dose-response curves. However, both dose-response curves were quasi-separated so the D_{50} values were indeterminant by their proposed method. Therefore, no RBE could be calculated. Nonetheless, the value of 1.1 is widely accepted as accurate, so all further discussions will use the dose in 4 MV equivalents as was done in the paper. Only two of the 10 dose-response data sets were quasi-separated.

This paper was re-analyzed in the same way as the Wong et al. and the Ruifrok et al. papers on retreatment of adult animals—with the intention of determining f, the fraction of damage repaired at the time of the second treatment, as measured by the dose (ERD) required to achieve the same effect as the retreatment dose given 1 day after the initial treatment. The only difference between those models and the one used for the weanling rats is that no delay is assumed, i.e., $t_d = 0$. Also since there were two different initial doses, there were correspondingly two different values of C (D_{init}) and t_0.

The results are given in Table 7.7.

Clearly $\alpha/\beta = 0$, in sharp contrast to the value of 3 Gy that was assumed, not derived, by the authors. The lower initial dose has a higher ultimate recovery, but a slightly slower recovery rate. The recovery curves and the individual data point as derived by including them individually in the ML estimation are shown in Fig. 7.11.

Whether the immediate initiation of recovery in the weanling rat is due to high cell cycle activity in a specific cell compartment is speculation and must be confirmed or refuted by histopathology studies. However, this study is one of few (if there are any others) that show that a lower initial dose is more fully recovered.

Knowles published dose-response curves for guinea pigs at age 1 day, 30 days, and 1 year (Knowles 1983). He also produced a dose-response curve for animals treated 1 year following 10-Gy treatment at age 1 day. The dose-response curves for 30-day weanlings and 1-year-old adults were not distinguishable with a D_{50} of about 20 Gy; however, the latent periods for the adult animals were longer. A D_{50} for the 1-day-old animals could not be obtained because the data were quasi-separated. However, he quoted a value of 14.8 Gy, and it does seem that the D_{50} for the newborn animals is roughly 25% lower than the adults. The data for animals irradiated at 1 day and 1 year were also quasi-separated, but the analysis indicated that the dose response was not different from the 30-day-old and adult animals. Thus, it appears that there was 100% recovery from the 10-Gy initial dose. The latent periods for animals treated at 1 day were shorter than for the other animals, but the latent periods for animals treated at 1 day and 1 year were not significantly different. Care was taken to irradiate the same spinal cord segments at ages 1 day and 1 year.

Re-analysis of the Knowles data was made uncertain because the number of animals at each dose point was not well specified. However, better estimates were possible for those whose responses were neither 0% nor 100%. Upon re-analysis, the D_{50} for the 30-day and 1-year animals was found to be 20.1 Gy. However, there was a large repair fraction for the animals given the initial dose of 10 Gy at age 1 day. The maximum likelihood estimate of the repaired fraction was 0.83 with a 95% confidence interval of 0.62 to 1.01. Because of the uncertainty in the number of animals irradiated, one may only suspect that less than 100% of the initial 10 Gy was repaired. Upon re-analysis, the equivalence of the dose response at 30 days and 1 year was confirmed. The χ^2 p-value for the re-analysis was an unimpressive 0.11, because of the noise in the sparse data. This means that the exact equivalence of the 30-day and one-year dose response is not conclusive.

Mason et al. investigated retreatment to the lumbar cord in young female guinea pigs. The animals were 9 weeks old at the time of treatment (Mason et al. 1993). As a reference, these sows (350 g) would be sexually mature at this age, but they are not bred until they attain a weight of about 500 g. Their lifespan is typically about 5 years. The experiment consisted of treatments using 4.5 Gy per fraction. Single course dose-response data were obtained for these and for 40-week-old animals. Retreatment of the young animals took place after 28 and 40 weeks after 9 fractions of 4.5 Gy with 3 and 4 dose points respectively, again using 4.5 Gy per fraction. An additional group received an initial dose of 12 fractions and a single retreatment dose point of 9 fractions. The treatment regimens are shown in Table 7.8. *The animals were irradiated 7 days per week with two treatments occurring Saturdays and Sundays for a total of 9 fractions per week.*

The author calculated D_{50}'s for the first two groups and combined the 28-week and 40-week data to calculate a third D_{50}. Their conclusions were based on these D_{50} values, but as stated previously it is statistically preferable to pool the data in a single model when possible. This was done using the following systematic component in a logistic regression:

$$z = \beta_0 + \beta_2^* \ln\left[f \cdot D_1 \cdot \mathrm{DMF}_{40-\text{week old}} + D_2 \right]$$

Table 7.8 Experimental design and results of Mason et al. (1993)

D_1	Time to retreatment	D_2	BID rx's	Age	r	m
54	—	—	2	8	0	8
58.5	—	—	2	8	0	8
63	—	—	2	8	10	24
67.5	—	—	2	8	7	15
72	—	—	3	8	13	16
76.5	—	—	3	8	6	7
81	—	—	4	8	7	7
85.5	—	—	4	8	7	7
58.5[a]	—	—	2	40	1	8
67.5[a]	—	—	2	40	2	8
72[a]	—	—	3	40	1	6
76.5[a]	—	—	3	40	4	8
85.5[a]	—	—	4	40	6	7
40.5	28	27	0	8	0	8
40.5	28	40.5	2	8	0	12
40.5	28	54	2	8	1	12
40.5	40	36	1	8	0	8
40.5	40	45	2	8	0	8
40.5	40	54	2	8	1	8
40.5	40	63	2	8	4	8
54	40	40.5	1	8	2	8

r number paretic, m number irradiated
[a]Data from animals treated at age 40 weeks

where f is the repair fraction, D_1 is the dose of the first course, and D_2 is the retreatment dose. An excellent fit to the data was obtained, and the results compare closely with the authors' conclusions: the D_{50} agree within 0.1 Gy (~66.2 Gy) and the repair fraction at 28 weeks agrees to within 1% (~92%). The $DMF_{40\text{-week old}} = 0.88$, indicating a 14% higher D_{50} for the 40-week-old animals. They did not determine the repair fraction for the initial dose of 12 × 4.5 Gy, which was determined by the above analysis to be about 60%. However, this is based on a single point.

Based on the single retreatment point 54 Gy 40 weeks after 40.5 Gy, the authors concluded that repair is decreased with increasing initial dose. This point was compared to 40.5 Gy initial dose followed 28 or 40 weeks by another 54 Gy, yielding the same total dose. The responses for the 40.5-Gy initial dose were combined for a total of 2/20 RM cases. Since this was about half of the 2/8 when the schedule was 54 Gy followed by 40.5 Gy, the authors concluded that the larger initial dose was more toxic. However, these ratios are not even close to being significantly different. Consequently this conclusion is not supported by the data. Nonetheless it has been repeatedly cited as evidence that higher initial doses are less efficiently repaired. This may very well be true, but the data in this paper cannot offer much support.

Important features of this paper include rapid and nearly complete repair of the initial 40.5 Gy dose, the shorter latency of myelopathy for animals irradiated at age 9 weeks versus 40 weeks, and the 14% higher D_{50} for the 40-week-old animals. Of course, all these facts relate to the young age of the guinea pigs at the time of initial treatment. These are important facts, but the applicability to adult animals is uncertain.

References

Ang KK, Price RE, Stephens LC, Jiang GL, Feng Y, Schultheiss TE, Peters LJ. The tolerance of primate spinal cord to re-irradiation. Int J Radiat Oncol Biol Phys. 1993;25(3):459–64. https://doi.org/10.1016/0360-3016(93)90067-6.

Ang KK, Jiang GL, Feng Y, Stephens LC, Tucker SL, Price RE. Extent and kinetics of recovery of occult spinal cord injury. Int J Radiat Oncol Biol Phys. 2001;50(4):1013–20. https://doi.org/10.1016/S0360-3016(01)01599-1.

Barendsen GW. Dose fractionation, dose rate and iso-effect relationships for normal tissue responses. Int J Radiat Oncol Biol Phys. 1982;8(11):1981–97.

Hubbard BM, Hopewell JW. The dose-latent period relationship in the irradiated cervical spinal cord in the rat. Radiology. 1978;128:779–81.

Knowles JF. The radiosensitivity of the Guinea-pig spinal cord to x-rays: the effect of retreatment at one year and the effect of age at the time of irradiation. Int J Radiat Biol. 1983;44(5):433–42. https://doi.org/10.1080/09553008314551411.

Mason KA, Rodney Withers H, Chiang CS. Late effects of radiation on the lumbar spinal cord of Guinea pigs: re-treatment tolerance. Int J Radiat Oncol Biol Phys. 1993;26(4):643–8. https://doi.org/10.1016/0360-3016(93)90282-Z.

Medin PM, Foster RD, van der Kogel AJ, Sayre JW, McBride WH, Solberg TD. Spinal cord tolerance to reirradiation with single-fraction radiosurgery: a swine model. Int J Radiat Oncol Biol Phys. 2012;83(3):1031–7. https://doi.org/10.1016/j.ijrobp.2011.08.030.

Nieder C, Grosu AL, Andratschke NH, Molls M. Proposal of human spinal cord reirradiation dose based on collection of data from 40 patients. Int J Radiat Oncol Biol Phys. 2005;61(3):851–5. https://doi.org/10.1016/j.ijrobp.2004.06.016.

Nieder C, Gaspar LE, Ruysscher DD, Guckenberger M, Mehta MP, Rusthoven CG, Sahgal A, Gkika E. Repeat reirradiation of the spinal cord: multi-national expert treatment recommendations. Strahlenther Onkol. 2018;194(5):365–74. https://doi.org/10.1007/s00066-018-1266-6.

Ruifrok ACC, Kleiboer BJ, van der Kogel AJ. Fractionation sensitivity of the rat cervical spinal cord during radiation retreatment. Radiother Oncol. 1992a;25(4):295–300. https://doi.org/10.1016/0167-8140(92)90250-X.

Ruifrok ACC, Kleiboer BJ, van der Kogel AJ. Radiation tolerance and fractionation sensitivity of the developing rat cervical spinal cord. Int J Radiat Oncol Biol Phys. 1992b;24(3):505–10. https://doi.org/10.1016/0360-3016(92)91066-V.

Ruifrok ACC, Kleiboer BJ, van der Kogel AJ. Reirradiation tolerance of the immature rat spinal cord. Radiother Oncol. 1992c;23(4):249–56. https://doi.org/10.1016/S0167-8140(92)80143-7.

Schultheiss TE, Higgins EM, El-Mahdi AM. The latent period in clinical radiation myelopathy. Int J Radiat Oncol Biol Phys. 1984;10(7):1109–15. https://doi.org/10.1016/0360-3016(84)90184-6.

Schultheiss TE, Stephens LC, Maor MH. Analysis of the histopathology of radiation myelopathy. Int J Radiat Oncol Biol Phys. 1988;14(1):27–32. https://doi.org/10.1016/0360-3016(88)90046-6.

Stewart FA, van der Kogel AJ. Retreatment tolerance of normal tissues. Semin Radiat Oncol. 1994;4(2):103–11. https://doi.org/10.1016/S1053-4296(05)80037-2.

Thames HD, Hendry JH. Fractionation in radiotherapy. London: Taylor & Francis; 1987.

van den Aardweg GJMJ, Hopewell JW, Whitehouse EM, Calvo W. A new model of radiation-induced myelopathy: a comparison of the response of mature and immature pigs. Int J Radiat Oncol Biol Phys. 1994;29(4):763–70. https://doi.org/10.1016/0360-3016(94)90564-9.

van der Kogel AJ. Radiation-induced damage in the central nervous system: an interpretation of target cell responses. Br J Cancer. 1986;53(Suppl. 7):207–17.

van der Kogel AJ. Central nervous system radiation injury in small animal models. In: Gutin PH, Leibel SA, Sheline GE, editors. Radiation injury to the nervous system. New York: Raven Press; 1991.

White A, Hornsey S. Radiation damage to the rat spinal cord: the effect of single and fractionated doses of X rays. Br J Radiol. 1978;51(607):515–23. https://doi.org/10.1259/0007-1285-51-607-515.

White A, Hornsey S. Time dependent repair of radiation damage in the rat spinal cord after X-rays and neutrons. Eur J Cancer. 1980;16(7):957–62. https://doi.org/10.1016/0014-2964(80)90335-7.

Wong CS, Hao Y. Long-term recovery kinetics of radiation damage in rat spinal cord. Int J Radiat Oncol Biol Phys. 1997;37(1):171–9. https://doi.org/10.1016/s0360-3016(96)00453-1.

Wong CS, Minkin S, Hill RP. Linear-quadratic model underestimates sparing effect of small doses per fraction in rat spinal cord. Radiother Oncol. 1992;23(3):176–84. https://doi.org/10.1016/0167-8140(92)90328-R.

Wong CS, Minkin S, Hill RP. Re-irradiation tolerance of rat spinal cord to fractionated X-ray doses. Radiother Oncol. 1993a;28(3):197–202. https://doi.org/10.1016/0167-8140(93)90058-G.

Wong CS, Poon JK, Hill RP. Re-irradiation tolerance in the rat spinal cord: influence of level of initial damage. Radiother Oncol. 1993b;26(2):132–8. https://doi.org/10.1016/0167-8140(93)90094-O.

Wong CS, Van Dyk J, Milosevic M, Laperriere NJ. Radiation myelopathy following single courses of radiotherapy and retreatment. Int J Radiat Oncol Biol Phys. 1994;30(3):575–81. https://doi.org/10.1016/0360-3016(92)90943-C.

Experimental Studies on Large Animals

8

8.1   Rhesus Monkeys

In 1984, the NCI sponsored a meeting in Philadelphia to chart new directions for radiation research, especially in the area of clinical radiobiology. Many of those attending were in the vanguard of radiation research at that time. Alfred Smith, Ph.D. was the NCI host. One of the results of the meeting was an NIH request for proposals (RFP) to perform large animal radiobiological experiments to determine clinically relevant dose-response information for major normal tissues. Emphasis was placed on volume effect studies.

In the late 1970s, M. D. Anderson Hospital's Division of Radiotherapy had initiated radiobiological studies on the rhesus monkey to establish the relative biological effectiveness (RBE) for neutron irradiation of the spinal cord. In answer to the NIH RFP, part of these studies would serve as preliminary data to their upcoming renewal of a long-standing program project grant. The first dose-response report from this NCI initiative was published in 1990. Considering the time it took for the RFP to be published, grants to be written in response, reviewed and rewarded, experiments to be performed with a two-year follow-up period after the spinal cord irradiation, results to be analyzed, manuscripts written, reviewed and published, the six-year period between the initial meeting and the first publication was reasonably brief.

The core group involved in the rhesus monkey radiation myelopathy studies at M.D. Anderson included Lester Peters (Chair of the Division of Radiotherapy), Kian Ang, M.D., Ph.D. (leader of the large animal project in the Program Project grant responding to the RFP), L. Clifton Stevens, D.V.M., Ph.D. (veterinary pathologist), and myself. Roger Price, D.V.M., Ph.D. (veterinary pathologist) soon joined the group also. One of the first fellows involved with this project was Guo-Liang Jiang, M.D., who later became President of Fudan University Cancer Hospital.

As stated above, the first data on rhesus monkey spinal cord tolerance had been collected in an experiment designed to explore the RBE for neutron irradiation (Hussey et al. 1982; Schultheiss et al. 1992; Stephens et al. 1983). Four groups of five

© Springer Nature Switzerland AG 2022

T. Schultheiss, *Radiation Myelopathy*,

https://doi.org/10.1007/978-3-030-94658-6_8

monkeys were treated to doses of 13 Gy to 15.5 Gy in 9 fractions with 50 MeV neutrons. We can now see that five subjects per dose group was too few to establish a dose-response function, but these were preliminary data collected to probe the dose axis for the location of the median tolerance dose. Subsequently six monkeys received 77 Gy in 35 fractions with Co-60. Of these six monkeys, three showed neurological deficits within the 2-year follow-up period. Based on this information, a series of experiments was designed in response to the RFP to determine the dose response for radiation myelopathy, the volume effect, and to explore hyperfractionation. Typically, small animal radiation dose-response curves were performed using eight animals per dose group. To establish the dose-response curve at 2.2 Gy per fraction, 16 animals were irradiated to 70.4 Gy and eight animals to 83.6 Gy. The reasons for using the larger number in the lower-dose group were to increase the accuracy of the estimate of D_{20} and because this group also served as one point in the volume effects study to follow. The D_{20} dose is important because the 20 and 80% response points are near optimal for estimating the slope of the dose-response curve. These were very expensive experiments to conduct so it was decided to use the animals that were free of complications at 2 years after 70.4 Gy in a retreatment study. The field length was 8 cm in all studies, except of course for the volume effect study.

Volume Effect Studies

The volume study consisted of the above 70.4 Gy group and two additional 70.4 Gy groups: 16 animals irradiated with a 16-cm field and 20 animals irradiated with a 4-cm field. It would have been preferable to get full dose-response curves at 2 or 3 different field sizes, but such experiments were prohibitively expensive. The large expense of these studies was because of the increased expense of purchase, housing and maintenance of primates and the need to follow the animals without complications for 2 years.

The experiments were designed using the linear logistic function in log (dose) as the dose-response model. The common statistical form of this function is

$$P = \frac{1}{1 + \left(\dfrac{D_{50}}{D}\right)^{k}} \tag{8.1}$$

with D_{50} is the median tolerance dose and k is slope parameter. Rewritten in standard form Eq. (7.1) becomes

$$P = 1 / \left(1 + \exp\left(\beta_0 - \beta_1 \ln\left(D\right)\right)\right) \tag{8.2}$$

where $\beta_0 = k \cdot \ln (D_{50})$ and $\beta_1 = k$. Based on the monkey preliminary data we used $D_{50} = 77$ Gy for the experimental design. Based on preliminary analysis of clinical data we estimated $k = 16$. To model the volume effect we used the probability model, aka the critical element model. This model does not require any additional parameters to model the volume effect.

The results of the dose-response and volume experiments are given in Table 8.1. The estimates of D_{50} and k were reported as 75.8 Gy and 18.7 (Schultheiss et al.

Table 8.1 Design and results of rhesus monkey dose/volume response experiments

Dose (Gy)	Field length (cm)	Number irradiated	Number of myelopathies	Incidence	Estimate
70.4	8	15	3	0.2	0.22
77	8	6	3	0.5	0.57
83.6	8	8	7	0.88	0.85
70.4	16	20	3	0.15	0.11
70.4	4	16	6	0.38	0.39

1990, 1994). (Re-analysis yields $k = 17.5$, which agrees with the stated values of χ^2 in the paper although 18.7 does not. The estimated response in the table is based on $k = 17.5$.) Obviously the agreement is good. In fact the model tends to overfit the data, but there is no way to reduce the number of parameters in the model.

The model for the volume effect was based on (1) the probability of producing a lesion in a volume that is large in comparison to the volume of the lesion itself and (2) the assumption that a complication arises as a result of a lesion being produced anywhere in an irradiated volume (Schultheiss et al. 1983). The mathematical form of the model results from the simple laws of adding probabilities. In this case, if one irradiated two subvolumes in an organ to some dose, the probability of not producing a lesion in the organ equals the probability of not producing a lesion in either of the subvolumes that were irradiated. The math is simple and leads to the following equation:

$$P(D,v) = 1 - \left(1 - P(D,1)\right)^{v} \tag{8.3}$$

where v is a unitless volume, which is the volume irradiated to dose D, divided by the reference volume; $P(D,1)$ is the dose-response function for the reference volume; and $P(D,v)$ is the dose-response function for volume v when uniformly irradiated to dose D. Estimating the probability of injury for an inhomogeneous dose distribution also follows from the same reasoning yielding

$$P(\{D\},v) = 1 - \prod_{i}\left(1 - P(D_i,1)\right)^{\Delta v_i} \tag{8.4}$$

where $\{D\}$ represents a nonuniform dose distribution and Δv_i is the volume element irradiated to dose D_i. The set of all pairs $(\Delta v_i, D_i)$ constitute the differential dose-volume histogram.

One must be careful not to interpret a subvolume of the spinal cord as a functional subunit since the spinal cord itself functions as a single unit and a lesion anywhere in the white matter will result in RM.

A consequence of this model for the volume effect is that the magnitude of the volume effect is determined by the steepness of the dose-response function. Therefore the very steep dose-response functions seen in RM and the small volume effect observed in most experimental studies are consistent with this probability model. (We are assessing the magnitude of the volume effect by the separation on the dose axis of two dose-response curves for different volumes.)

8.2   Pigs

Shortly after the publication of the field-size effect for RM in rhesus monkeys, van den Aardweg et al. published a similar paper using single doses to treat 2.5-, 5-, and 10-cm fields in large white pigs (van den Aardweg et al. 1994, 1995). Three to six animals were used for the dose groups. Because the dose was delivered in a single fraction, the dose response was very steep with $D_{50} = 26.9$ Gy and $k = 55.2$. The pathogenesis is similar in the pig model and the primate model, although the latency in the pigs was very short, varying from 7 to 16.5 weeks. Pigs that were held for 12 months before being killed showed glial scarring and changes in the main ventral artery. The latter is not reported in other species. Nonetheless, the probability model produces a good fit to the data, shown graphically in Fig. 8.1, with a χ^2 value of 4.1 with 12 degrees of freedom. One can appreciate the significant dispersion in the data, probably owing to the very small number of animals used at each dose.

In more recent volume effect studies of RM in large animals, Medin et al. compared the dose response in Yucatan mini pigs treated with a 5-cm long field with a uniform dose to pigs treated with a 5-cm but a nonuniform transverse dose (Medin et al. 2011, 2013b). For the nonuniform treatment, the dose distribution was arranged such that the spinal cord was in the sharp lateral gradient of the distribution with the center of the cord receiving 50% and the lateral edges receiving about 90% and 10% of the plan maximum dose. This dose distribution was designed to simulate a SBRT treatment. There were four to six animals per dose group. The dose response for the two experiments did not differ statistically. When re-analyzed for this work, a dose-response function was obtained with $D_{50} = 20.1$ Gy and $k = 37.0$ when a relative volume of 1 ($v = 1$, Eq. 8.3) was assumed for both the uniform and nonuniform dose distribution. Using $v = 1$ for the uniform dose and $v = 0.5$ for the nonuniform dose distribution, $D_{50} = 19.8$ and $k = 48.6$ was obtained. Although using $v = 1$ for all data points fits the data better, the assumption of 50% relative volume for the second group also yielded an excellent fit. Whereas this experiment does not conclusively

Fig. 8.1 Field-size effect in large white pigs fitted using the probability model (Eq. 7.3) (van den Aardweg et al. 1994, 1995).

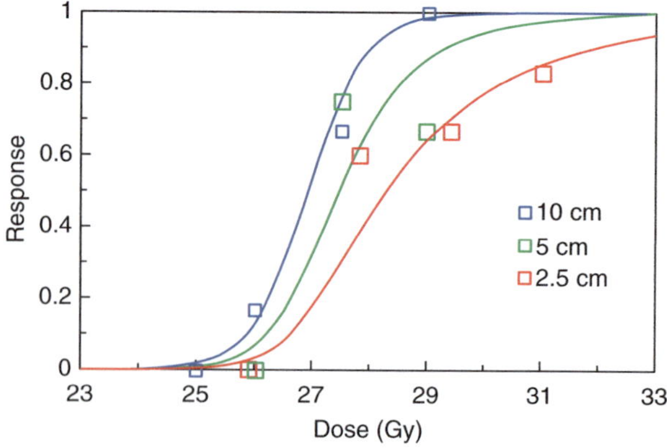

prove the lack of a volume response for lateral dose gradients, it indicates the lack of a detectable volume response in these data.

It is possible that under the conditions in the Medin experiments, the biological response for lateral hemi-cord irradiation and full cord irradiation are the same. However if this is not the case, then there are a number of possibilities that may explain why these experiments showed no difference between hemi-cord and full cord irradiation. First, very few animals per dose point were used. Comparing *frequencies* of 1/6 versus 6/6 in a 2×2 contingency table does not give a significant difference even when a 1-sided test is used. A one-sided test is appropriate in this case since a larger volume cannot reasonably be protective. Thus using only 6 and fewer animals per dose group could hardly be expected to produce statistically different outcomes. Furthermore, the individual doses varied by about ±0.2 Gy within any given dose group, and the animals were treated on different equipment. Given some additional inherent inaccuracies related to dose delivery added to the extreme steepness of the dose-response functions, dramatic differences in observed response rates could occur as a result of slight variations in dose within the range of experimental error. Also the experiments were published about 2 years apart. If these experiments were in fact performed 2 years apart, having identical cohorts could not be assured. Finally, the latencies were statistically different for the two groups. Whether this was due to exogenous differences in the two cohorts or due to the different dose regimens is a matter of speculation. Thus the Medin studies serve primarily to show that these experiments were not capable of detecting a volume effect given the experiment design. In fact, because of the steepness of the dose-response function in these animals, the experimental design virtually precluded a volume effect being detected.

An interesting and somewhat disconcerting note to the pig experiments was published by Medin et al. following the initial publications of data on radiation myelopathy. In a companion study, vertebral bodies of pigs were irradiated to determine the dose response of the vertebrae (Medin et al. 2013a). Animals in this study became paretic at doses above about 24 Gy. It was subsequently determined that the etiology of this injury was damage to the spinal nerve. Moreover, the latency for the injury, the estimated ED_{50}, and the symptoms were all "consistent with the spinal cord as determined in [their] companion studies" on radiation myelopathy. Since there was no histological examination of the spinal nerves in those "companion studies," one cannot be certain that the observed endpoints in those studies were not contaminated by the response of the spinal nerves. In fact, this might apply to the van den Aardweg studies as well and we have seen that the dog studies also elicited changes in the spinal nerve.

To conclude, none of the large animal studies contradict the probability model (Eq. 8.3) of the volume effect, although not all of the studies are edifying. The clinical implication is that the probability of RM is approximately linear in the field length at clinically relevant doses. However, the likelihood of RM at these doses is so small that the volume effect is fundamentally unobservable.

8.3   Beagles

The next set of experiments on field-size effects in RM published on large animals came from two papers by Powers et al. using the beagle model, although the first was a pathology report without different field sizes (Powers et al. 1992a, 1998). A large portion of the data for these papers came initially from the dissertation of Beck (Beck 1990). The data in the Beck dissertation comprised five dose groups from 44 to 76 Gy, each with nine animals. The 76-Gy group in fact had 16 animals, some of which were killed early to assess the progression of radiation lesions. About half of the other animals in the lower-dose groups were killed at 11–13 months (nominally 12 months) or sooner if symptomatic progression necessitated it. All animals underwent evoked spinal cord potentials at least four times during the first year (if they survived), and all animals had a necropsy performed. The field length was 17.5–24 cm and was delivered using lateral opposed portals of a 4 MV beam. The dose was delivered in evenly spaced fractions of 4 Gy. The responses of the animals at 1 year are given in Table 8.2. It was stated that "dogs were replaced if they proved unsuitable or were lost from anesthesia or other causes unrelated to this study." Thus not all groups contain nine animals. Dogs occasionally exhibited permanent neurological deficits following the procedure for evoked spinal cord potentials and were excluded from the analysis.

The 1992 paper categorized the pathological responses in the spinal cord (as well as the meninges and dorsal root ganglia). The five most significant lesion types were multifocal white matter necrosis, mass hemorrhage, segmental parenchymal atrophy, focal fiber loss, and scattered white matter vacuolation. The 76-Gy dose group was deleted from the 1992 publication, and the animals were grouped according to when they were killed: less than 1 year as a result of early symptoms, at 1 year, and at 2 years. A further detail given was that animals were originally scheduled as follows: four to be killed at the end of year one, and five at the end of year two. The number surviving year one corresponds accurately to the data given in the Beck dissertation. However, the 1992 (and 1998) paper's use of a full nine animals per dose group does not agree with the dissertation because only seven or eight animals

Table 8.2 Clinical responses of the 4- and 20-cm experiments as reported in different publications

Beck dissertation (12-month follow-up)			Powers et al. (1992a) (12- or 24-month follow-up)	Powers et al. (1998) (18-month follow-up)	
Dose (Gy)	Field size (cm)	Clinical Response	Number of animals in Powers, 1992	Clinical Response 20-cm	Clinical Response 4-cm
44		0/9	9	2/12	0/6
52	17.5–24	2/7	9	6/12	0/6
60	Nominally	5/8	9	7/12	1/6
68	20	8/8	9	16/17	4/6
72		10/10	–	–	–

were available for evaluation at 12 months in the dissertation. This discrepancy is possibly due to the use of animals in the publications that were excluded in the dissertation.

One may infer from the introduction that the use of pathological endpoints for the dose response in the 1992 paper was done to clarify the pathogenesis of RM. However, all other dose-response papers, whether experimental animal or human, have at least included the clinical responses and have generally used *only* the clinical response in the dose-response analysis. Although the clinical responses were readily available to the investigators, they were not included in this paper. Interestingly, seven animals were said to have clinical symptoms whose only pathological lesions were focal fiber loss or scattered white matter vacuolation. Clinical symptoms without more severe lesions had not been previously reported.

Additionally, Powers et al. published a paper in 1998 that supplemented the data from the 1992 report and added two additional groups: a group of animal irradiated with a 4-cm field to doses of 48–80 Gy at 4 Gy per fraction, and a group irradiated with a 20-cm field to doses of 60–84 Gy with 2 Gy per fraction. The 2 Gy per fraction animals *sometimes* were given 2 fractions per day (at least 6 h apart), making their doses uninterpretable. Three additional animals supplemented the 44- to 60-Gy 1992-dose groups and eight animals were added to the 68-Gy dose group. Clinical responses were given in graphical form for the 20-cm and 4-cm groups who received 4 Gy per fraction. All animals from the Beck dissertation are now attributed to the 20-cm dose group. The responses, read from a graph, are given in Table 8.2 for the 20-cm and 4-cm dose groups who received 4 Gy per fraction. Again, the excluded animals from the Beck dissertation were apparently included in the 20-cm dose group.

The authors state that the assessment of dose response is more accurate using pathological endpoints "because of the inherent subjectivity in evaluating clinical signs." However, it is only because of clinical signs in patients that RM is studied at all, so clinical signs are of paramount importance. Furthermore, it is not at all certain that histological evaluations are inherently less subjective than clinical endpoints. In both papers, the histological endpoint preferred was any combination of massive hemorrhage, white matter necrosis, or parenchymal atrophy, as it was felt that together this set of histologies correlated best with the clinical endpoint. However using data from the 1992 paper, Table 8.3 can be generated. The data from this table show a sensitivity of 70% for the histological surrogate of radiation myelopathy, a specificity of 95%, a positive predictive value of 94%, and a negative predictive value of 72%. Whereas these figures may be excellent with regard to a clinical test, they are not sufficient replacing an endpoint in statistical analysis.

Table 8.3 Comparison of histological and clinical endpoints in dog studies

Histological surrogate	Clinical signs (true myelopathy)	
	+	−
+	16	1
−	7	18

In fact, in the 1998 study they state there was a "lack of correlation between lesions and neurological signs" in the 4-cm group of animals. Thus based on endpoint alone, these studies in beagles cannot be compared directly to any other large animal or human studies where the endpoint was a clinical one. Finally, one may question the relevance of isolated, asymptomatic histological lesions compared to clinical signs.

Two additional issues with these studies are the duration of follow-up and the other procedures to which these animals were subjected. The follow-up periods were 12, 18, or 24 months for animals not killed as a result of advanced symptoms. This would certainly result in biased results given that the latency for RM in dogs was reported to be as long as 19 months for the 20-cm field. In addition, the animals from the Beck thesis, in which spinal cord evoked potentials were performed, were all included in the published studies. In some cases needle tracks were observed in the spinal cord parenchyma post mortem, and in some cases paresis was attributed to the procedure. Therefore one cannot assume that the clinical responses of these animals would be identical to the animals in the 1998 study, which were not subject to this procedure.

Finally, the histology reported by these investigators had features not observed previously associated with RM (Schultheiss and Stephens 1992; Powers et al. 1992b). Meningeal thickening and adhesion to the cord were observed as well as a dose-related increase in neurons with large intracytoplasmic vacuoles and reduced number of satellite cells in the dorsal root ganglia. Abnormal gait and posture were included in indications for neurologic changes indicating radiation myelopathy and these symptoms could result from the dorsal root ganglia changes. Also animals with only focal fiber loss and white matter vacuolation nonetheless had clinical symptoms of RM. Thus it is plausible that clinical symptoms assigned to RM were in fact a result of changes in the peripheral nervous system (nerve roots).

With the issues of variable follow-up and paresis from evoked potentials and/or damage to the dorsal root ganglia, the 20-cm (17.5–24) and the 4-cm group cannot be assumed to have identical tolerance profiles. In fact, the Beck dissertation data and the 1998 paper's supplementary 20-cm data cannot be assumed to have identical inherent responses. Thus, we are left with only the 4-cm data that may have the appropriate clinical responses for analysis and no clinical information on the volume effect in dogs.

8.4   Fractionation

As part of M. D. Anderson's large animal study, an attempt was made to explore the effect of hyperfractionated doses on the expression of RM. Rhesus monkeys were treated twice per day with 1.2 Gy per fraction. Eleven animals were treated to a total dose of 84 Gy and 15 received 98.4 Gy. The twice daily anesthesia administration resulted in weight loss and a general malaise, so treatment breaks were given such that the 84-Gy group had an average treatment duration of 57.5 days, and the 98.4-Gy group had an average of 65.8 days. The response rates are given in Table 8.4.

Table 8.4 Response rates in rhesus monkeys treated BID with 1.2 Gy

Dose	84 Gy	98.4 Gy
Incidence:		
Clinical	6/11 (55%)	8/15 (53%)
Histological	6/11 (55%)	11/17 (65%)
Treatment time (days)	57.5	65.8
Latency (months)	14.8	11.6

The investigators were surprised by the result. The experiment was designed to achieve a response of approximately 20% at the lower dose and approximately 50–75% at the higher dose. Of course these predictions depended upon both the α/β ratio for RM in rhesus monkeys and the half-time of repair of sublethal damage. It seems that the lower dose is less compatible with the predictions than the higher dose. The investigation into the histological response was an attempt to salvage the unexpected outcome, but it did not seem particularly informative. All other studies performed by the group using the rhesus used the clinical outcome exclusively. The explanation for the lack of a dose effect eluded the investigators, and the results were never published although they were presented in various public venues.

As stated above, because some of the 2-Gy-per-fraction dogs in Powers et al. (1998) were irradiated twice per day, no fractionation comparison can be made with animals irradiated with 4-Gy per day QD.

References

Beck ER. Radiation response of the canine spinal cord. Thesis. Colorado State University, Fort Collins Colorado USA; 1990.

Hussey DH, Jardine JH, Raulston GL, Stephens LC, Gray KN, Maor MH, Rodney Withers H. 50-MeVd→Be neutrons: a comparison of normal tissue tolerance in animals with clinical observations in patients. Int J Radiat Oncol Biol Phys. 1982;8(12):2083–8. https://doi.org/10.1016/0360-3016(82)90549-1.

Medin PM, Foster RD, van der Kogel AJ, Sayre JW, McBride WH, Solberg TD. Spinal cord tolerance to single-fraction partial-volume irradiation: a swine model. Int J Radiat Oncol Biol Phys. 2011;79(1):226–32. https://doi.org/10.1016/j.ijrobp.2010.07.1979.

Medin PM, Foster RD, van der Kogel AJ, Meyer J, Sayre JW, Huang H, Öz OK. Paralysis following stereotactic spinal irradiation in pigs suggests a tolerance constraint for single-session irradiation of the spinal nerve. Radiother Oncol. 2013a;109(1):107–11. https://doi.org/10.1016/j.radonc.2013.08.025.

Medin PM, Foster RD, van der Kogel AJ, Sayre JW, McBride WH, Solberg TD. Spinal cord tolerance to single-session uniform irradiation in pigs: implications for a dose-volume effect. Radiother Oncol. 2013b;106(1):101–5. https://doi.org/10.1016/j.radonc.2012.08.007.

Powers BE, Beck ER, Gillette EL, Gould DH, LeCouter RA. Pathology of radiation injury to the canine spinal cord. Int J Radiat Oncol Biol Phys. 1992a;23(3):539–49. https://doi.org/10.1016/0360-3016(92)90009-7.

Powers BE, Gillette EL, Gould DH. Response to Drs. Schultheiss and Stephens. Int J Radiat Oncol Biol Phys. 1992b;23(5):1093–4. https://doi.org/10.1016/0360-3016(92)90922-5.

Powers BE, Thames HD, Gillette SM, Smith C, Beck ER, Gillette EL. Volume effects in the irradiated canine spinal cord: do they exist when the probability of injury is low? Radiother Oncol. 1998;46(3):297–306. https://doi.org/10.1016/S0167-8140(97)00213-2.

Schultheiss TE, Stephens LC. The pathogenesis of radiation myelopathy: widening the circle. Int J Radiat Oncol Biol Phys. 1992;23(5):1089–91. https://doi.org/10.1016/0360-3016(92)90920-D.

Schultheiss TE, Orton CG, Peck RA. Models in radiotherapy: volume effects. Med Phys. 1983;10(4):410–5. https://doi.org/10.1118/1.595312.

Schultheiss TE, Stephens LC, Jiang GL, Ang KK, Peters LJ. Radiation myelopathy in primates treated with conventional fractionation. Int J Radiat Oncol Biol Phys. 1990;19(4):935–40. https://doi.org/10.1016/0360-3016(90)90015-C.

Schultheiss TE, Stephens LC, Ang KK, Jardine JH, Peters LJ. Neutron RBE for primate spinal cord treated with clinical regimens. Radiat Res. 1992;129(2):212–7. https://doi.org/10.2307/3578159.

Schultheiss TE, Stephens LC, Ang KK, Price RE, Peters LJ. Volume effects in rhesus monkey spinal cord. Int J Radiat Oncol Biol Phys. 1994;29(1):67–72. https://doi.org/10.1016/0360-3016(94)90227-5.

Stephens LC, Hussey DH, Raulston GL, Jardine JH, Gray KN, Almond PR. Late effects of 50 MeV d→Be neutron and cobalt-60 irradiation of rhesus monkey cervical spinal cord. Int J Radiat Oncol Biol Phys. 1983;9(6):859–64. https://doi.org/10.1016/0360-3016(83)90012-3.

van den Aardweg GJMJ, Hopewell JW, Whitehouse EM, Calvo W. A new model of radiation-induced myelopathy: a comparison of the response of mature and immature pigs. Int J Radiat Oncol Biol Phys. 1994;29(4):763–70. https://doi.org/10.1016/0360-3016(94)90564-9.

van den Aardweg GJMJ, Hopewell JW, Whitehouse EM. The radiation response of the cervical spinal cord of the pig: effects of changing the irradiated volume. Int J Radiat Oncol Biol Phys. 1995;31(1):51–5. https://doi.org/10.1016/0360-3016(94)E0306-5.

Multispecies biomathematical models are most commonly deployed to describe multispecies fish populations or more generally predator/prey biodynamical populations. In the attempt to devise a multispecies dose-response model for radiation myelopathy, no interspecies interactions occur, so the model simply needs to describe the dose response for various species with as few species specific parameters as possible.

Despite being able to avoid RM in nearly all conventional clinical treatments, the dose and volume response of the spinal cord remain of major importance in a number of radiation therapy applications. Currently, these include stereotactic body radiation therapy (SBRT), craniospinal irradiation and other large-field intensity modulated radiation treatments (IMRT) that include the spinal cord, dose escalation, hypofractionated IMRT, and retreatment. Many of these techniques have been addressed in experimental animals and to some extent in the human data. However, it has been unclear how to apply animal data to the human situation, or how human models of dose and volume response may be validated by experimental settings.

9.1 Biostatistical Models

The probability model (or critical element model) for the volume effect was originally developed as a method to calculate a value for the normal tissue complication probability (NTCP) for an organ irradiated with an inhomogeneous dose when the dose-response function for the homogeneously irradiated organ was known (Schultheiss et al. 1983). It is given by the simple Eq. (8.3).

$$P(D,v) = 1 - \left(1 - P(D,1)\right)^{v} \tag{8.3}$$

where $P(D,v)$ is the dose-response function for the organ when the relative volume v is homogeneously irradiated to dose D. When the organ is inhomogeneously irradiated, Eq. (8.3) becomes

© Springer Nature Switzerland AG 2022

T. Schultheiss, *Radiation Myelopathy*,

https://doi.org/10.1007/978-3-030-94658-6_9

$$P(\{D\}, v) = 1 - \prod_i \left(1 - P(D_i, 1)\right)^{\Delta v_i} \qquad (8.4)$$

where Δv_i is the volume element irradiated to dose D_i. The set of ordered pairs (Δv_i, D_i) comprise the differential dose-volume histogram (DVH).

The dose-response function, $P(D,1)$, most commonly used in these equations is the logistic function in dose or log dose, so $z = \beta_0 + \beta_1 \times$ [dose or ln (dose)] and $P = $ logit (z). Thus for a given species, the dose-volume response function for a single fractionation schedule is determined by only two parameters. How these two parameters may vary across species is speculative. To include fractionation effects requires at least one more parameter, viz. α/β from the LQ model, which adds a term $\beta_2 \times$ dose $\times$ dose-per-fraction to z. (Thus $\alpha/\beta = \beta_1/\beta_2$.) The probability model is the simplest dose-volume response model, and in other models, the most popular of which is the LBK model (Lyman and Wolbarst 1989), an additional parameter is added specifically to address the volume effect thereby achieving a four-parameter model. These parameters are a location parameter (D_{50}), a slope parameter (k for the probability model, γ, or σ for the LBK model), a fractionation parameter, and a volume parameter. When fitting a model to single species dose-volume response data, a 4-parameter model will generally result in an overfit for well-behaved data, often making the results difficult to interpret. Of course, the above discussion applies only to daily fractionation. Furthermore, to address the issue of retreatment requires an additional set of assumptions and leads to other biostatistical models and additional parameters.

In 1990s, the pathogenesis of radiation myelopathy became much better resolved. As stated earlier, the initiating event appears to be radiation death of the endothelial cell lining of the venous microvasculature leading to increased vascular permeability, edema, release of cytokines, breakdown of the BBB, and impaired ability of the toxic byproducts in the interstitial fluid to be removed by paravenous efflux. Because histopathological analysis shows that demyelination can occur in a diffuse pattern in animals without symptoms, it can be reasonably assumed that *symptomatic* radiation myelopathy occurs only when a white matter lesion reaches some minimum size, and that this would occur only with a sufficiently large amount of local endothelial damage.

Following this reasoning, we deploy the k-out-of-N system to model the radiation response of the spinal cord. This is in fact a family of systems used in reliability testing. The simplest k-out-of-N system is one that is made of N identical components; the system fails when k or more of the components fail. A *consecutive* k-out-of-N:F system continues to function until k consecutive components fail (F). A k-out-of-N:G system functions as long as k components still function, i.e., are good (G). The mathematics for the nonconsecutive k-out-of-N system are relatively simple (binomial statistics), but the consecutive system becomes very complex, especially at large values of N. A consecutive k-out-of-N:F system is a series system if $k = 1$ and a parallel system when $k = N$ (Kuo and Zuo 2003).

k-out-of-N systems have a pedigree in biological modeling for this field. The probability model is a 1-out-of-N:F system (Schultheiss et al. 1983). Jackson et al. modeled radiation injury to the liver using a nonconsecutive k-out-of-N:F system (Jackson and Kutcher 1993; Yorke et al. 1993; Jackson et al. 1995). Stavreva et al. have used a consecutive k-out-of-N system, augmented with further assumptions (requiring more parameters) and applied to the spinal cord injury in the dog (Stavreva et al. 2001).

For the large animal data in this study, we model the spinal cord as a consecutive k-out-of-N:F system, and we will suppress the notation in F. Thus we consider the spinal cord to contain N unspecified tissue injury units (Hopewell et al. 1989); a myelopathy occurs when a failure run (spatially sequential failures) of k or larger occurs. Each hypothetical unit is identical in its dose response and therefore has the same probability of failure when it is irradiated to dose D. Then the probability of myelopathy is the probability, P, that the maximum run length is k or larger, given a failure probability, p, of a single unit. It can be shown that the probability of failure, P, is

$$P = \sum_{i=0}^{\left\lfloor \frac{N+1}{k+1} \right\rfloor} (-1)^i \, p^{i-1} \left(1-p\right)^{ik} \left(\binom{N-ik}{i-1} + p \binom{N-ik}{i} \right) \tag{9.1}$$

$$\cong 1 - e^{-N(1-p)p^k} \tag{9.2}$$

The sum becomes inconvenient for recursive algorithms at large N and the approximation (9.2) fails at values of p very close to 1. Therefore we use a formulation based on the upper and lower limits of P given by Muselli and crosschecked against the exact Eq. (9.1) (Muselli 2000).

$$P \cong 1 - e^{\left(\frac{(N-0.72k)(1-p)}{1-p^k} + 1 \right) \ln\left(1-p^k\right)} \tag{9.3}$$

The hypothetical units we are considering are not cells. They are not functional subunits, but they may be related to tissue injury units or tissue rescue units if such things actually exist. In other words, we would reject the idea that they are a specific biological entity until their identity can be verified. For now, we consider them as a mathematical device, and that they are related to the endothelial cell (or cells) and its environment.

The form of p should have no more than 2 parameters. Typically these parameters are a location parameter and a slope parameter. However, it must also be possible to convert from one fractionation schedule to another. Finally, in analogy to cell survival, we expect that for a given fractionation scheme, the log of the survival of the units should be linear in total dose (when it is given in constant fraction sizes). A function that satisfies these requirements is

$$p = 1 - e^{-\beta_1 D - \beta_2 Dd} \tag{9.4}$$

where D is the total dose given in fraction sizes of d. Perhaps more familiarly, we could use

$$p = 1 - e^{-D_e / D_0} \tag{9.5}$$

where

$$D_e = D_1 \frac{\dfrac{\alpha}{\beta} + d_1}{\dfrac{\alpha}{\beta} + d_e} \tag{9.6}$$

where D_e is the dose given in fractions of d_e that is equivalent to D_1 given in fractions of d_1. In fact, we have found that Eq. (9.5) performs better in the data fitting process than Eq. (9.4). Thus rather than β_1 and β_2 as variates, we have α/β and D_0.

Thus we have a 4-parameter model: k, N, D_0, and α/β. In general, this is more parameters than are needed in dose-response analysis. However as noted above, even the most popular NTCP model, the LBK model, has 4 parameters. If the model deploys a separate parameter to account for the volume effect, a minimum of 4 parameters are necessary per species. We now describe how we reduce the number of parameters when considering multiple species.

Using the consecutive k-out-of-N system, we will model the dose response for large animals and humans, keeping k, N, and α/β *constant across species*, varying only the value of D_0 for each species. The large animal data come from studies by Ang et al. (rhesus monkeys), van den Aardweg et al. (pigs), Medin et al. (pigs), Powers et al. (beagles), and Schultheiss et al. (rhesus monkeys). For these studies, we normalize the value of N to a field size of 8 cm (used in rhesus monkeys) and scale N linearly with other field sizes. This process fails to account for the dimensions of the spinal cord other than length. The human data come from new findings presented in Chap. 4. There is no field-size information in Chap. 4, which contains summaries of literature reports, so we simply assume that the value for N for 8 cm applies. Since the monkey data were all generated using 2.2 Gy per fraction, no estimate of α/β is possible for these data. Accordingly, we converted all other doses to equivalent doses in 2.2 Gy fractions using Eq. 9.6.

Equations (9.5) and (9.6) deploy the clinical LQ model. However, this model is used only to convert from one dose per fraction regimen to another. *It should not be interpreted in the cell survival context.*

There are more complex k-out-of-N models. There are two- and three-dimensional models, where instead of a single sequence of elements there are two- and three-dimensional arrays of elements. Although such models would appear to be closer to the biological situation, they introduce both unnecessary complexity and additional but unrequired parameters.

Parameter values were estimated using the maximum likelihood method applied to all the large animal data in a single likelihood equation. Goodness of fit was assessed using the pearson χ^2 statistic and the pearson residual was used for residual analysis. Low-dose consecutive responses of 0% and high-dose consecutive responses of 100% were not included in the goodness-of-fit analysis.

9.2   The Data

All the published, useable x-ray data for large animals and humans are shown in Tables 9.1 and 9.2. Their origins and the rationale for editing, if any, are briefly given below as the papers have been discussed in greater detail elsewhere in this volume.

Monkeys Schultheiss et al. published these data on rhesus monkeys treated with Co-60 with the dose calculated at the depth of mid spinal cord from a single posterior field (Schultheiss et al. 1990, 1994). The fields extended from the top

Table 9.1 Data from large animal radiation myelopathy used in analysis

Reference; animal; follow-up time (mo)	Dose (Gy)	Dose/fx (Gy)	Field length (cm)	r	N[a]
Schultheiss et al. (1994, 1990); rhesus monkey; 24	70.4	2.2	8	3	15
	77	2.2	8	3	6
	83.6	2.2	8	7	8
	70.4	2.2	16	6	16
	70.4	2.2	4	3	20
	44	2.2	8	0	70
van den Aardweg et al. (1995); large white pig; 16–25	25	25	10	0	6
	26	26	10	1	6
	27.5	27.5	10	4	6
	29	29	10	3	3
	26	26	5	0	5
	27.5	27.5	5	3	4
	29	29	5	2	3
	25.9	25.9	2.5	0	3
	27.8	27.8	2.5	3	5
	29.4	29.4	2.5	4	6
	31	31	2.5	5	6
Medin et al. (2011, 2013b); Yucatan mini pig; 12	17.5	17.5	7	0	5
	19.5	19.5	5.1	1	6
	22	22	5.1	5	5
	24.1	24.1	5.1	4	4
	16.9	16.9	5[b]	0	5
	18.9	18.9	5[b]	1	5
	21	21	5[b]	4	5
	23	23	5[b]	4	4
	25.3	25.3	5[b]	4	4
Powers et al. (1998); beagle; 18	48	4	4	0	6
	56	4	4	0	6
	60	4	4	0	6
	64	4	4	0	6
	72	4	4	1	6
	80	4	4	4	6
	68	4	20	7	8

[a]N is the number of animals followed for the full follow-up period including those killed early because of symptoms. It excludes symptom-free animals that did not survive the full follow-up. For the human data, N is the number of patients who survived the minimum follow-up.
[b]In the 5-cm group, a 90% to 10% dose gradient was deployed laterally across the cord.

of C1 to approximately T2 for the 8-cm field and to T11 for the 16-cm field. The inferior border of the 4-cm field was placed at T1-T2 and extended in the cephalad direction. The original design called for a 20-cm field rather than a 16-cm field, but it was felt that the larger field involved too much esophagus. The longest observed latent period was 20 months (24 months in retreated animals (Ang et al. 2001)) and all animals were held for at least 24 months after the final treatment.

Table 9.2 Human data from retrospective studies (Chap. 4)

Spinal cord level	Dose (Gy)	Dose/fx (Gy)	Field length[a] (cm)	r	N[b]
Cervical	60	2	8	1	12
	65	1.63	8	0	24
	54	3	8	7	15
	19	9.5	8	4	13
	47.5	1.85	8	0	211
	52.5	1.85	8	0	22
	60	2	8	2	19
	65	1.63	8	0	19
	62.8	1.93	8	3	109
Thoracic	37.8	3.15		0	86
	34.1	5.68		8	87
	34.9	5.82		4	31
	38	8, 4	Effective fields length for the thoracic data is a fitted parameter	8	157
	34	8, 4, 2		9	230
	17.9	8.83		3	524
	41	3.15		2	153
	31.5	5.25		0	36
	36	6		1	5
	41.2	4.12		6	200
	42	4.2		4	97

[a]A field length of 8 cm was assumed for the cervical cord and the field length for the thoracic cord was included in the data fitting. The sources of the data are given in Table 4.1
[b]N is the number of patients in the cohort

Pigs van den Aardweg et al. published the dose response in large white pigs irradiated with parallel opposed lateral fields of Co-60 to 2.5, 5, and 10 cm field lengths (van den Aardweg et al. 1995). The center of each field was located at the C3–C4 space. The animals were treated with single doses. The longest latent period was 16.5 weeks and the animals were held for a minimum of 70 weeks.

Medin et al. published two separate papers on radiation myelopathy in Yucatan mini pigs treated with single doses of 6MV photons (Medin et al. 2011, 2013b). The first paper used dynamically shaped arcs to generate a dose distribution with a lateral dose gradient from about 90% to 10% across the cord. The field length was 5 cm, extending from mid C4 to mid C7. The pigs were held for a minimum of 12 months with the longest latency being 23 weeks. The second paper treated the pigs with uniform irradiation to field sizes of about 5.1 cm.

An interesting and somewhat disconcerting note to the pig experiments was published by Medin et al. after their initial publications of data on radiation myelopathy (Medin et al. 2013a). In this third study, vertebral bodies of pigs were irradiated to determine the dose response of the vertebrae. Animals in this study became paretic at doses above about 24 Gy. It was subsequently determined that the etiology of this injury was damage to the spinal nerve. Moreover, the latency for the injury, the estimated ED_{50}, and the symptoms were all "consistent with the spinal cord as determined in companion studies" on radiation myelopathy above. Since there was some histological examination of the spinal cord but not the spinal nerves in those original studies, one cannot be certain that the observed endpoints in those studies were not conflated with the response of the spinal nerves. Paralysis resulting from spinal nerve injury in addition to myelopathy could at least partially explain the very similar results of homogeneous and lateral gradient dose responses. In fact, this effect could have occurred in the van den Aardweg studies as well.

Beagles Powers et al. published two papers using the beagle model with much of the data for these papers coming from the dissertation of Elsa Beck (Beck 1990; Powers et al. 1992, 1998). The experimental design in Beck's dissertation included five dose groups, 4 Gy per fraction, from 44 to 76 Gy, with nine animals in each group. Half the animals were to be followed only for 12 months, but some were killed sooner owing to symptoms of RM. All animals underwent evoked spinal cord potentials at least four times during the first year after irradiation and all animals had a necropsy performed. The fields encompassed T1 to T13 (17.5 to 24 cm) and were delivered using lateral opposed portals. Dogs that exhibited permanent neurological deficits after the procedure for evoked spinal cord potentials were excluded from the analysis in the thesis but probably not from the published papers.

The 1992 publication highlighted the pathological responses in the 44 to 68 Gy animals. The meninges and dorsal root ganglia were also studied. The five most common spinal cord lesions were multifocal white matter necrosis, massive hemorrhage, segmental parenchymal atrophy, focal fiber loss, and scattered white matter vacuolation. Animals were originally scheduled to be killed as follows: four at the end of year one, and five at the end of year two. The animals killed with symptoms at less than a year comprised another group. The number alive at 1 year corresponds exactly to the data in the Beck dissertation. An important statement in the dissertation is that all 68-Gy animals exhibited clinical symptoms.

The objectives of the 1992 study were "to identify the pathologic lesions" and "to correlate these lesions with clinical neurologic findings." No pathognomonic lesion nor combinations of lesions for RM were found. Although available to the investigators, the clinical responses were not included in this paper. In fact, the analysis of binary data depends upon a quantal response. Converting a continuous histological response into a quantal response is much more difficult than determining if an animal is exhibiting paresis. Interestingly, seven animals were said to have clinical symptoms whose only pathological lesions were focal fiber loss or scattered white matter vacuolation. Clinical symptoms without more severe lesions had not been previously reported.

Unfortunately, the data from the 1992 paper cannot be used directly in this analysis for a number of reasons. Of course, a major reason is that the clinical responses were not reported by dose. Another major factor is that the spinal evoked potential introduced additional traumatic events directly to the spinal cord. Also, the reported pathologic lesions did not have a one-to-one correspondence with the clinical outcome, which is the foundation for radiation myelopathy dose-response analysis.

In a 1998 paper, Power et al. supplemented the data from the 1992 report and added two additional groups: a group of animals irradiated with a 4-cm field to doses of 48 to 80 Gy at 4 Gy per fraction, and groups irradiated with a 20-cm field to doses of 60 to 84 Gy with 2 Gy per fraction. Three additional animals supplemented the 44- to 60-Gy 1992 dose groups and eight animals were added to the 68-Gy dose group.

The animals in the 2-Gy-per-fraction group were given an unspecified number of BID treatments at 6-hr intervals. Without more specific dosimetric information, no dose-response analysis can be performed due to an unknown amount of incomplete repair and it cannot be used for determining the α/β value. Therefore the 2-Gy groups were not used in this analysis.

The clinical responses were given in graphical form for the 20-cm and 4-cm groups, all of whom received 4 Gy per fraction. In the graph for the 20-cm group, it is possible to determine that at 68 Gy, 16 of the total 17 animals were paretic. Since all nine animals from Beck's dissertation showed clinical symptoms, this means that seven of the additional eight animals at this dose were positive. Using some of the information from the *Clinical signs* section of the paper, it is possible to determine that only one other animal in the 20-cm group had clinical signs, but its dose was not stated.

There is some indication that the beagles presented in the Powers et al. (1998) paper were also subjected to spinal evoked potentials, which would make them also ineligible for inclusion in the analysis and therefore totally eliminate all data from this species. In a 1991 abstract, Gillette et al. discuss the volume study that included 20- and 4-cm irradiations. Quoting from that abstract "The functional response was evaluated … by electrophysiologic studies. Electrophysiology was done at the end of irradiation and at 6 months and 1 year after irradiation." There was no mention of limiting the electrophysiologic studies to the dogs treated to 20 cm. Furthermore, Beck, who performed the electrophysiologic studies on the 20-cm dogs as part of her dissertation, was not a coauthor on the abstract, possibly indicating that the animals who were the subject of the abstract were not part of the Beck dissertation. As a result of this ambiguity, the statistical analysis was done with and without dogs from the 1998 publication. There was no material difference in the result.

Retreatment The spinal cord has been shown to be capable of significant long-term repair. Assuming the damage model implied by the k-out-of-N model, there are a number of potential repair models. (Only for monkeys are there dose-response data for retreatment in large animals.) Using the value of N determined from single course treatments, we create a Monte Carlo model. This model consists of h histories, each having N units. The probability of deactivating each of the N units is based on the dose schedule of the first course. From these probabilities, a series of failure runs is created for each history. Thus each history represents a treated spinal cord. This is illustrated in Fig. 9.1. To model the long-term repair, we assume that during the interval between treatments, repair or recovery of deactivated units is possible. There are a number of ways this repair can be envisioned, but most repair mechanism rely on the end units of a failure run to be repaired before the interior units.

Fig. 9.1 A representation of (a portion) the spinal cord as a consecutive k-out-of-N system. The system fails (clinical neurological damage) when the number of units in a failure run exceeds k. This figure depicts two failure runs—one of length 4 and a second of length 6. After treatment, recovery is modeled by creating an algorithm that describes the stochastic repair of the failure runs inward from the end units.

First, units on the end of a failure run can be repaired with a fixed probability, then the next unit is available for repair. When an end unit goes unrepaired, as determined randomly by the fixed probability, repair on that end of the failure run ceases. Both ends of all failure runs are subject to the repair process. One could also model the repair of a fixed number of units. Another alternative is to model a minimum run length that is completely repaired and run lengths larger are repaired with a probability that depends on the number of remaining unrepaired units. After the repair process is completed, re-irradiation is modeled by determining the probability of deactivation for all remaining viable units, based on the re-irradiation dose schedule and the probability given by Eq. (9.5). We assume that at the time of re-irradiation, all viable units, whether they were repaired or were undamaged after the initial radiation, have the same dose response as in they did *de novo*. The units that failed as a result of the re-irradiation are determined and the run lengths are recounted. If a failure run length exceeds k units, then the history is counted as a myelopathy case. In the case where we use the nonconsecutive k-out-of-N system (rodents), it is not necessary to keep track of individual run lengths for retreatment. In this case we are concerned only if the total number of deactivated cells exceeds k.

It was demonstrated in Chap. 6 that in rodents, there is a hybrid volume effect. The probability (critical element) model holds for field lengths greater than 10 mm, and the critical volume model holds for lengths less than 8 mm. The dose-response function is a product of these probabilities. Rodents are not addressed further in this chapter.

9.3   Results of Large Animal Data Analysis

De Novo Irradiation in Large Animals   Table 9.3 gives the maximum likelihood estimates (MLE) for the free parameters for the large animals and human. Figure 9.2a–e show the dose-volume responses. Good fits were obtained for all spe-

Table 9.3 Parameter values for the model defined by Eqs. (9.1), (9.5), and (9.6) fitted to the data in Tables 9.1 and 9.2

N	k	α/β (Gy)	v_{thor} (cm)	D_0 (Gy) human	D_0 (Gy) monkey	D_0 (Gy) pig[a]	D_0 (Gy) pig[b]	D_0 (Gy) dog
4000	28	0.63	2.09	46.6	51.6	180	92.4	78.0

[a]van den Aardweg et al.
[b]Medin et al.

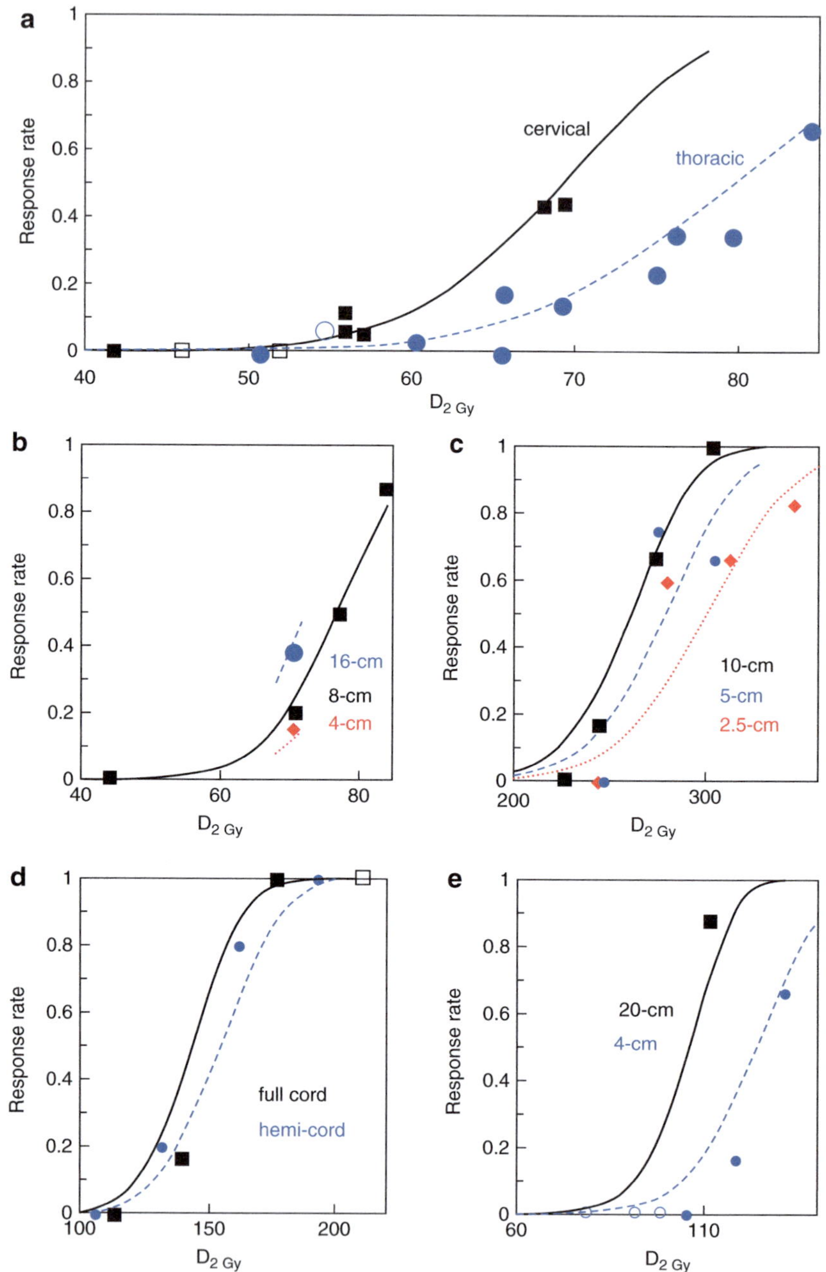

Fig. 9.2 Dose-response curves for humans and 4 animal models using Eq. (9.1), (9.5), and (9.6) and maximum likelihood estimates in Table 9.3. (**a**) humans with black points being cervical data and blue points being thoracic data. The open squares are consecutive 0% data points used in the analysis but not in the goodness-of-fit statistic. The blue open circle in (**a**) represents a data point from Macbeth et al. (1996) that was included in the analysis but could be considered an outlier. (**b**) rhesus monkeys. Black data points are 8-cm data, blue is the 16-cm data point, and red is the 4-cm data point. (**c**) pig data of van den Aardweg et al. (**d**) pig data of Medin et al. (**e**) dog data of Powers et al.

cies and field sizes. The MLE estimate for α/β in this study was 0.63 Gy whereas the analysis of human data in Chap. 4, Model 4 found $\alpha/\beta = 0.56$ Gy.

The goodness-of-fit statistics is a surprisingly high value of p (χ^2) = 0.75. An examination of the residuals shows only one out of 48 residuals used in the χ^2 is greater than 2, but it is from the reports by Macbeth of MRC clinical trials and does not present any concerns regarding its validity. However, the distribution of the residuals is not consistent with the normal distribution, so the p-value of χ^2 may be suspect.

Of greater concern is the highly correlated estimates of k and N. Figure 9.3 shows values of k and N optimized using MLE with the maximum likelihood varying only from -299.8 to -299.3 when N varies from 1000 to 16,000 and k from 52 to 20. This correlation is more than merely statistical. For given values of k and p, the probability of failure of a single unit, the value of N is fixed exactly. Of course, p is determined by the dose regimen, of which there are 53. We should expect D_0 and α/β also to be correlated since we can see from Eqs. (9.4) and (9.5) that $D_0 = \alpha + 2.2\beta$. This will result in large joint confidence intervals.

It can be seen that the highest values of D_0 are found in the pig model and the lowest values in primates, both human and non. Given this, in Fig. 9.4 we plotted

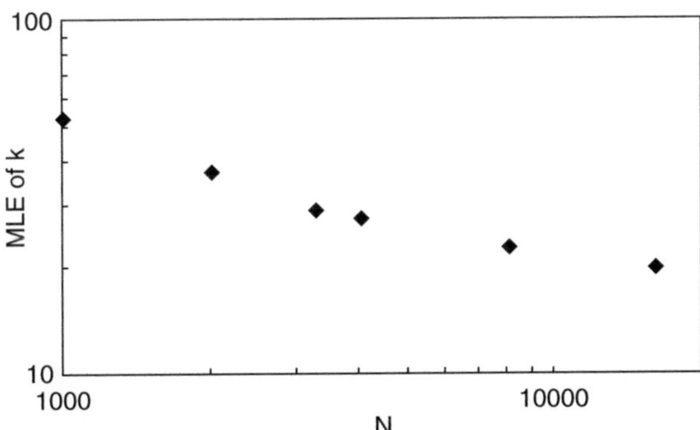

Fig. 9.3 Maximum likelihood estimate of k given N.

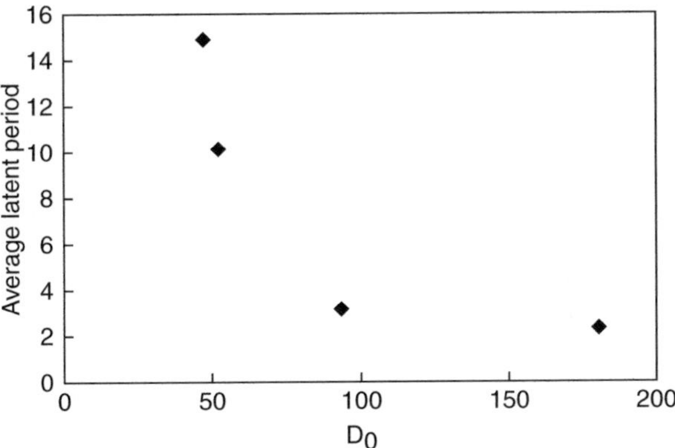

Fig. 9.4 Average latent period for myelopathy cases versus D_0 of the species. Data from the dog experiment were not fully available.

the latent periods versus D_0 for the animal models. A fairly dramatic dependence is seen. Although this may be merely coincidental, there is a potential biological explanation. The smaller the value of D_0 is, the more sensitive the tissue is to radiation. Presumably with N, k, and α/β constant, it would be reasonable to predict that greater target damage would manifest itself sooner. This is certainly congruent with the dose dependency of latency observed within a species. Although the D_0 values can vary widely because of correlations with other parameters, the relative magnitude remains fairly stable.

Many dose-response models have been applied to radiation myelopathy data, but always to one species and usually one experiment at a time. The consistent histopathological picture and the increased understanding of the pathogenesis prompted this attempt to create a biologically based model that could be applied across species.

Merely using the same model with different parameters for each species does not constitute a cross-species model. Any empirical model, including a logistic model, would be a cross-species model if this were the case. A biologically based cross-species model should contain both species dependent and species independent parameters. The k-out-of-N model adapted here is a simple version of such a model. k-out-of-N models can be very complex, even in the one dimensional version. Clearly, a higher dimensional version would be more appropriate to the biological situation we are modeling. However, given the paucity of data from any given experiment, the high correlation of parameters estimates, and the complexity of the mathematics, a higher dimension model would not be better supported by the data and would be mathematically intractable. Furthermore, the 3-dimensional models recently explored in reliability studies are generally limited to N-values of one or two digits and cannot easily accommodate the large values of N found here.

In broad terms, two of the parameters of our model, k and N, are related to the formation of the lesion, and two, α/β and D_0, are related to the response of the individual targets. *In this context, "target" refers to an unknown number of cells that constitute or support the endothelium in the white matter of the spinal cord.* These targets should not be construed as individual cells. According to this model, the loss of a sufficient number of these targets in contiguity results in a lesion that ultimately leads to radiation myelopathy. We have tried to reflect the biology elucidated by a number of investigators who have provided convincing evidence that it is damage to the vascular endothelium that results in radiation myelopathy. Because the estimates of k and N are highly correlated, their specific values reported here may not reflect underlying biology. However, the ratio of k/N may be indicative of the minimum size of the lesion that can produce RM. We standardized our model on the volume of rhesus monkey cervical spinal cord, which in cross section is approximately 5 mm by 8 mm. Assuming that the cord is approximately two thirds white matter, and with $k = 44$ and $N = 4000$, the minimum clinically detectable lesion size is approximately 17 mm^3 or 2.6 mm on a side. This may be considered too large to constitute the minimum clinically detectable lesion, and we would not attempt further justification. However, this result is more accurate than one might expect, given that this prediction was not intentionally built into the model.

The volume effect in this study was modeled by scaling the value of N with field size, using 8-cm as the standard. The value of k was held constant across species and therefore, in large animals, the number of contiguous targets that constituted a lesion of minimum size was also constant. This does not take account of the fact that the cross section of the spinal cord would vary by species. If this were taken into account, the estimates would change somewhat, but it is doubtful that the overall fit would improve.

The value of D_0 reflects the sensitivity of the specific animal model to radiation. It reflects the position of the dose-response curve on the dose axis.

The value of α/β reflects the fractionation sensitivity of the endpoint. It is generally accepted that the value of this parameter for RM is below 3 Gy. This parameter is also responsible for how large the separation is between dose-responses curves generated from different doses per fraction.

Although many aspects of this k-out-of-N model suggest its general applicability to RM, it is presented only as an example of and a preliminary effort in modeling multiple species data. Its actual implementation has a number of difficulties. Chief among these are the problems that arise as a result of not being able to write the model in closed form. Its mathematical complexity makes its applications to arduous and virtually impossible in some cases. For example applying it to an inhomogeneous dose distribution requires Monte Carlo coding. Furthermore, it cannot be recast in a generalized linear form and consequently the regression diagnostics are much more difficult and limited.

If one were to pursue this line of analysis, the next step would be to eliminate the parameter N. The strength of this model is that it highlights the utility of including the concept of a minimum lesion size. However, there should not be a correlation between minimum lesion size and the irradiated volume, even though the probability of injury is related to each of these. Another benefit of this model is that it clearly distinguishes between dose and damage, and it shows a methodology to analyze retreatment in terms of damage repaired and redelivered that is separate from dose.

For retreatment of rhesus monkeys, first we need to consider the distribution of the lengths of failure runs. This distribution is the geometric distribution, which is the discrete analog of the exponential distribution. Figure 9.5a shows an example of this distribution and Fig. 9.5b shows what it might look like after repair, depending on the exact algorithm used to model the repair. From this figure, one can see that this model has successfully depicted the separation between dose and damage. Because the remaining failure runs (damage) shown in Fig. 9.5b do not have the same distribution shape as the initial damage (Fig. 9.5a), there is no way to equate the remaining damage with a dose equivalent.

Because most repair mechanism would tend to repair fully more short failure runs than longer failure runs, the average run length after repair will increase. But because the average run length also increases with dose, the effect of repair is to make the damage distribution look more like a small volume where a higher dose remained unrepaired, rather than the original volume where a lower dose remained unrepaired. Realizing this, a simple model was deployed to approximate repair in the k-out-of-N system. The repaired-volume model of Chap. 7 was deployed. In this

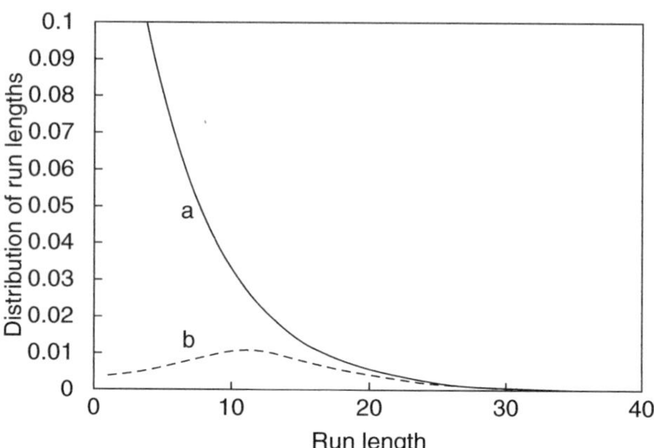

Fig. 9.5 Damage distribution in the k-out-of-N model. The solid curve (**a**) represents the distribution of run lengths after the initial radiation. The dashed curve (**b**) represents the distribution after repair has occurred. This figure depicts why dose and damage cannot be equated since the dashed curve could not be achieved by a single dose.

model the retreated spinal cord is modeled as comprising two components—a fraction, v_r, with the damage completely repaired and a complementary fraction, 1- v_r, with the damage completely unrepaired. Thus upon treatment, the fraction v_r behaves as having received only the retreatment dose, and the fraction 1- v_r behaves as having received the original plus the retreatment dose. This essentially turns the retreatment effect into a volume effect.

Applying this model to the monkey retreatment data, we get the results shown in Fig. 9.6. The data on this graph are depicted with the retreatment dose on the abscissa, except on the far right is shown the original *de novo* data and model fit (in black). Also falling on this line is the single data point with retreatment of 66 Gy 3 years after the initial 44 Gy. The best fit value for value of v_r above is 0, indicating complete repair of all damage. The three data points on the next graph to the left show the data 2 years following the initial 44 Gy. The curve for the model was extended to show that it does not cross the *de novo* graph but approaches it asymptotically. The next curve to the left and the single point on it represents 1 year following 44 Gy. Finally, on the far left are the data from the preliminary data where four animals in each group were retreated following the initial 70.4 Gy, never before shown with the full retreatment series. Clearly the model is challenged by these data that do not exhibit a dose response. However, since there are only 4 animals per group, the residuals for these three points for this model are acceptable—less than 1.6. Furthermore, the current model flattens the retreatment dose response more than any other models explored (except those that include a different steepness parameter for retreatment, and those models then intersect and cross the *de novo* response). Another possible scenario, which is mentioned in Chap. 7, is that the two myelopathy cases that appear after the second treatment probably were simply very late events occurring as a result of the initial 70.4 Gy.

The question of whether the retreatment dose-response curve has a lower slope than the *de novo* curve remains unresolved. There simply is not enough data to clarify the answer. In pigs, rats, and guinea pigs, it seems that when the initial damage is completely repaired, the retreatment dose response is as steep as the initial dose response. Further complicating the issue is the fact that the steepness parameter in dose-response

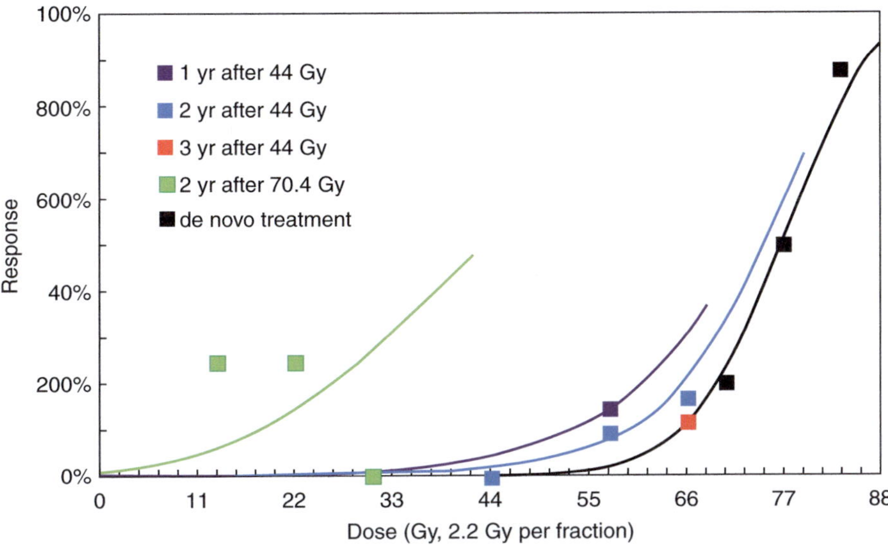

Fig. 9.6 Dose response for retreatment of rhesus monkey spinal cord. In black is the *de novo* treatment response. The retreatment for various times after treatment is modeled by a fraction, v_r, of the volume in which the damage is completely repaired and a complementary fraction, $1\text{-}v_r$, in which no repair of damage occurs.

Table 9.4 Fraction of volume completely repaired after 44 Gy

Time after 44 Gy	Fractional volume completely repaired	Repaired fraction of 44 Gy to achieve 10% response
1	0.98	0.720
2	0.989	0.867
3	0.999	1

analysis innately has a larger variance than the location parameter. However, as always the clinical situation makes the concerns more moderate since radiation myelopathies in humans rarely result from planned treatment doses. They are more likely to result in the most sensitive individuals in the population or as a result of unplanned events.

Finally we return to the issue of damage versus dose (or "remembered" dose). Table 9.4 and Fig. 9.7 illustrate long-term repairs in two ways. Table 9.4 gives the fractional volume in which the damage is fully repaired, according to the model, versus time after treatment. Also given is the total dose causing a 10% myelopathy rate minus the *de novo* dose causing a 10% rate all divided by the initial dose (44 Gy). These fractions represent the amount of recovered dose using the 10% response as the criterion and are very close to the 76%, 85%, and 101% of dose recovered given in Ang 2001. (The 10% response rate was chosen as being close to the retreatment responses for the 2-year retreatment data.) Obviously a relatively small amount of unrepaired volume must be compensated by a much larger portion of dose.

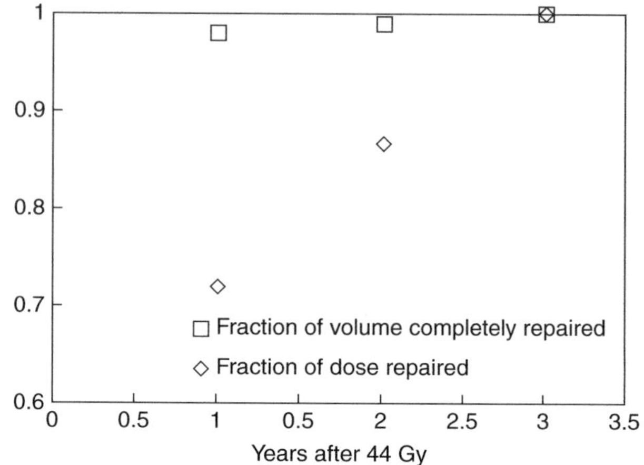

Fig. 9.7 Fraction of dose and fraction of volume repair versus time after treatment. It is striking that a very small volume in which the damage goes completely unrepaired can have such a significant effect on the dose response.

The Transition from Large Animals to Rodents It was not possible to achieve an acceptable fit with the consecutive k-out-of-N model for the rodent data of Bijl et al. and Hopewell et al. However, as addressed in Chap. 6, both the Bijl et al. and the Hopewell et al. data are well fitted by the hybrid probability/critical volume model. The data from Philippens et al. were not analyzed because different volume groups had different histological endpoints, i.e., some of the endpoints represented radiculopathy and not myelopathy.

As stated above, the probability model (aka critical element model) is a 1-out-of-N model and the critical volume model is a nonconsecutive k-out-of-N model. Thus the critical volume model depends on a minimum number of targets being sterilized before a clinical endpoint is observable. This does not exactly match the general picture of injury at the microscopic level. Even in rats, the white matter lesions are discrete foci of necrosis and not characterized by diffuse damage. On the other hand, this depiction of the damage that causes RM may be too literally related to what is seen at the time of the onset of symptoms. The latent periods is months after treatment, but the initial damage presumably occurs within hours, or at most days, of the final dose. It may be that during the latent period some level of the initial diffuse damage ultimately becomes manifest as discrete lesions.

No matter what the exact mechanism is, the k-out-of-N model systems serve as a prototype of biomathematical model that requires a minimum size of damage. These models have been shown in this volume to fit experimental data whose behavior are poorly described by other model systems.

References

Ang KK, Jiang GL, Feng Y, Stephens LC, Tucker SL, Price RE. Extent and kinetics of recovery of occult spinal cord injury. Int J Radiat Oncol Biol Phys. 2001;50(4):1013–20. https://doi. org/10.1016/S0360-3016(01)01599-1.

Beck ER. Radiation response of the canine spinal cord. Thesis. Colorado State University, Fort Collins Colorado USA; 1990.

Hopewell JW, Calvo W, Campling D, Reinhold HS, Rezvani M, Yeung TK. Effects of radiation on the microvasculature. Front Radiat Ther Oncol. 1989:85–95. https://doi.org/10.1159/000416573.

Jackson A, Kutcher GJ. Probability of radiation-induced complications for normal tissues with parallel architecture subject to non-uniform irradiation. Med Phys. 1993;20(3):613–25. https://doi.org/10.1118/1.597056.

Jackson A, Ten Haken RK, Robertson JM, Kessler ML, Kutcher GJ, Lawrence TS. Analysis of clinical complication data for radiation hepatitis using a parallel architecture model. Int J Radiat Oncol Biol Phys. 1995;31(4):883–91. https://doi.org/10.1016/0360-3016(94)00471-4.

Kuo W, Zuo MJ. Optimal reliability modeling: principles and applications. Wiley; 2003.

Lyman JT, Wolbarst AB. Optimization of radiation therapy, IV: a dose-volume histogram reduction algorithm. Int J Radiat Oncol Biol Phys. 1989;17(2):433–6. https://doi.org/10.1016/0360-3016(89)90462-8.

Macbeth FR, Wheldon TE, Girling DJ, Stephens RJ, Machin D, Bleehen NM, Lamont A, Radstone DJ, Reed NS, Bolger JJ, Clark PI, Connolly CK, Hasleton PS, Hopwood P, Moghissi K, Saunders MI, Thatcher N, White RJ. Radiation myelopathy: estimates of risk in 1048 patients in three randomized trials of palliative radiotherapy for non-small cell lung cancer. Clin Oncol. 1996;8(3):176–81. https://doi.org/10.1016/S0936-6555(96)80042-2.

Medin PM, Foster RD, van der Kogel AJ, Sayre JW, McBride WH, Solberg TD. Spinal cord tolerance to single-fraction partial-volume irradiation: a swine model. Int J Radiat Oncol Biol Phys. 2011;79(1):226–32. https://doi.org/10.1016/j.ijrobp.2010.07.1979.

Medin PM, Foster RD, van der Kogel AJ, Meyer J, Sayre JW, Huang H, Öz OK. Paralysis following stereotactic spinal irradiation in pigs suggests a tolerance constraint for single-session irradiation of the spinal nerve. Radiother Oncol. 2013a;109(1):107–11. https://doi.org/10.1016/j.radonc.2013.08.025.

Medin PM, Foster RD, van der Kogel AJ, Sayre JW, McBride WH, Solberg TD. Spinal cord tolerance to single-session uniform irradiation in pigs: implications for a dose-volume effect. Radiother Oncol. 2013b;106(1):101–5. https://doi.org/10.1016/j.radonc.2012.08.007.

Muselli M. New improved bounds for reliability of consecutive-k-out-of-n: F systems. J Appl Probab. 2000;37(4):1164–70.

Powers BE, Beck ER, Gillette EL, Gould DH, LeCouter RA. Pathology of radiation injury to the canine spinal cord. Int J Radiat Oncol Biol Phys. 1992;23(3):539–49. https://doi.org/10.1016/0360-3016(92)90009-7.

Powers BE, Thames HD, Gillette SM, Smith C, Beck ER, Gillette EL. Volume effects in the irradiated canine spinal cord: do they exist when the probability of injury is low? Radiother Oncol. 1998;46(3):297–306. https://doi.org/10.1016/S0167-8140(97)00213-2.

Schultheiss TE, Orton CG, Peck RA. Models in radiotherapy: volume effects. Med Phys. 1983;10(4):410–5. https://doi.org/10.1118/1.595312.

Schultheiss TE, Stephens LC, Jiang GL, Ang KK, Peters LJ. Radiation myelopathy in primates treated with conventional fractionation. Int J Radiat Oncol Biol Phys. 1990;19(4):935–40. https://doi.org/10.1016/0360-3016(90)90015-C.

Schultheiss TE, Stephens LC, Ang KK, Price RE, Peters LJ. Volume effects in rhesus monkey spinal cord. Int J Radiat Oncol Biol Phys. 1994;29(1):67–72. https://doi.org/10.1016/0360-3016(94)90227-5.

Stavreva N, Niemierko A, Stavrev P, Goitein M. Modelling the dose-volume response of the spinal cord, based on the idea of damage to contiguous functional subunits. Int J Radiat Biol. 2001;77(6):695–702. https://doi.org/10.1080/09553000110047555.

van den Aardweg GJMJ, Hopewell JW, Whitehouse EM. The radiation response of the cervical spinal cord of the pig: effects of changing the irradiated volume. Int J Radiat Oncol Biol Phys. 1995;31(1):51–5. https://doi.org/10.1016/0360-3016(94)E0306-5.

Yorke ED, Kutcher GJ, Jackson A, Ling CC. Probability of radiation-induced complications in normal tissues with parallel architecture under conditions of uniform whole or partial organ irradiation. Radiother Oncol. 1993;26(3):226–37. https://doi.org/10.1016/0167-8140(93)90264-9.

Epilogue

Chapter 1

- The dose limit of 45 Gy in 25 fractions was reached by an evolutionary process that involved statistical errors and poor understanding of fractionation effects. It works as an "avoidance" target, but absolute adherence to this limit can be unwise.
- More than two fractions per day should be avoided until a thorough understanding of the CHART experience is attained.

Chapter 2

- Radiation damage to the white matter is initiated by damage to the vascular endothelium. It is probable that the venous endothelium is of primary importance.
- Early apoptosis is not predictive of late injury.
- The final stages of white matter necrosis are endothelial cell damage, activation of astrocytes, breakdown of the blood-brain barrier, repetition of this cycle, and buildup of toxins in the peri-venous space.
- This process applies almost exclusively to the cervical and thoracic levels.
- Paresis of the lumbar cord is more likely a result of demyelination and necrosis of nerve roots.
- Radiation oncologists should be alert to the possible increase in late effects from radiation in patients with or who have recovered from Covid-19.

Chapter 3

- The majority of rodent experiments in RM have an excessive number of extreme (0 or 100%) responses. Thus they are more representative of preliminary data and a final design.

© Springer Nature Switzerland AG 2022
T. Schultheiss, *Radiation Myelopathy*,
https://doi.org/10.1007/978-3-030-94658-6

- To support the validity of model assessment, no more than one low dose and one high dose extreme response should be used in goodness-of-fit tests. The degrees of freedom should be adjusted accordingly.
- Journals should use reviewers with statistical training in every submission that reaches conclusions based on data.

Chapter 4

- The thoracic cord is less radiosensitive than the cervical cord.
- The α/β ratio is probably less than 1 Gy in humans. There are no human data indicating that it is as high as 2 Gy.
- The difference in the thoracic cord and the cervical cord dose response can be successfully modeled as a volume effect or an odds ratio.
- The volume effect is undetectable at clinical doses, as is the difference between the response of the cervical and thoracic levels.
- Extreme care should be taken in relying upon the volume effect to increase the tolerance to SBRT.
- Assuming $\alpha/\beta = 2$ Gy results in overestimating the single dose equivalent of a fractionated dose, which would be anticonservative when estimating the equivalent SBRT dose of a fractionated dose.

Chapter 5

- Latency is correlated with dose, but it is not indicative of the level of response.
- Nearly all data after 1990 were analyzed using the LQ model. Re-analysis found that the LQ model fits about half the time. α/β values vary widely. Thus the conclusion can be reached that the LQ model does not reliably fit the data.
- The FE-plot should never be used to assess the adequacy of the LQ model.
- Two studies show that the placement of the top-up dose at the end of treatment is more toxic than at the beginning. This calls into question the assumptions of equal effects from equal fractions.
- The LEM and BIANCA models of RBE do not fit the carbon ion data.
- A new LQ RBE model is proposed that fits the carbon ion data.
- Rodent data on the RBE of protons are not analyzable.

Chapter 6

- At field lengths below 8 mm, the radiation tolerance in rodents increases rapidly.
- The data do not verify that the bath-and-shower effect exists.

- A volume effects model that combines the probability model (critical element model) and the critical volume model fits the field length dose-response data. This is supportive of a minimum lesion size for RM.
- It would be interesting to repeat bath-and-shower experiments with more animals and orthovoltage x-rays to get a sharper beam profile.

Chapter 7

- Analysis of retreatment experiments generally assumes that dose can be used as a surrogate for damage.
- It is generally assumed that the amount of dose that is recovered decreases with increasing dose. This was not seen in the several experiments re-analyzed here.
- The amount of dose recovered as a function of time behaves according to first order kinetics, usually with an initial delay.
- For large animals, repair of 100% of the damage seems possible for initial doses not close to the D_{50}.
- A new model (the volume-repaired model) based on using volume rather than dose as the surrogate for repair fits the rhesus monkey data where other models have failed.
- For humans, the algorithm of Nieder for determining retreatment risk is endorsed.
- Nearly 100% of repair of is seen in immature animals. Repair commences without a time delay.
- In immature rodents, a higher amount of repair with a lower dose was seen in one experiment.
- Because of the difference in the absolute size of the spinal cord in adults and children for the same number of levels, extrapolation is not possible from the adult to the pediatric response.

Chapter 8

- Of the two groups using the pig model, one reported significant involvement of nerve roots after irradiation and one did not.
- The pig field length data did not reject the probability model.
- Although the beagle data was said to reject the probability model, the data were reported using histological endpoints rather than clinical one, variable follow-up times were used, comorbid procedures were done, and the histological endpoint was not an accurate representation of the clinical endpoint.
- Rhesus monkey data were well described by the logistical model with the probability model fitting the field-size data.
- The rhesus monkey response at 2.2 Gy per fraction is close to the human response at 2 Gy per fraction.

- The smallest field size used in a large animal volume effects experiment is 2.5 cm. Below that there are no data, making extrapolation to SBRT treatments somewhat speculative.

Chapter 9

- A dose-response model based on reliability theory is developed to model all large animal data simultaneously. Three parameters apply to all animals and one additional parameter is animal specific. The human data are included.
- This model highlights the difference between damage and dose.
- The volume-repaired model (Chap. 7) is successfully applied to the monkey retreatment data.

Appendix: Medico-Legal Issues

Although legal issues in medicine include much more than medical malpractice, this chapter addresses only the latter. In the USA, malpractice laws are governed by the state. However, there are a few common principals that apply generally. There are a number of somewhat overlapping criteria that must be met, but how they are parsed and in what order they are addressed is not important. They must all be met. This Appendix addresses malpractice issues specific to cases involving radiation myelopathy, and the radiation professional that may be contemplating consulting in such a case.

Basic Elements of Medical Malpractice

Like so many of the subjects that writers cover, the writers describing medical malpractice like to divide it into three necessary components. Interestingly, these three components are not the same for every writer and therefore one can infer that there are in fact more than three. However, the subjects that most frequently make to top three are duty of care, harm, and causation.

Duty of Care

The duty of care arises from the doctor–patient relationship. This is often inverted by saying that the doctor–patient relationship, which naturally occurs first, implies a duty of care. Therefore, the first consideration that is listed is that a doctor–patient relationship must have existed. While the existence of this relationship is rarely in question in radiation oncology, it is also possible to arise unexpectedly. In radiation oncology, it is not uncommon for physicians to seek advice informally from other physicians regarding patient's treatment. Depending upon state laws, the information that is provided by the original doctor, and the nature of the advice, it is possible for liability to attach to the adviser who has never seen the patient. A doctor–patient relationship does not normally arise from advice given in a quality assurance setting, such as a Tumor Board, but guidance should be sought from legal counsel if there is any doubt. The duty of care arising from the doctor–patient relationship means that the physician must provide medical care that meets the applicable standard of care at the time. That is, a reasonable and prudent practitioner acting under the same or similar circumstances during the same time period would have behaved

© Springer Nature Switzerland AG 2022
T. Schultheiss, *Radiation Myelopathy*,
https://doi.org/10.1007/978-3-030-94658-6

similarly. Expert witnesses are called by both sides to produce opinions regarding the standard of care at the time of treatment and whether the treatment given the patient was within that standard. These opinions may ultimately take the form of affidavits, depositions, or court appearances. The standard of care does not require that the patient receive the highest possible medical care.

Another aspect of the duty of care is the responsibility of the physician to provide the patient with sufficient information from which they can determine the course of their medical care. Thus, a patient's consent to receive treatment can only be given if they are adequately informed of the implications of and alternatives to a treatment, although every conceivable alternative does not have to be covered. Obtaining informed consent is not possible in every situation, but these rarely arise in radiation treatments. It is said that informed consent is a process and not merely a document, but obviously this consent should be carefully documented. Failure to receive informed consent from a patient may be judged negligence in itself. Obtaining an informed consent cannot protect a practitioner if the treatment is otherwise malpractice, but failure to obtain it can subject the practitioner to malpractice liability even if the treatment was within the standard of care.

Other providers of patient care also have varying duties of care to the patient. Healthcare institutions also have specific duties related to the patient care provided under their authority.

Negligence is a factor that is often listed in the necessary components of medical malpractice. It is listed here under duty to care because negligence can be understood as a failure to perform that duty.

Harm

For malpractice to exist, the patient must have suffered harm. This is not limited to physical injury in general, but in the current context it is. In radiation oncology and certainly in the case of radiation myelopathy, injury is the readily assessed. In radiation oncology, it is not uncommon for the patient to experience direct side effects of tumor progression. In fact, this is much more common than serious late effect of treatment. For malpractice to have occurred, the patient's injury must be a consequence of the treatment. Of course, harm alone does not indicate malpractice, even if it was a direct result of the treatment.

Fear of doing harm or of legal action may lead to conservative treatment. The standard of care nearly always covers a range from cautious to aggressive. It is outside this range where negligence and liability are to be found.

Causation

The foundation of a malpractice claim is negligence. To establish a cause of action, the plaintiff must show that the defendant had and failed to fulfill a duty of care owed to the patient, that the patient suffered harm as a result of this failure—the negligent treatment caused the harm. Negligence may be an act or a failure to act when there is a duty to do so.

Finally physicians can do everything exactly according to current standards and yet be involved in malpractice litigation as a result of the errors of those they supervise. In radiation oncology, the physician is responsible for directing all aspects of the treatment, and each component of the treatment should be approved by the physician. Even a predictable complication of a treatment within the standard of care can become the genesis of a malpractice claim. An adverse outcome alone *does not* constitute malpractice but that does not mean that all plaintiff attorneys would reject the case nor that jurors would be unsympathetic to the plaintiff.

Malpractice Related to Product Liability

Radiation oncology is arguably the most technologically complex field in medicine. It was probably the first field to deploy computers in its daily applications. It would be impossible to plan and deliver treatment today without advanced computers systems as well as computer control and monitoring of treatment machines. Because treatment planning, treatment delivery, and treatment record-and-verify systems were often developed by different companies often from different countries, these systems did not always operate using the same metrics and conventions. Consequently, manufacturers were at risk for using different systems to express directions and distances. (Think of the Mars Climate Orbiter, built at a cost of $125 million, that crashed into the Martian surface because Lockheed Martin Astronautics supplied acceleration data in English units but the JPL assumed they were working with metric units.) Furthermore, manufacturers could not always control what other equipment might be interfaced with their own. As a result, it has happened that well-known corrections to convert the system of measures of one system to that of another were not securely applied and the radiation dose was delivered in the wrong configuration and to the wrong location in the patient. Other similar problems can result at the point of connection between disparate systems from two manufacturers. A connection is made but the correct information may not be properly transferred from one system to another. Although such errors almost always included a component of operator error, the manufacturers found themselves as defendants in product liability litigation. The treatment or treatment planning equipment operators and those who supervised or employed them were likewise subject to medical malpractice suits in that the duty of care includes assurance of the safe and secure delivery of treatment.

The Expert Witness

Potential Responsibilities

Expert witnesses are asked to apply their expertise to determine relevant facts in a case. In a case of radiation myelopathy, this may include examining, evaluating, or assessing that:

- The defendant had the appropriate training and qualifications.
- The patient's symptoms and their presentation are consistent with radiation myelopathy.

- The Radiation Oncology Department's QA program is adequate and whether it is accredited by a professional organization.
- The radiation oncologist received or sought the necessary medical information on the patient to plan an appropriate treatment.
- The consulting, work-up, and planning process met the standard of care with special attention to the informed consent.
- The proposed treatment met the standard of care.
- The treatment delivery process met the standard of care.
- The spinal cord dose was properly monitored.
- The spinal cord dose was consistent with the patient's subsequent neurological symptoms.

The expert in a radiation myelopathy case should also determine if there were pre-existing neurological problems or comorbidities that explain the symptoms fully or in context with the spinal cord dose and if there were other neurotoxic therapies administered.

In addition, the expert must have access to the affidavits and depositions of the patient, the defendant, and other experts. The expert must be prepared to demonstrate that they have evaluated all relevant information. Meticulous attention to all details is essential if the expert is to render a competent and informed opinion.

How the expert reports their findings depends on the legal processes defined by the state. An affidavit summarizing their findings may be required. They may be asked to be deposed by the opposing counsel. In this deposition, they will be required to testify under oath as if in a court room. The deposition may be videotaped, and under some circumstances the tape can be shown to the jury. Ultimately, the expert may be expected to testify in person.

American College of Radiology (ACR) Guidelines

Both the American College of Radiology and the American Association of Medical Physicists have guidelines for how their members who perform as expert witnesses should conduct themselves.

ACR Practice parameter on the physician expert witness in radiology and radiation oncology:

1. Although the nature of legal proceedings is adversarial, the expert witness must be as impartial and objective as possible.
2. In a medical liability case, the expert witness should be familiar with the relevant standard of care. Care must be taken to distinguish between the expert's personal opinion and the standard of care.
3. The expert witness should review all relevant material and information in order to assure an informed and fair opinion. Images and other relevant materials reviewed by the expert witness should be the original images and other relevant materials used by the interpreting or treating physician in the case. If original images or other relevant materials are not available, good-quality copies of the originals may be acceptable. In cases involving images originally interpreted

using a picture archiving and communication system (PACS), the expert witness review should consider the original algorithm and format (PACS or hard copy) used by the interpreting physician.

4. The expert witness should be prepared to explain the basis of an opinion and should take care that proffered testimony will be scientifically valid and applicable to the facts at issue, can be or has been tested, and has withstood or reasonably could withstand a peer review. The expert witness should be familiar with and be prepared to address the known or potential limitations regarding an opinion, as well as the degree to which that opinion is accepted in the medical community.

5. Compensation of the expert witness should reflect the time and effort involved. Linking compensation for expert testimony to the outcome of the case (contingency fee) is unethical.

6. The expert witness should strive to minimize all potential sources of conscious and subconscious bias when reviewing case materials. Images and other relevant material presented in a blinded fashion to the expert in a malpractice lawsuit strengthens the credibility of the opinion rendered by the expert (https://www.acr.org/-/media/ACR/Files/Practice-Parameters/ExpertWitness.pdf).

And from the ACR Bylaws ARTICLE XI – Ethics and Discipline
Section 3. Rules of Ethics.

In providing expert medical testimony, members should exercise extreme caution to ensure that the testimony provided is non-partisan, scientifically correct, and clinically accurate. The diagnostic radiologist, radiation oncologist, interventional radiologist, nuclear medicine physician, or medical physicist shall not accept compensation that is contingent upon the outcome of litigation (https://www.acr.org/-/media/ACR/Files/Governance/Bylaws.pdf).

American Association of Physicists in Medicine and the ACR for Medical Physicists

Likewise the American Association of Physicists in Medicine and the ACR have jointly published similar guidelines for the medical physicist which state:

The Qualified Medical Physicist expert witness should:

1. Be impartial and should not adopt a position as an advocate or partisan in the legal proceedings.

2. Review the facts in the case and testify to the contents of the case fairly and impartially.

3. Review the standards and the state of the art of the subject matter prevailing at the time of the occurrence.

4. Be prepared to state the basis of the testimony presented and whether it is based on personal experience, specific references, or generally accepted opinion in the areas of specialization of the Qualified Medical Physicist expert witness.

5. Be aware that transcripts and courtroom testimony are public records, subject to independent peer review.

6. Not knowingly provide testimony that is false.
7. Not misinterpret his or her credentials, qualifications, experience, or background. Furthermore, a Qualified Medical Physicist expert witness should not engage in advertising or solicit employment as an expert witness where such advertising or solicitation contains false or deceptive representations about the Qualified Medical Physicist's qualifications, experience, titles, or background (https:// www.acr.org/-/media/ACR/Files/Practice-Parameters/ExpertWitnessMP.pdf).

Misleading, incompetent, or false statements by the physicist or physician expert may constitute ethics violations for which the expert may be disciplined by the professional organization.

Specific Issues in Radiation Oncology

Quality Assurance (QA) in Treatment Planning and Delivery

The majority of radiation physicists in clinical practice in the USA is certified by the American Board of Radiology, the same board that certifies radiologists and radiation oncologists. This is the only certification of non-M.D.'s that is recognized by the American Board of Medical Specialists. Because of their expertise and credentials, radiation physicists along with dosimetrists and radiation therapists also have a duty of care and can become defendants in malpractice cases.

It is the responsibility of the physicist to perform or oversee the output calibration of all treatment machines and equipment that may impact dose delivery. It is now extremely rare for a linear accelerator to be sufficiently out of calibration that its miscalibrated output could possibly contribute to a complication. Redundant systems, secondary output checks, external checks of output measures, and safeguards built into the linac operating systems prevent the output from straying too far without being discovered. This was not true in the past, especially when treatment was delivered by cobalt units. In fact, radiation overdose incidents have been the primary catalyst to encourage the Nuclear Regulatory Agency and state radiation regulatory bodies to write more meaningful radiation regulations.

In addition to machine calibration, the physicist is responsible for ensuring the accurate planning and secure delivery of the treatment prescribed by the radiation oncologist in their written directive.

Although there have been dramatic instances of miscalibration causing radiation injury, advances in treatment delivery, technology, dose monitoring equipments, and quality assurance processes and vigilance has dramatically diminished the likelihood of that occurrence. It is far more likely that somewhere in the complicated chain of simulation, imaging data transfer, treatment planning, treatment data transfer, and treatment delivery, the accurate and secure transfer of information breaks down. Even if that occurs, the most likely outcome is that the treatment process would simply come to an abrupt halt with control software not allowing the next step to proceed. However, mistakes are possible that allow the further progression of treatment but significantly alter the planned dose distribution. It is these sorts of errors that the department's QA program is designed to detect. At a higher level, the accreditation bodies are quite explicit in what such a QA program should include, and important among the elements is an end-to-end testing of the secure transfer of data affecting the final delivery of treatment. These highly technological aspects of radiation treatments distinguish the treatment process in radiation oncology from most other medical specialties.

In some cases where a serious error that could cause harm to the patient has been made, a settlement might be reached before symptoms are experienced. In nearly all states, there are statutory error-reporting mechanisms. These generally are triggered when the prescribed dose was exceeded by a certain amount, the dose per fraction was in error by more than a certain amount and for more than a given number of fractions, the incorrect area was treated, or of course if the wrong patient was treated. The parties to be notified include the state, the referring physician, and the patient, but not necessarily the latter two if the error was sufficiently small. Certain types of treatments are also regulated by the Nuclear Regulatory Commission and these too have reporting conditions if the treatment was not delivered as planned.

Outdated Treatment Techniques

As illustrated earlier in this book, there have been radiation treatments that were at one time considered to be standard of care and that had the potential to induce radiation myelopathy. An easily illustrated example of this is the now discredited lung cancer treatment of 40 Gy in 10 fractions. This treatment was commonly given in a split-dose fashion with the initial 20 Gy given every other day, followed by a 2- to 3-week break before repeating the first course. (The break was intended to allow some normal tissue repair but was found to yield no beneficial effects.) This technique produced radiation myelopathy with a crude frequency of about 5%, which was published by multiple authors. However, as late as the 1990s some radiation oncologists were still using this technique, by which time it was outside the standard of care. No informed consent would have saved a defendant who had prescribed this treatment that was known to cause radiation myelopathy. Likewise during this same time period, it became well established that using higher doses per fraction was much more likely to result in late complications than had been previously appreciated and that the overall treatment time was much less important with respect to late complications. The clinical implication was that it was much better to treat with every field every day rather using a single field per day of two parallel treatment fields or two fields of a four-field treatment. Treating lung cancer with parallel opposed fields to 50 Gy in 25 fractions would be very unlikely to result in radiation myelopathy if both fields were treated in each session, but would definitely put the patient at risk if only a single field per day were used. Using the latter technique in the 1990s would certainly not be considered standard of care and would expose the radiation oncologist to negligence claims.

Literature Reports of Misdiagnosis

Misdiagnosis I

(R. Abadir, Radiation myelitis: Can diagnosis be unequivocal without histological evidence? International Journal of Radiation Oncology, Biology, Physics 6:649–650, 1980.)

A 21-year-old man presented with difficulty breathing, choking, bilateral supra-clavicular adenopathy, and a mediastinal mass. He was diagnosed with Hodgkin's lymphoma and was treated with an aggressive course of radiation therapy typical of the early 1970s. After radiation he had numbness and tingling in the right arm that resolved after 8 months. This was likely a L'hermitte's sign.

Five years after treatment he developed weakness in the lower extremities, and tingling in the fingers of the left hand. Two medical centers diagnosed him with radiation myelopathy. In 7 months the symptoms progressed to paralysis in all limbs, difficulty in bowel and bladder control, and left Horner's syndrome. A myelogram showed a complete block from C4 to T2. It was followed by a laminectomy. The patient died 10 weeks after surgery. The surgical and autopsy findings were both pleomorphic glioma.

The dose of 4965 cGy in 16 fractions was certainly consistent with a risk of radiation myelopathy. However, the clinical presumptive diagnosis was erroneous.

Misdiagnosis II

(KJ Zülch and H Oeser H (1974) Delayed spinal radionecrosis-a juridical error? Neuroradiology 8:173–176.)

A 47-year-old patient presented with paresthesias, motor palsy, and difficulty with autonomic functions of one-year duration. An epidural tumor from T5 to T7 was diagnosed and partially resected. Three pathologists reviewed the slides, and although no consensus was reached regarding the histology, it was agreed that the tumor was malignant. The patient received postoperative radiation "with a conventional dose," "more than 5000 r" in an unspecified fractionation but "according to the technique conventional for the 1950s." After some improvement, the symptoms recurred and a second surgery ensued. No tumor was found, but the "spinal cord was found to be embedded in scar tissue and was described as 'whitish'." The case went to court as a case of radionecrosis, where the plaintiff lost, but an out-of-court settlement was reached (all of which occurred in Frankfurt, Germany). After the patient died, 12 years later, an autopsy was performed that showed only scarring, probably related to the tumor regression after the radiation, but no evidence of radiation damage to the spinal cord. Professors K.J. Zulch and H. Oser, who presented this case, cogently observed "Not every transverse lesion following radiation must necessarily be caused by delayed radionecrosis."

Acknowledgement I would like to thank Larry Brisbee, Esq for his generous advice on this Appendix.

Index

© Springer Nature Switzerland AG 2022
T. Schultheiss, *Radiation Myelopathy*,
https://doi.org/10.1007/978-3-030-94658-6